The Antikythera Mechanism

The Antikythera Mechanism

The Story Behind the Genius of the Greek Computer and its Demise

Evaggelos G. Vallianatos, PhD

Universal-Publishers
Irvine • Boca Raton

The Antikythera Mechanism:
The Story Behind the Genius of the Greek Computer and its Demise

Universal Publishers, Inc.
Irvine • Boca Raton
USA • 2021
www.Universal-Publishers.com

ISBN: 978-1-62734-358-9 (pbk.)
ISBN: 978-1-62734-359-6 (ebk.)

Front cover painting of the Antikythera Mechanism by Evi Sarantea

Painting of the Antikythera Mechanism by the Greek artist Evi Sarantea. The painting is a panorama of the artist's vision of what the Antikythera Mechanism was at the time of its creation in the second century BCE and what it became in the twenty-first century: an emblem in a galactic temple decorated by a Renaissance clock and, on the left side, surrounded by the Wheel of the Sun god Helios depicting the Cosmos of the Sun, the Moon, and the planets, all bathed in stars and constellations, including the interlocking gears below the Cosmos. The right side of the painting depicts two interconnected spirals, the upper one showing a nineteen-year calendar and the lower depicting the eclipse-predicting eighteen-year calendar. Below them there's a painted representation of fragment A, Wheel of the Sun. And at the center lower section, one sees hanging flames representing the unknown scientists who created the Greek computer of genius. Courtesy Evi Sarantea.

Typeset by Medlar Publishing Solutions Pvt Ltd, India
Cover design by Ivan Popov

Library of Congress Cataloging-in-Publication Data

Names: Vallianatos, Evaggelos G., author.
Title: The antikythera mechanism : the story behind the genius of the Greek computer and its demise / Evaggelos G. Vallianatos, PHD.
Description: Irvine : Universal Publishers, 2021. | Includes bibliographical references.
Identifiers: LCCN 2021028486 (print) | LCCN 2021028487 (ebook) | ISBN 9781627343589 (paperback) | ISBN 9781627343596 (ebook)
Subjects: LCSH: Antikythera mechanism (Ancient calculator) | Astronomical clocks--Greece--History.
Classification: LCC QB107 .V36 2021 (print) | LCC QB107 (ebook) | DDC 681.1/11--dc23
LC record available at https://lccn.loc.gov/2021028486
LC ebook record available at https://lccn.loc.gov/2021028487

For Mark Andreas, Corinna Lia, Ellis (Herakles),
Katya (Athena) and Argo

Contents

Preface

Suppose you could travel back 2,300 years in time to visit Alexandria, the capital city of the Greek kingdom of Egypt. This was the most enlightened, wealthy, and powerful of all the Greek states that flourished after the death of Alexander the Great.

Alexandria was famous for its Mouseion (University-Institute of Advanced Studies) and the great library, which held the collected science, philosophy, and literature of the Greek and Mediterranean world. The great library was rather like merging the Library of Congress with all the resources of Stanford, Harvard, and MIT. Its collections included the epics of Homer; the books of early natural philosophers like Anaximander; the plays of Aeschylus, Sophocles, Euripides, and Aristophanes; the atomic theory of Leucippus and Democritus; the engineering of Ktesibios, Archimedes, Philon of Byzantium, and Heron of Alexandria; the history of Herodotus and Thucydides; the medical knowledge of Hippocrates and Galen; the philosophy of Plato; the philosophy and science of Aristotle; the scholarship and geography of Eratosthenes and Ptolemaios; the astronomy of Eudoxos, Aristarchos of Samos, Hipparchos, and Ptolemaios; and the mathematics of Eudoxos, Euclid, Archimedes, and Apollonius of Perga. Homer and this small sample of scientist-philosophers defined Greek civilization.

However, if you were to revisit Alexandria 800 years later, you would find no trace of the great Library of Alexandria, or of its Mouseion University. These had been destroyed and wiped from the map by Christian fanatics.

In 415 of our era, a mob of Christian monks set out to expunge all traces of "pagan" Greece from history. To make their point, they murdered the mathematician Hypatia and tore her limb from limb in the public marketplace;

she was the head of a prestigious school of philosophy in Alexandria, but to the monks she was "anathema."

When the Christian faith was finally established as the official religion of the Roman Empire in the fourth century, an all-out Holy War was launched against the pagan unbelievers and all aspects of their culture. In the sixth century, Justinian, the Emperor of the Eastern Roman Empire, suppressed and destroyed the Platonic Academy in Athens, which had flourished as a university for about 900 years. This war against the science and culture of ancient Greece unleashed a wave of darkness and ignorance for both East and West. Christian fanaticism plunged Europe into the Dark Ages for almost a thousand years.

These imperially sanctioned attacks on Greek culture gave Christian monks tremendous power. They physically destroyed priceless texts of science by scraping the very ink from the parchments. Then they over-wrote the empty parchments with religious texts.

In 1998, an American paid $2 million for a 769-year-old Christian prayer book, known as a "palimpsest" (a manuscript from which the original text had been scraped off, enabling the blank parchment to be overwritten again). Christian prayer-texts were written over the erased scripts of some of the books of Archimedes.

This philanthropist was willing to pay such a vast sum of money for the "ghost" of a vanished book because of the slim hope that the content of Archimedes's original manuscript might be resurrected from oblivion.

In 2005, seven years after the Archimedes palimpsest surfaced in New York, an even greater drama was unfolding in the National Archaeological Museum in Athens.

A team of international scientists and engineers were struggling to "decode" the secrets of an ancient mechanical computer: the Antikythera Mechanism. In 1900, sponge divers had found this artifact among the broken amphorae and statues of an ancient shipwreck, on the seabed of the tiny Aegean island of Antikythera.

To the divers, it just looked like a lump of shell-encrusted bronze and it was impossible to see what it really was. In fact, the technology needed to reveal the innermost workings of this apparent lump of concrete did not even exist in 1900. This went hand in hand with the prejudices of most scientists, who believed they had a monopoly on science and technology. Only after decades of technological advances and the appearance in the early 1970s of a virtuous scientist interested in the origins of the mechanical device was science able to penetrate this complex metal artifact and illuminate its innermost secrets.

Later still, scientists used MRI scanners and X-rays to probe this object. They were shocked to discover an interior mechanism of astonishing complexity: finely meshed cogs and toothed wheels. The scans revealed an instrument of mathematical exactitude, fabricated by advanced engineering. Inscriptions on the face and back of the computer indicate it was manufactured 2,200 years ago.

It is instructive to ask what most of the world was like 2,200 years ago. America, from the Andes to Alaska, was inhabited by Stone Age tribal cultures, to whom metalworking was largely unknown. Writing existed only as the stone-carved glyphs of the Maya, or the painted hieroglyphs of the Aztecs. With the possible exception of Rome, Europe was much the same: warrior tribes of sheep and cattle herders mingled with settled agriculturalists, while mountainous regions were still populated by hunter-gatherers. Writing only existed as stone-carved images or runes. Literacy did not exist; there were no books, no sciences, no architecture, no physics, no medical science, no engineering, no history, and no political theory.

Metalworking was limited to bronze swords and axe-heads, sickles and farm implements; war chariots represented the pinnacle of engineering.

Yet astonishingly, in that very same era, Greece had produced architects, artists, scientists, and engineers who built the Parthenon, great amphitheaters, harbors, and bridges. Mathematicians such as Eratosthenes had used geometry to estimate the circumference of the Earth. Democritus had produced the atomic theory of matter. Aristarchos of Samos invented the heliocentric theory of the Cosmos. The Sun rather than the Earth was at the center of the universe. Aristarchos also wrote a book on the measurement of the distances of the Sun and the Moon from the Earth, including measuring the sizes of the Sun and the Moon. The list of Archimedes's achievements is astonishing: he expounded the physics of levers and forces; laid the foundations of hydraulics and hydrostatics; calculated the value of Pi (Π), and invented calculus. In his work *On the Circle and the Sphere*, he derived the formula to calculate the surface area and volume of any sphere. In his book on *Parabolas*, he was the first to use the concept of "infinity." On a practical engineering level, he invented the Archimedes screw, used to raise water from rivers to irrigate fields, and in military technology he invented the "Claw of Archimedes"—a giant crane, which could grasp ships attacking a city, to destroy them. He also wrote a book on *Spheres*, now lost, in which he described the construction of a mechanical universe. A model of the spherical mechanical universe of Archimedes survived until the first century BCE.

These thinkers link us to the sale of the Archimedes palimpsest and the X-ray investigation of the Antikythera device. These are parallel threads of the same narrative, which illuminate the fate of Greek science and civilization.

What happened to the advanced work of scientists like Eudoxos, Euclid, Aristarchos, Archimedes and Eratosthenes during the Dark Ages? How could the ancient Greeks have developed such a sophisticated science and technology more than 2,000 years ago, when most of the world was scientifically primitive? Is the Antikythera Mechanism the proof that other advanced technologies existed?

This book attempts to answer these questions.

My thesis contends that the ancient Greeks did indeed develop advanced science and engineering technologies, which enabled them to build the Antikythera astronomical computer and similar devices.

This is a crucial issue because many scholars and historians still believe that modern science is entirely the product of the fifteenth-century Renaissance and the eighteenth-century Enlightenment. Moreover, they cannot believe that the ancient Greeks ever developed advanced technologies.

So, who cares! Why should we be interested in the ancient Greeks, much less their technology? After all, modern science and technology has given us nuclear weapons and taken us to the Moon. How can the ancient Greeks possibly be compared to us?

"We are all Greeks." These were the heartfelt words of the English poet Percy Bysshe Shelley in 1821: "Our laws, our literature, our religion, our arts, have their roots in Greece."[1] Shelley was one of many philhellenes who fought alongside the Greeks during their War of Independence, in the heroic struggle to expel the Turkish occupiers of their homeland.

Jacob Burckhardt, the famous Swiss cultural historian, was grateful to the Greeks for laying the foundations of Western culture, enabling humans to become civilized. In 1872, he wrote that the Europeans see the world through the eyes of the Greeks and to abandon them would be to accept their own decline.[2]

---

[1] Percy Bysshe Shelley, *The Complete Poems* (New York: The Modern Library, 1994) 501.

[2] Jacob Burckhardt, *The Greeks and Greek Civilization*, tr. Sheila Stern, ed. Oswyn Murray (New York: St. Martin Press, 1998) 12.

Another European scholar from England, W. R. Inge, wrote in 1921 that there was no way the Europeans could go on without understanding the debt they owed to Greek civilization and culture. "Without what we call 'our debt to Greece'," he says, "we should have neither our religion nor our philosophy nor our science nor our literature nor our education nor our politics. We should be mere barbarians. Our civilization is a tree which has its roots in Greece ... [our civilization] is a river ... but its headwaters are Greek."[3]

In 1948, the English poet W. H. Auden suggested that the people of the West owe their very existence to the Greeks. He said the Greeks taught us to think about thinking, that is, to ask questions. Without the Greeks, he said, "We would never have become fully conscious, which is to say that we would never have become, for better or worse, fully human."[4]

In 1949, the famous British classical scholar Gilbert Highet added Greek logic and metaphysics to building reason among the people of the West.[5]

In 1951, another British scholar, H. D. F. Kitto, gave a brilliant and insightful perspective on the Greeks. The Greeks, he wrote, were "not very numerous, not very powerful, not very well organized." They had "a totally new conception of what human life was for and showed for the first time what the human mind was for." Comparing the historical chronicles of barbarians to the work of Thucydides, he said, is "the difference between a child and a man who cannot only understand, but also make his understanding available to others. Epic poetry, history and drama; philosophy in all its branches, from metaphysics to economics; mathematics and many of the natural sciences—all these begin with the Greeks."[6]

In the late 1950s, E. J. Dijksterhuis, Dutch historian of mathematics and natural sciences, captured the Western scientific esteem of the work of the Greeks. He said that the origins of present-day knowledge, especially in mathematics and natural sciences, go straight to ancient Greece.[7]

---

[3] W. R. Inge, "Religion" in *The Legacy of Greece*, ed. Richard Livingstone (New York: Oxford University Press, 1969) 28.

[4] W. H. Auden, *Forewords and Afterwords*, ed. Edward Mendelson (New York: Random House, 1973) 32.

[5] Gilbert Highet, *The Classical Tradition: Greek and Roman Influences on Western Literature* (New York: Oxford University Press, 1949) 542.

[6] H. D. F. Kitto, *The Greeks* (Penguin Books, 1967) 7, 9.

[7] E. J. Dijksterhuis, "The Origins of Classical Mechanics from Aristotle to Newton" in *Critical Problems in the History of Science*, ed. Marshall Clagett (Madison, WI: University of Wisconsin Press, 1969) 164.

In 1993, the American historian of astronomy Anthony Aveni reported that theory, logic, and models are gifts of the Greeks: "Indelibly written on the faces of all our modern texts about the natural world. If we can clearly trace one branch of the taproot of modern science back to the ancient Greece, it is this habit of formulating practical, non-human-centered, mechanical models of how nature works," he wrote.[8]

In 1999, Charles Freeman, a classical scholar, summed up the Greek influence with the chromosome metaphor. The chromosome was one of the greatest discoveries in biology in the twentieth century: the genetic material in every living thing responsible for passing one generation's biological characteristics to the next. The Greeks, he said, "provided the chromosomes of Western civilization."[9] The chromosome is an idea and metaphor of science; knowledge of where the Greeks have had their greatest lasting influence on the West, the theme of this book.

First, this book is about the greatest technological achievement of ancient Greek civilization: the Antikythera computer. It describes the political, scientific, and technological revolutions that enabled the Greeks to build it. The book explains how and why they were able to manufacture the first astronomical computer in the world.

However, the book is more than a biography of a computer. Indeed, that biography occupies but one chapter in a study tracing the history of the Greeks from the seventh century BCE to the second century of our era. In addition, two chapters delve into the fortunes of Greek civilization after the triumph of Christianity. Second, this book goes beyond technology, to investigate the ethics, which went along with the scientific advances. The Greeks believed that science and technology had to be used for the good of society. Technology does not exist in a vacuum or for selfish advancement and profit; it must contribute to "the good."

My research is founded upon the use of primary sources. I trace the history of science and technology from original texts. This helped me understand the thought and labor that went into the building of a scientific computer and calendar that synchronized the movements of the heavens and the works of human beings by advising farmers when to plant for a good harvest, and when

[8] Anthony F. Aveni, *Ancient Astronomers* (Washington, DC: Smithsonian Books, 1993) 53–54.
[9] Charles Freeman, *The Greek Achievement* (Penguin Books, 2000) 434.

to expect the Olympic games. I have linked Archimedes and the Antikythera device to science and culture in ways that have never been done before.

Third, this book illuminates the relevance of Greek thought and engineering to our own times, building a case for why we should care about the world's oldest computer, which the ancient Greeks called Tablet. This cosmological and Promethean computer of genius has a history full of insights for today's computer geeks, high-tech executives, and discerning readers interested in Greek and Western civilization. This book documents and explains this history—a cautionary tale of what serves and what hurts the public good. It's a warning: that in their rush to make money, developers of artificial intelligence, machine learning, and driverless cars are opening another Pandora's box of uncontrollable violence and possible extinction.

The lesson from the Greeks is this: If technology is not used for the public good, it undermines society, as we saw in the global financial crash of 2008, or in Volkswagen's lies about their cars' poisonous emissions, or the deleterious and plague-like consequences of industrialized farming. The harm from modern farmers is so widespread that their animal feeding operations have become sources for pestilence. Young mothers are finding pesticides in their breast milk. And even worse than that, the United States and Russia are on a hair-trigger alert in unleashing a catastrophic nuclear war.

Galen, the Greek medical genius of the second century, put it bluntly: "If wealth is valued above Virtue, it spoils and corrupts science and society."[10] In other words, if we ignore the good and the beautiful, we are doomed.

In marked contrast, the Antikythera computer—Tablet—was built to serve the public good. (In chapter 6, I will explain why "Tablet" is the name ancient Greeks gave astronomical devices like the Antikythera Mechanism.) The Antikythera Tablet brought the heavens within the reach of human understanding. It was an astronomical compendium of astonishing complexity and accuracy. It provided accurate tracking of the movement and position of the planets, the Sun, the Moon, the constellations, and prominent stars. It mechanized the predictions of scientific theories, including the eclipses of the Sun and the Moon, as well as the phases of the Moon. It was a celestial calendar, a moving map of the constellations, and a mirror of nature and the heavens.

[10] Galen, "The Best Doctor is also a Philosopher" 57–61, in *Selected Works*, tr. P. N. Singer (New York: Oxford University Press, 1997).

Finally, this book sets this ancient history against the modern drama of the resurrection of the vanished texts of Archimedes from the palimpsest, and the decoding of the secret interior of an ancient Antikythera computer. Science gave us satisfactory answers to these enigmas. Indeed, science is our greatest connection with the Greeks. We acknowledge their invention of democracy. We also celebrate their achievements in laying the foundations of mathematics, physics, biology, and astronomy, which eventually took us to the Moon. This is what should make their story so important to the modern reader.

That is why the discovery of a vandalized book of Archimedes is big news and the decoding of a 2,200-year-old mechanical computer is of extraordinary importance. This computer was at the cutting edge of an evolving technology, which revealed the heavens, and connected the Greeks to nature, to their athletic and religious festivals, and to the gods. This book relates the dramatic story of those who built the Tablet and those who, more than two millennia later, unearthed its secrets from within an encrusted lump of marine debris and deciphered the corroded inscriptions.

Introduction

Science and the Public Good

Science was the greatest Greek achievement and the Antikythera computer its most brilliant expression. Scientists and engineers translated natural philosophy, astronomy, mathematics, physics, and engineering into the bronze gears of the mechanical universe of the computer.

But behind this celestial gear machine, there was a great metaphysical virtue, just as fundamentally important as the gears: the promise that science would rarely, if ever, be separated from the public good and the beautiful. This restraint, inherent in Greek thinking about science and technology, kept the potential excesses of scientific knowledge in check.

The successors of Alexander the Great, especially the kings of Egypt, adopted Aristotle's model of science: lavishing it with money, a great library, a university, and other support.

However, the Christianization of Greece and the West in the fourth century led, first, to the near disappearance of science during several centuries of darkness, usually described as the Middle Ages (chapter 9). Then, starting with the Renaissance in the fifteenth century, science revived in the West primarily because Greek scholars in Padua, Venice, Florence, Rome, and other cities in Western Europe translated ancient Greek scientific and philosophical texts into Latin. Western scientists studied those texts and absorbed their science, triggering additional insights that became modern science. But without the virtues science had in its Greek origins and development, science in the West got into trouble. It had its numerous flashes of genius, but it almost lost its integrity, becoming by the twentieth and twenty-first centuries primarily a factory for nuclear weapons, intercontinental missiles, nuclear power plants, satellites, pesticides, genetic engineering, high-tech computers, artificial intelligence, iPhones, pharmaceuticals, countless cars run on petroleum, and countless chemicals and machines.

This metamorphosis of science, among other things, explains why modern scientists took more than a hundred years to understand the science and purpose of the Antikythera computer. It did not meet their expectations. It was too much like the science they practiced, a reality that shocked them. They could not answer a question that arose immediately when they began to understand the complexity of the Greek computer: How could Greeks, 2,200 years ago, do things we do now? The Antikythera device challenged their understanding and metaphysics.

The contrast between ancient Greek science and modern science is necessary if only in helping us appreciate the enormous potential that the Greek science model presents for our science and civilization in the twenty-first century. In addition, people who understand the limitations of what science has become are likely to judge fairly and appreciate the Greek computer.

Even a cursory review of a few facts in the history of modern science might help place the ancient device in proper perspective. The Antikythera computer, after all, is more than an interesting if spectacular curiosity and a fantastic example of genius. It seamlessly weaves scientific technology and the Cosmos around a portable and small metal device that reveals the secrets of the heavens to farmers, priests, and astronomers. Benefiting all, as the Antikythera computer did, is not exactly an easy theoretical or practical achievement. For this reason, the Antikythera computer may well prove essential for our survival.

Here's why.

Science for the Greeks was *episteme (ἐπιστήμη)*, expert knowledge of something like archery and *techne (τέχνη)* or cunning craftsmanship, technology. Aristotle said that *episteme* comes with certain knowledge of a thing. When we are certain we know the cause behind a fact, we are doing science.[11]

[11] Aristotle, *Posterior Analytics* 71b8–12, ed. Richard McKeon, *The Basic Works of Aristotle* (New York: The Modern Library, 2001). In most of my citations of ancient Greek authors, I use the Greek texts of the Loeb Library or the Oxford Classical Texts. The English translations in the Loeb Library are for the most part obsolete. So, with some exceptions, all translations in this book are my own.

Knowledge for the Good of Society

Like technology and culture, *episteme* came to mean primarily knowledge for the good of society. This idea of the good, almost always connected with the beautiful, was pervasive in Greek thought and society.

For example, Xenophon, c. 428–354 BCE, an Athenian writer, student of Socrates, and military man, tells the story of Herakles's choice by Prodikos, a fifth-century BCE sophist. Young Herakles, the greatest hero of the Greeks, is reflecting on what he should be doing in life: What path should he follow? At that moment, two goddesses approach Herakles. One of them, Vice, offers Herakles a career full of pleasures. The other goddess, Virtue, tells Herakles that a life merely full of pleasures promised by Vice is meaningless and empty; she advises Herakles to work for a living, insisting that only honest labor makes life truly good and happy:

"The gods give nothing to man without effort and work," says the goddess of Virtue. "If you expect the favor of the gods, worship them: if you want your friends to respect you, be good to them: if you want a polis to honor you, help that polis: and if you are thinking of winning the admiration of all Hellas, you need to do things that benefit Hellas."[12]

For example, the founder of the Olympics deserved public recognition and honor all over Hellas. A legend has it that Herakles founded the Olympics.

Plato warned us about the misuse of knowledge.[13] Plato, 427–347 BCE, was a great philosopher (chapter 3). His original name was Aristokles. The epigram from his grave praises him to the heavens, calling him divine, and extolling his virtues of temperance, justice, and wisdom.[14] But Plato grew up in a country, Athens and Greece, torn apart by war. Plato witnessed Athens humbled by Sparta. He also lived through the wrongful death of his teacher Socrates by democratic Athens.

About 800 years after Plato, the Platonic philosopher Proklos, 412–485, denounced a group of philosophers calling themselves Skeptics who refused to accept that knowledge was possible. The Skeptics, Proklos said, "would do away

[12] Xenophon, *Memorabilia* 2.1.28.

[13] Plato, *The Republic* 7.519a, ed. Ioannes Burnet (Platonis Opera, Oxford Classical Texts, Oxford: Oxford University Press, 1902, 1978). With some exceptions, all my citations of Plato come from the Oxford Classical Texts. The text edited by Burnet was revised by later editors.

[14] *The Greek Anthology*, 7.60, tr. W. R. Paton (Loeb, 1960–1968), 5 vols.

with all knowledge, like enemy troops destroying the crops of a foreign country, in this case a country that has produced philosophy [Hellas/Greece]."[15]

Despite these rare excesses, science served the public interest. First of all, science was not, strictly speaking, a technical process or enterprise. It was much more than using reason and experiment for understanding society, nature, and the universe. Science was asking questions. It was philosophy. It included the appreciation and understanding of the beauty and workings of nature and the Cosmos, which, in Greek, means an ordered universe that even the gods dare not interfere with.

In fact, according to fifth-century BCE historian Herodotus, c. 490–c. 425 BCE, the Greeks called their divinities gods because they believed those divinities had set all things in order and assigned a place to everything.[16] In other words, the gods opened the path of organized knowledge, order, and reflection. According to Aristotle, those cultivating reason and doing science were dear to the gods.[17] (I examine Aristotle and science in chapter 3.) The most outstanding example of that divine order was the Cosmos, which, in the vision of the Greeks, was also a world made beautiful, equitable, and just as well as having the appearance of an ornament. Greek science was the rigorous search for the truth in society, the natural world, and the Cosmos. No wonder the Cosmos found its place within and without the computer Greeks invented and manufactured in the second century BCE.

In the fifth century BCE, Hippocrates, c. 460–370 BCE, exemplified science and craftsmanship. An unknown commentator alleged Hippocrates was the son of *Phoibos* (Shining) Apollo. In addition, we learn Hippocrates was from Thessaly but was born in the Aegean island of Kos. The commentator hastened to explain that Hippocrates triumphed over many diseases not by accident but by the application of *techne*.[18]

Hippocrates expected the Greek physicians not to harm those seeking their help. Greek doctors took an oath to heal. In that sense, Hippocrates mirrored the main purpose of Greek science, which was to improve the human condition, not to injure men and women or nature. Aristotle searched for the truth in the natural world, human society, and the Cosmos.

---

[15] Proklos, *A Commentary on the First Book of Euclid's Elements* 199, tr. and ed., Glenn R. Morrow (Princeton, NJ: Princeton University Press, 1992).

[16] Herodotos, *The Histories* 2.52.

[17] Aristotle, *Nicomachean Ethics* 1177b31–1179a33.

[18] *The Greek Anthology* 7.135.

The biology of Aristotle (chapter 3) and the mathematical physics of Archimedes (chapter 4) remain the same to this day; in fact, they triggered "modern" science, the science we have today in the third decade of the twenty-first century. But they are different from our science in the sense they stood by their principles of the discovery of truth and the application of that knowledge for the protection of nature and the good of society, not its harm.

In the history of science, the technical component of science keeps becoming more complicated but science remains pretty much the same. What has become alienated in modern times is the purpose of science.

Who Invented Science?

The Greeks and the Babylonians were stargazers. The Greeks decoded the heavens because they sought to understand the working of the Cosmos, home of the gods. After all, Ouranos, the sky, was a god who hosted other gods, the stars.

According to Edward Rosen, distinguished professor of the history of modern science at the City University of New York, the Babylonians had no interest in "the inner workings of the cosmos. Their aim was to master those cosmic features which they believed controlled human destiny."[19] I will discuss the differences between Greek science and non-Greek science in chapter 8. But at this point I am suggesting that Greek science triggered Western science; and to some degree, Western science became a continuation of Greek science. This is true provided one also takes into consideration that modern times altered our relations with the ancient Greeks in a lasting fashion. To the invisible but real influence Christianity exercises over science, we must add the overwhelming power and influence nuclear bombs have been having on science and civilization.

Modern Science

This is not the place for a history, even a sketchy one, of modern science. I will, instead, summarize a few key developments that highlight the new nature of science and technology in the West. It's always important to know where we

[19] Edward Rosen, *Copernicus and the Scientific Revolution* (Malabar, FL: Krieger Publishing, 1984) 15.

come from and what we have created. Is present knowledge, what we think we know and what has shaped what we know, an appropriate channel or environment for understanding something like the Antikythera computer? A clue for answering this question comes straight from the Dark Ages.

Lynn White, an American historian who spent about fifty years studying Europe after it became Christian, makes huge claims for the technology of that continent. Historians call the first millennium of Christian Europe Middle Ages or Dark Ages. First, White says, "Modern technology is the extrapolation of that of the Western Middle Ages not merely in detail but also in the spirit that infuses it." And, second, he argues, medieval technology has been the essential ingredient, indeed the reason behind the domination of the world by Europe and America. The intriguing element in the argument of White is his emphasis on Christianity inspiring medieval and, hence, modern technology. At about 830 of our era, White says, Benedictine monks advanced technology as part of Christian virtue. And because medieval people dominated nature for the glory of their Christian god, White concedes, technological aggression becomes "the normal Western Christian attitude toward nature."[20]

There's an element of truth in that assertion. Christian monks had no idea what science was, but they knew how to dominate the natural world by sheer force and practical machines they inherited from the Greeks. Their urge to control the wild received theological blessings, so aggressiveness became natural to them. Their technological aggression became part and parcel of modern dealings with nature. It still continues. Technological aggression is a fact. However, technological aggression leaves a tragic footprint in the mind of the scientist-engineers, as well as society and the natural world.

A novelist, Mary Shelley, captured the aggression of modern science. Shelley was also prophetic of the misuse of technology in the West. Her 1818 novel, *Frankenstein*, captures the arrogance of engineers giving birth to monsters. Victor Frankenstein created a terrifying creature that, in good time, turns against him.

Our engineers keep following Frankenstein, but never have the conversation Frankenstein had with his creation. "If I cannot inspire love, I will cause fear; and chiefly towards you my archenemy, because my creator, do I swear inextinguishable hatred," the monster warns Frankenstein.

[20] Lynn White, Jr., *Medieval Religion and Technology* (Berkeley: University of California Press, 1978) xxiv, 219, 250–251.

"Have a care: I will work at your destruction, nor finish until I desolate your heart, so that you shall curse the hour of your birth.... Remember that I have power; you believe yourself miserable, but I can make you so wretched that the light of day will be hateful to you. You are my creator, but I am your master;—obey!"[21]

Shelley spoke to the winds. A hundred and twenty-seven years after the novel *Frankenstein* saw the light of day, scientists and engineers gave birth to a gigantic Frankenstein monster: the nuclear bomb. This bomb is so powerful that it has reorganized the world.

From Know-How to Nowhere

In 1974, Elting E. Morison, an MIT professor, explained the confusion between science and the public good. "It is not so clear ... that the problems solved by modern science and engineering coincide nicely with the fulfillment of needs felt or unfelt in men and women," he wrote.

"The chances seem to be," he warned, "that the stream of new artifacts ... will not fit ... with the nature of human being.... There seems ... to be a developing mismatch between our extending knowledge of what we can do with the materials and forces in the world around us and our older, but less certain, understanding of what we have to do to be ourselves. And in this mismatching—such is the power in our machinery and such is the confusion about our real needs—we are likely to come away losers—ground down, blown up, twisted out of shape, crammed into computer-designed compartments, bored to death."[22]

Science in America shaped philosophy as well. That was inevitable. Christianity, wars, industrialization, and the nuclear bomb demanded obedience. Philosophy followed science in specialization. It shut down its probing questions and cleansed its broad ethical mantle, becoming neutral to good and bad. It had nothing to do with society. That's why modern philosophers lose no sleep over nuclear bombs or pesticides or political corruption or environmental destruction. They know, or should know, the knowledge they produce is divorced from goodness and virtue. Modern philosophy is no longer

[21] Mary Shelley, *Frankenstein or the Modern Prometheus*, ed. M. K. Joseph (New York: Oxford University Press, 1998) 145, 167.

[22] Elting E. Morison, *From Know-How to Nowhere: The Development of American Technology* (New York: New American Library, 1977) 137.

philosophy. "The point of philosophy now [in 2016]," say two American professors of philosophy, Robert Frodeman and Adam Briggle, "is to be smart, not good. It has been the heart of our undoing."[23]

The Religion of Modern Science and Technology

David F. Noble, who taught history at York University in Toronto, Canada, was aware of this dramatic change of values. Like Lynn White, he sought explanations from the dark past of Europe. He links the American technological enthusiasm to monastic otherworldliness, transcendence, all in an effort of recovering the "lost divinity of man."

This makes sense not because of any lost divinity of man/woman, but because Christianity and technology are intimately related in America. Blind faith in technological gadgetry—what Noble describes as "an unrivaled popular enchantment with technological advance"—parallels a dangerous superstitious belief in the return of Jesus.

"If we look closely at some of the hallmark technological enterprises of our day," Noble wrote in 1997, "we see the devout not only in the ranks but at the helm. Religious preoccupations pervade the space program at every level and constitute a major motivation behind extraterrestrial travel and exploration. Artificial intelligence advocates wax eloquent about the possibilities of machine-based immortality and resurrection, and their disciples, the architects of virtual reality and cyberspace, exult in their expectation of God-like omnipresence and disembodied perfection. Genetic engineers imagine themselves divinely inspired participants in a new creation. All of these technological pioneers harbor deep-seated beliefs which are variations upon familiar religious themes."

The boosters of space travel, says Noble, "brought the world but minutes away from mutually assured annihilation." The civilian leaders of NASA have not abandoned the military ends of their agency, advancing the "militarization" of space in terms of surveillance and the use of weapons.

The same thing happened to the advocates of artificial intelligence searching for the "immortal mind." They are on the payroll of the military, and in

[23] Robert Frodeman and Adam Briggle, "When Philosophy Lost Its Way," *The New York Times*, January 11, 2016.

return, they are arming the military with even more lethal weapons and technics of surveillance and control.

They also transfer their technologies to corporations, "which have deployed them the world over to discipline, deskill, and displace untold millions of people, while concentrating global power and wealth into fewer and fewer hands." As for genetic engineers, Noble sees them, rightly, as laying the "foundations for an Orwellian future."[24]

This preposterous vision was still at the heart of Google—in 2016. Google is America's giant information technology company. Its director of engineering, Ray Kurzweil, speaks eloquently about human immortality. He translates what he calls "exponential growth of information technology" into "expanding our life expectancy for thousands of years."

How does he propose to bring about such a miraculous metamorphosis? He bets that by 2030, we will have extremely powerful computers the size of blood cells. These "little robotic devices," he says, will flow through the bloodstream, eliminating diseases and aging. He also asserts people will be mostly machines, which will do most of their "nonbiological" thinking for them. With such premises, Kurzweil has no doubt Google engineers will extend the lives of very wealthy people "indefinitely."[25]

In my view, ideas about human immortality and humans made mostly of computer hardware are products of science fiction. They mirror the nightmares of an amoral oligarchy of high tech and wealth married to militarism. Their immortality schemes are likely to produce more Frankenstein monsters—and worse.

Stephen Hawking, late British physicist; Elon Musk, CEO, Tesla Motors; and Nick Bostrom, director, Future of Humanity Institute, Oxford University, equate artificial intelligence to a life-and-death threat to human civilization.[26]

Next to Christianity and, probably, connected to it, the second force shaping the increasingly hazardous nature of modern science is money. Most of that money for scientific research comes from governments and global corporations. Petroleum, natural gas and coal companies, responsible for global warming, have been funding scientists spreading doubt about climate change.[27]

---

[24] David F. Noble, *The Religion of Technology* (New York: Alfred A. Knopf, 1997) 5, 205–206.

[25] "Inventor Ray Kurzweil sees immortality in our future," *PBS Newshour*, March 24, 2016.

[26] "Will we be wiped out by machine overlords?", *PBS Newshour*, July 20, 2017.

[27] Naomi Oreskes and Erik M. Conway, *Merchants of Doubt* (New York: Bloomsbury Press, 2010).

The same corporate strategy is polluting science not merely in support for agribusiness and war, but for additional life-threatening activities with deleterious consequences for human survival.

Three examples:

In 2005, John Ioannidis, professor of medicine at Stanford, argued that "most published research findings are false." Thirteen years later, in 2018, he equated ignoring of science by the Donald Trump administration to a "nightmare."[28]

On April 11, 2015, Richard Horton, editor of the preeminent medical journal, *The Lancet*, issued the following warning about modern science:

"The case against science is straightforward: much of the scientific literature, perhaps half, may simply be untrue. Afflicted by studies with small sample sizes, tiny effects, invalid exploratory analyses, and flagrant conflicts of interest, together with an obsession for pursuing fashionable trends of dubious importance, science has taken a turn towards darkness."[29]

On June 21, 2015, Ulrich E. Loening, professor of zoology, University of Edinburgh, accused the owners and scientists of crop genetic engineering with corruption. They do everything in their power, he said, to create blind consensus among gene scientists about the "safety" of their products—genetically engineered food—while vigorously rejecting "any contradictory findings."[30] Another byproduct of crop genetic engineering is biological warfare. The mission of this war biology is plagues. The 2020 corona virus plague could have been a product of biological warfare.

With such metaphysics and corrupt practices and politics, it is easier to understand why Westerners rushed to deny the Greek origins of science and technology. Greek science reminds them of what they should be doing: searching for truth and without any strings attached to corporate money or political influence—virtues in decline in our times. Indeed, the unpleasant record of modern science presented here, though peripheral to the purpose of this book, explains the difficulties as well as opportunities of understanding and learning from the lessons of the Antikythera computer.

---

[28] John P. A. Ioannidis, "Why most published research findings are false," *PLOS Med* 2(8): e124, August 30, 2005; "All science should inform policy and regulation," *PLOS Med* 15 (5): e1002576, May 3, 2018.

[29] Richard Horton, "What is medicine's 5 sigma?" *The Lancet*, Vol. 385, April 11, 2015, 1380.

[30] Ulrich E. Loening, "A Challenge to Scientific Integrity," *Environmental Science Europe*, 21 June 2015.

History of Science

I did my postdoctoral studies in the history of science at Harvard. I also worked for twenty-seven years with science issues on Capitol Hill (Congressional Office of Technology Assessment and the Office of Congressman Clarence Long of Maryland) and the US Environmental Protection Agency. I saw that under both Republican and Democratic administrations, the EPA prefers to side with polluters, in a sense becoming a polluters' protection agency. This reached heights of criminal negligence under the Trump administration.

I observed how the state, politics, and money corrupt science and restrict its benefits to a select few. I could and did analyze the rhetoric and policies behind the mega enterprise science has become in America and the West. I wrote four books warning of the dangerous path of modern science.[31] Indeed, the fourth book, *Poison Spring*, revealed the dramatic consequences of having industry in charge of public policy.

I also connected my studies of science in America with my Greek interests, studying the behavior of scientists and scholars interpreting the history of Greek science. It was this interest that brought me to the writing of this book. I kept up with my readings in classical Greek studies, never ceasing to admire how original the ancient Greeks were, especially in science. I noticed, however, that many writers and classical scholars in America and elsewhere were not merely critical of the Greeks but biased against them. They charged the Greeks with all kinds of mischief, from having slaves to being misogynists, to appropriating knowledge from other people, to having contributed very little to science, which they consider a strictly modern achievement, to having done practically nothing in mechanics and technology.

For example, in 2009, a British researcher questioned the legacy of the Greeks. This was Patricia Fara. She treats science as an exclusive product of war, politics, and business. She picked the year 1833 to say that was when the word "scientist" was invented; hence there was no science before 1833. She is scornful of the Greeks' "grandiose systems that aimed to explain the cosmos." She denied Archimedes was more than an "armchair philosopher," certainly not a "scientist" or a "technologist."

---

[31] Evaggelos Vallianatos, *Fear in the Countryside* (Cambridge, MA: Ballinger, 1976); *Harvest of Devastation* (New York: The Apex Press, 1994); *This Land is Their Land* (Monroe, ME: Common Courage Press, 2006); and, with McKay Jenkins, *Poison Spring* (New York: Bloomsbury Press, 2014).

Fara repeats the conventional argument about Greek class warfare when, supposedly, the few rich theorized about the Cosmos and the many did dirty work. She argues that Greek scientists like the second-century Ptolemaios failed to "mention not only the skills of Greek artisans, but also their dependence on older techniques originating in Mesopotamia and Egypt." Fara even went so far as to question whether the murder of the Greek mathematician Hypatia by Christian monks in 415 actually happened. She says that Hypatia "was (allegedly) torn to bits by an angry mob."[32] I discuss Hypatia in chapter 9.

With this kind of talk, Fara's credibility drops to zero. Greece gave birth to science and advanced technology. For about nine centuries, from the seventh century BCE to the second century of our era, Greece was a science powerhouse, giving birth to what we call science, including the scientific method, and the infrastructure of both theoretical and applied science. Greek thinkers asked fundamental questions about how the world works. They came up with theories that explained natural phenomena. They invented science.

Here's a synopsis of Greek scientific and technological achievements explored in this book:

Thales was the first Greek natural philosopher who thought of water as the foundation of all things. He flourished in Miletus in late seventh century BCE. His student, Anaximander, spoke of the abounded or *apeiron* as the maker of nature and the Cosmos. He was born in Miletus in late seventh century BCE.

The sixth century BCE gave birth to Pythagoras of Samos. Pythagoras used the number and mathematics in understanding the Cosmos. Herakleitos of Ephesos (late sixth century BCE) said fire was logos and logos was fire. The natural world and the Cosmos were perpetually changing. In contrast, Parmenides (late sixth century BCE) from Elea, southern Italy, found truth only in existing things. What "is" defines his philosophy.

In the fifth century BCE, Empedocles, healer, statesman, and philosopher from Acragas (Agrigento), Sicily, spoke about the four elements making up everything: water, air, Earth and fire. He also added love and strife in the making and governing of the Cosmos. Leucippus and Democritus (fifth century BCE) invented the atomic theory. Hippocrates (fifth century BCE) was the father of scientific medicine.

---

[32] Patricia Fara, *Science: A Four Thousand Year History* (Oxford, 2009) xiii–xiv, 1–39.

Eudoxos of Knidos (first half of the fourth century BCE) was a mathematical genius who created the first geometrical model of the Cosmos. I already said Aristotle was a philosopher of science. He invented biology. He was also the father of scientific method. His student, Theophrastos (early fourth to early third centuries BCE), invented botany. Autolykos of Pitane (fourth century BCE) made contributions to spherical astronomy; Euclid (probably late fourth century BCE), was a genius in geometry. His *Elements* summarizes the Greeks' mathematical knowledge.

Ktesibios of Alexandria, third century BCE, invented geared mechanisms, including sundials. Most sundials were clocks for "reckoning the passage of time and its relationship to the larger universe."[33] Other sundials were as simple as the face of the Sun and its rays of light on the land.

Ktesibios probably inspired Archimedes, a mathematical and engineering genius of the third century BCE. In fact, Archimedes became the greatest scientist the world has ever seen. His mathematical physics opened the eyes of Galileo Galilei and Isaac Newton. Archimedes, in fact, informed the world that Aristarchos of Samos (third century BCE) invented the heliocentric theory of the universe.[34] Herophilos and Erasistratos (turn of the fourth and third centuries BCE) left their mark on medicine, especially human anatomy. They dissected humans and discovered the nervous system. Erasistratos came very close to revealing the circulation of the blood.

Apollonios of Perga (second half of the third century BCE) advocated the epicycle for explaining the spherical movements of the planets. He was also a great mathematician.

Philon of Byzantium, 240–200 BCE, and Heron of Alexandria, c. 62 of our era, followed in the engineering footprints of Archimedes. Hipparchos, the greatest Greek astronomer, flourished in the second century BCE. He set the mathematical foundations of astronomy. Indeed, his astronomy made the Antikythera computer possible. Poseidonios (late second century BCE) and Geminos (first century BCE), working in Rhodes, continued the technological revolution of Ktesibios, Archimedes, Hipparchos, and Poseidonios.

Eratosthenes, c. 275–195 BCE, and Strabo, c. 64 BCE–c. 24 CE, set the foundations of scientific geography. In fact, Eratosthenes calculated the circumference of the Earth and drew the first scientific map of the world.

---

[33] John Noble Wilford, "Antiquity on the Clock," *The New York Times*, October 25, 2016.

[34] Archimedes, *The Sand-Reckoner* 135.8–23.

Pedanios Dioskourides (first century) was the father of pharmacology and herbalism. In the second century of our era, Ptolemaios codified astronomy for the next millennium and a half. Galen of the second century expanded Hippocrates's medicine. The works of Galen taught the Arabs and Europeans medicine for centuries.

Out of the work of Ktesibios, Archimedes, Hipparchos, Poseidonios, Geminos, and other Greek scientists and engineers came advanced engineering and computers. The Antikythera Mechanism of this book comes straight from that tradition.

The works of the Greek scientists educated and inspired Renaissance scientists to create modern science. I learned my mathematics from the *Elements* of Euclid.

I still remember the axiomatic foundations of mathematical theories and the marvel of deductive reasoning and economy of the Greek text: nothing superfluous; all accepted knowledge guiding me to the right conclusion.

The early twentieth-century editor and translator of Euclid, Thomas Heath, was right in saying the *Elements* is "one of the noblest monuments of antiquity," an "immortal work" that is "the greatest elementary textbook in mathematics that the world is privileged to possess."[35]

The *Elements* of Euclid is still at the heart of mathematics, just as the *Almagest* of Ptolemaios was the astronomy textbook of Europe until the 1600s.

Writing in the late twentieth century, Benno Artmann, professor of mathematics at the University of Gottingen, Germany, underlines his enthusiasm for the beauty and enormous value he found in the *Elements*. He says the manner in which Euclid presents his material "has set the standard for more than two thousand years." Just like an architect forms his aesthetic judgment from studying Greek architecture, so does a student of mathematics learn the language of mathematics from knowing the *Elements*. Euclid created the "famous algorithm that bears his name for the solution of specific problems in arithmetic, and he shows us how to master the infinite in its various manifestations."[36]

[35] Euclid, *The Thirteen Books of the Elements*, tr. and ed. Thomas Heath, second edition, 3 vols. (New York: Dover Publications, 1965) I, vii, ix.

[36] Benno Artmann, *Euclid – The Creation of Mathematics* (New York: Springer, 1999) v–vi.

I also remember that in my study of medieval Greek history, we learned of the automata, robot-like devices serving the palaces of the emperors in order to impress or frighten foreign diplomats.

There was also "Greek fire," a military weapon the Greeks used effectively for centuries against their enemies.[37]

Archimedes enriched the stock of military weapons. He even invented solar weapons to burn Roman ships. So, I concluded quickly the Greeks did have a technological culture but it was hidden from view either because the subject was too esoteric and complex or at the hands of biased scholars. But probably the most compelling reason for the cover-up of Greek technology was the ego of Westerners claiming primacy in the development of science and technology.

Then, suddenly, two discoveries, in 1998 and 2005, changed the Greek technological picture and shattered the delusions of those laboring against the Greek technological achievements. As I said in the Preface, the first had to do with the recovery of some lost works of Archimedes and the second was about the "decoding" of the Antikythera Mechanism. This mechanical device is the star example of the Greek golden age of science and technology.

Gearing technology is fundamental. Gears, after all, move power from source to application. In 1985, three scientists working for the National Aeronautics and Space Administration defined the modern importance of gearing this way:

"Gearing and geared transmissions drive the machines of modern industry. Gears move the wheels and propellers that transport us over the sea, on the land, and in the air. A sizable section of industry and commerce in today's world depends on gearing for its economy, production, and livelihood."[38]

These NASA scientists admitted that gearing started with the Greeks, citing the gears of the Antikythera computer. This explains the excitement from its discovery.

I will demonstrate that mechanical models with gears came out of Syracuse, Rhodes, and Alexandria, three great poleis that became the foundation of the Greek golden age of science inspired by Alexander the Great. Yet of all Greek poleis, Athens deserved and received the greatest credit for the achievements

[37] J. R. Partington, *A History of Greek Fire and Gunpowder* (New York: Barnes and Noble, 1960).

[38] John J. Coy, Dennis P. Townsend and Erwin V. Zaretsky, "Gearing" (NASA Reference Publication 1152, AVSCOM Technical Report 84-C-15, Lewis Research Center, Cleveland, Ohio, 1985) 1.

of the Greeks. In 1998, Peter Hall summarized the genius of Athens, which he rightly called the "fountainhead" of Greek civilization:

Athens was first in several arts of civilization. It gave us democracy, "in a form as pure as we are likely to see." It gave us philosophy, "in a form so rounded, so complete, that hardly anyone added anything of moment to it for well over a millennium." Athens also gave us "the world's first systematic written history. It systematized medical and scientific knowledge, and for the first time began to base them on generalizations from empirical observation. It gave us the first lyric poetry and then comedy and tragedy, all again at so completely an extraordinary pitch of sophistication and maturity, such that they might have been germinating under the Greek Sun for hundreds of years. It left us the first naturalistic art; for the first time, human beings caught and registered forever the breath of a wind, the quality of a smile. It single-handedly invented the principles and the norms of architecture, which all its western practitioners have learned and followed (even where they were consciously rejecting them) for the next twenty-five centuries. These achievements have formed a central part of European life in all the centuries since," he wrote.[39]

The culture of Athens influenced the rest of Greece, including Macedonia and the kingdoms that came out of Alexander's empire. Science and technology linked the old and the new. *Techne*, craftsmanship, science, and wisdom, was the linchpin of both, especially the new in the Alexandrian Era. The Antikythera computer comes from this technological culture. I am grateful to Derek de Solla Price, 1922–1983, the first modern scientist and historian of science who opened our eyes to the advanced technology of the ancient Greeks. In 1959 and 1974, he correctly identified the ancient Antikythera computer as a product of scientific technology.

Price urged scholars to use the Antikythera computer as a new text for the reinterpretation of the Greek past. But the Antikythera computer is also a model of high tech serving the public good.

I wrote this book in that spirit. The fragments of the Antikythera astronomical computer inspired me. Fragments A to G, shown below, are the most complex and important in making the Antikythera Mechanism the marvel of technology and astounding compendium of Greek civilization.

[39] Peter Hall, *Cities in Civilization* (New York: Pantheon Books, 1998) 24.

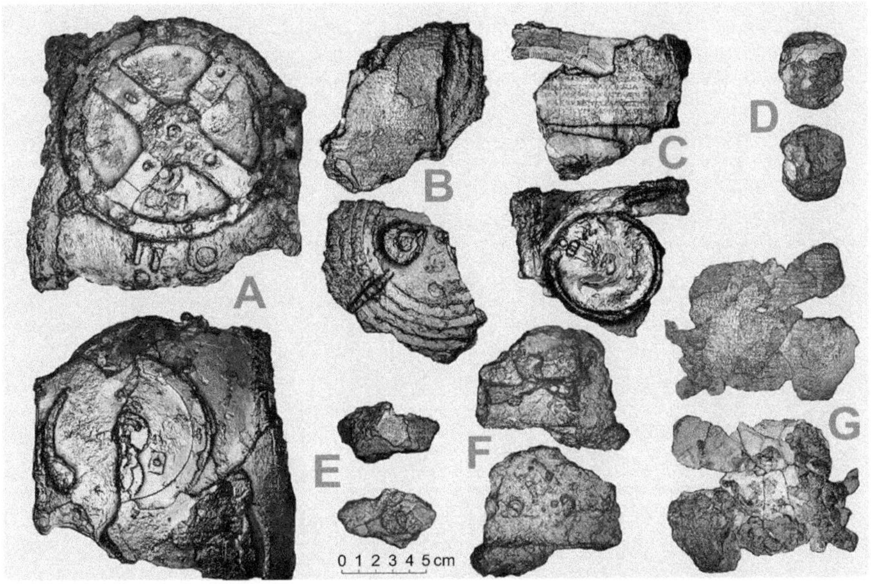

Main fragments of the Antikythera Mechanism. They were imaged by means of Polynomial Texture Mapping. Each fragment, A to G, is seen from the front and back sides. Courtesy Tom Malzbender and Hewlett Packard.

I wanted to know how the scientists of the second century BCE in Rhodes or Alexandria or Epiros in northern Greece translated their beautiful and advanced engineering conception of the Antikythera Mechanism into a computer of genius. That is, how they employed an almost flawless technology to the construction of the bronze-geared device, the first of its kind in the world. My story explains that nearly magical moment by highlighting the technological flowering of the Alexandrian Era. My hope is that the scientific technology of the Antikythera computer puts to rest doubts whether or not Greeks contributed to the history of technology. They did. In fact, they sowed the seeds for the scientific and technological revolutions of the sixteenth and seventeenth centuries. Second, modern science can still model itself on its Greek beginnings and start serving the public good.

Chapter 1
Myths and the Beginnings of Craftsmanship

The Greeks called a skilled workman or handicraftsman *demiourgos*, one who works for people. Homer says that a man who is a master of his craft benefits everyone. Men like that are welcomed everywhere in the world. Among these craftsmen, he includes seers, medical healers, master builders, and singers of tales giving pleasure with their songs and stories.[40]

The Fire of Knowledge

If you enter the large lecture room of the Academy of Athens, you are dazzled by a series of murals on the ceiling and walls. One of those beautiful murals shows Athena and Prometheus creating man.

Creating humans was not enough. Prometheus could see things were not going as he had hoped. He put to use his "fire"—knowledge for the benefit of humanity. This Titan, god of forethought, found the Greeks and other humans living like ants in sunless caves.

Aeschylus, 525–456 BCE, an Athenian tragic poet who reports on Prometheus, says nothing about the time Prometheus took pity on humanity. He did not know. Such time must be counted in hundreds of thousands of years or millennia before Homer.

Aeschylus says that those first Greeks did not know how to build homes facing the Sun or to work with wood. They grew no food and did not recognize

[40] Homer, *Odyssey* 17.382–386, ed. Kaktos Philological Team (Athens: Kaktos, 1992).

the seasons. Prometheus saved them from extinction, giving them the use of their wits, making them masters of their minds. Prometheus showed them the risings and the settings of the stars, teaching them how to count and write; how to take drugs for fighting diseases; how to interpret dreams; and how to yoke and harness animals for hard work or for riding horses, "the crowning pride of the rich man's luxury." Prometheus also instructed the Greeks in the worship of the gods and in setting oracles throughout the land. He showed them how to craft and sail ships and, in general, taught them mining and all the mechanical and other arts and crafts.[41]

Plato also reports that gods came to the assistance of the early Greeks: the fire of knowledge and crafts from Prometheus, more crafts from the god Hephaistos, and seeds and plants from other gods. This vital divine aid enabled the Greeks to stand on their own feet.[42]

Metis, Athena, and Hephaistos: Gods of Science and Technology

The Greeks also had an additional god of technology. This was Athena, daughter of Zeus, who had assisted Prometheus in bringing men into being. She was goddess of war, wisdom, and the arts and crafts of civilization. Her mother was the water goddess Metis, which, in ancient Greek, means counsel, advising, wisdom, cunning, and craft. Metis helped Zeus overcome his father, Kronos. Zeus married her, though he had difficulty in embracing her. She was like water, instantly changing form. Nevertheless, Zeus got her pregnant with Athena. However, his grandparents Ouranos (Sky, Heavens) and Gaia (Earth) warned him that Metis was destined to also have a son who would dethrone him. Zeus immediately swallowed pregnant Metis. That way he safeguarded his supreme position among gods and humans while he had at his disposal the extraordinary intelligence of Metis.

Athena became like her mother, a mighty goddess of intelligence. Like Prometheus, she spurred men in the development of science and technology, especially technical skills and handicrafts for meeting their own needs and improving their welfare. She invented the bridle for the domestication of the

[41] Aeschylus, *Prometheus Bound* 441–504.

[42] Plato, *Statesman* 274b–e.

horse, cattle, donkeys, and mules. This invention enabled humans to feed themselves and create civilization.

Athena also fostered the cultivation of the olive tree, and the teaching and practice of carpentry, metalworking, and pottery while she taught women spinning and weaving.[43]

Writing in the first century BCE, the Greek historian Diodoros of Sicily (Siculus) reported that the Greeks honored Athena and Hephaistos. He explained that the Greeks were grateful to Athena because she helped them domesticate and cultivate the olive tree. This included her instruction on the preparation of the olive fruit.

Diodoros said that before the birth of Athena, the olive tree was wild. Athena taught the Greeks how to cultivate and care for the olive tree. She also taught them to make clothes, work with wood, build homes, as well as master other useful arts and crafts. In addition, she was the inventor of music pipes. In short, Athena invented many sophisticated and cunning devices useful to life.

During the thirteenth day of Pyanopsion, October–November, the Athenians honored Athena and Hephaistos. In this instance, Athena was Athena *Ergane*, the Worker, patron of artisans and craftsmen. Hephaistos was the patron of smiths. Bronze workers, *Chalkeis*, and craftsmen celebrated these two gods in a festival known as *Chalkeia*. The participating craftsmen, engineers, and smiths put their best creations to a competition. Those that won offered their creations to Athena Ergane.

During the Alexandrian Age, the workers who worshipped Dionysos—comic poets and artists in literature, music, dance, and theater—had their own guild. They, too, called themselves craftsmen (*technitai*) but *technitai* for Dionysos. They performed at religious festivals, traveling from city to city where they enjoyed a good living and protection.[44]

Diodoros said Hephaistos was the inventor of metallurgy, teaching human beings the uses of fire for working iron, copper, gold, silver, and everything else requiring fire. He also invented the remaining uses of fire for the benefit of all people, especially those working in the crafts like pottery. As a result, Diodoros says, "The workers who are skilled in these crafts offer prayers and sacrifices to

---

[43] Hesiod, *Theogony* 886–900, 924–926, tr. Apostolos N. Athanassakis (Baltimore, MD: The Johns Hopkins University Press, 1983); *The Homeric Hymns: To Athena* 1–18, tr. Apostolos N. Athanassakis (Baltimore, MD: The Johns Hopkins University Press, 1976).

[44] Christian Habicht, *Athens from Alexander to Antony*, tr. D. L. Schneider (Cambridge, MA: Harvard University Press, 1997) 103.

Hephaistos before all other gods." Workers and all humankind preserve the memory of this god forever and honor him by calling fire "Hephaistos."[45]

As we already explained, intelligence, the foundation of science, was a gift of two goddesses, Metis and her daughter, Athena. Because the word "metis" comes from Metis, goddess of intelligence, it means skill in counsel or device, astuteness, shrewdness, contrivance or scheming, as well as what one plans to do and what one has in mind to do. Thus, metis is deliberation and weighing in the mind, necessary steps in doing science.

Keld Zeruneith, member of the Royal Danish Academy of Sciences and Letters, explained the role of metis in Greek culture:

"Metis lies in the individual's ability to consider and choose lucidly in a critical situation. In this sense, metis is a power of thought and a form of intelligence based on rapidity, acuteness, the ability to change one's mind, to find new paths, to be directed, precise, pensive, strategic, cunning and wise," he wrote.

"It is crucial," he continued, "that [goddess] Metis derives fundamentally from water. This explains why she can change her nature and transform herself like the waves. Since these metamorphoses contain all possible life forms, it provides her with insight into life's vicissitudes, an understanding of the beginning and cohesion in past, present, and future. This is her true wisdom."[46]

Athena's brother, Hephaistos, was older than Athena. He was fire and technology. Homer sings about the modern-like engineering works of Hephaistos, especially the golden women robots serving him at his workshop on Olympus and the extraordinary shield he designed and built for Achilles.[47]

Hephaistos married Kabeiro, daughter of the sea god Proteus. Hephaistos and Kabeiro had the twin sons known to us by their mother's name: Kabeiroi. Like their father, Hephaistos, these gods thrived in works of metallurgy and craftsmanship. They were primarily known in the Aegean islands of Lemnos, Syros, Delos, Chios, Paros, and Samothrace as well as in the coastal regions of Thrace, Macedonia, and Thebes in Central Greece. Their works were so

---

45 Diodorus Siculus, *The Library of History* 5.73.7–8; 5.74.2–3. As I will do to other texts, I modified the translation.

46 Keld Zeruneith, *The Wooden Horse: The Liberation of the Western Mind from Odysseus to Socrates*, tr. from the Danish by Russell L. Dees, ed. W. Glyn Jones (New York: Overlook Duckworth, 2007) 126–127.

47 Homer, *Iliad* 18.368–379, 417–421, 478–617. See Stephanos A. Paipetis, *The Unknown Technology in Homer* (New York: Springer, 2010).

beneficial, Greeks celebrated and worshipped them with mysteries, only second in national importance to the mysteries and worship of the goddess Demeter in Eleusis.

Demeter exerted her influence in the religion of the Kabeiroi, who, in addition to their promotion of craftsmanship, were gods of the natural world.

Handsome silver coin of the mid-second century BCE from the Cycladic Aegean island of Syros shows the Kabeiroi gods. Courtesy Numismatic Collection. Alpha Bank, Athens.

The Great Craftsman and Artist Daidalos

The next great technological innovator in Greek culture was Daidalos (cunning worker). Plato considers him a great craftsman and artist.[48] Aristotle does not deny he existed.[49] In the third chapter, I will return to Plato, who discusses the connection between Daidalos and Socrates. Now, however, I rely on the ancient historian Diodoros of Sicily. Diodoros did not dispute that Daidalos

[48] Plato, *Republic* 7.529d.
[49] Aristotle, *Politics* 1253b23–37.

lived more than a millennium before his time. His only reservation was the story about the "flying" of Daidalos out of Crete. He did not reject the story but explained this was part of what he had heard and read about Daidalos. Diodoros reported the following about Daidalos:

Daidalos was an Athenian who flourished during the second millennium BCE, long before the Trojan War. He was related to the Athenian royal house of Erechtheus. He was so outstanding in everything he did, especially in architecture and sculpture, that his works were masterpieces of art and technology. His statues looked like humans made of stone or marble. However, Daidalos killed Talos, a child inventor, out of jealousy. Talos was his nephew. He invented the potter's wheel, the iron saw, and a kind of lathe for the carpenter to draw circles. These great tools and other "cunningly contrived devices" promised fame to the young inventor. Daidalos could not stand that Talos was becoming such a genius and he killed him.

To avoid the Athenians who charged him with murder, he left for Crete, where he put his masterly art to work for the benefit of King Minos, son of Zeus and great legislator. At that time, Minos used to sacrifice his most handsome bull to Poseidon. However, one of those bulls was so beautiful that Minos replaced him with one of inferior looks.

Poseidon became furious and took revenge. He made Minos's wife, Pasiphae, fall in love with the beautiful bull. Pasiphae turned to Daidalos, who built a cow-like machine that allowed her to have sex with the bull. Pasiphae gave birth to a monster: the Minotaur, up to his shoulders a man, with the head of a bull. Daidalos then built the labyrinth, a building with interlaced corridors, to keep the monster in prison. And as long as Athens paid tribute to Crete, seven young men and seven young women every year, the keepers of the Minotaur fed him Athenian youth.

Eventually, Daidalos learned that helping Pasiphae to mate with the bull was not such a good idea. Minos wanted his head. Minos offered money for his capture while his troops were searching any departing ship from Crete. Daidalos faced his opponent's overwhelming power with another one of his spectacular inventions. This time he took himself and his son Ikaros to the air. He made wings for both of them and flew away from Crete.

Ikaros, thoughtlessly, flew too high and the Sun melted the wax holding his wings together. He fell to the sea and drowned. Daidalos made it to Sicily where, like in Crete, he became the chief engineer to the king, Kokalos. He built a river dam; an acropolis on a rock that was impregnable to attack; he turned a cave into a steam healing facility; and he constructed a golden ram that had the "perfect image of a real ram."[50]

Palamedes: Inventor of Writing and the Arts of Civilization

Daidalos had a worthy successor named Palamedes. Aeschylus, Sophocles, c. 496–405 BCE, and Euripides, c. 485–406 BCE, the greatest dramatic poets of Athens, wrote Palamedes plays, all lost. From the surviving fragments of those works, we know the dramatic poets speak highly of Palamedes, another Daidalos-like inventor. Indeed, his name, Palamedes (Παλαμήδης), means the inventor.[51] If a man invented something, he became worthy of Palamedes. In Greek: Palamedeios.

Palamedes was the son of Nauplios, who was the son of Poseidon. Nauplios was one of the Argonauts who lived sometime during the late thirteenth century BCE. His son, Palamedes, was a soldier in the Trojan War. He came into conflict with Odysseus. Indeed, Odysseus convinced Diomedes, another Greek hero of the Trojan War, to assist him in murdering Palamedes. One story says Odysseus hated Palamedes because Palamedes uncovered his scheme not to join the Greeks for the Trojan War. An oracle had informed Odysseus that joining the Trojan War would cost him twenty years of his life. So, Odysseus pretended madness to avoid induction to fighting for Helen against the Trojans. But Palamedes, representing Agamemnon and Menelaus, the two brother kings responsible for the Trojan War, discovered Odysseus's trick and embarrassed him into joining the Greek armies against the Trojans. Odysseus never forgave Palamedes and, at the appropriate time, killed him.[52]

Another story reports Odysseus accused Palamedes of treason. In this instance, Odysseus faked a letter from the king of Troy, Priam, to Palamedes. Priam offered gold to Palamedes for spying for him. Meanwhile, Odysseus had hidden gold in the very tent of Palamedes. The discovery of that gold convinced Agamemnon that Palamedes was a spy. The penalty for treason

[50] Diodorus Siculus, *The Library of History* 4.76–78, tr. C. H. Oldfather (Loeb, 1933–1967).

[51] Pausanias, *Guide to Greece* 2.20.3.

[52] Pausanias, *Guide to Greece* 10.31.1.

was death. Greek soldiers stoned Palamedes to death.[53] The Roman poet Virgil says Palamedes was a victim of the "guile and envy" of Odysseus.[54] Despite the importance of Palamedes, Homer says nothing about him.

The fame of Palamedes came not from his conflict with Odysseus but from his technological ingenuity and achievements. Sophocles credits him with building a fortification wall and establishing communications by fiery beacons for the Argive army. He invented the battle array, writing, astronomy, numbers, counting, weights and measures, board games, and battle strategies, and assisted the Greeks in the management of their food. According to Sophocles, Palamedes decoded the heavens, explaining the periods and revolutions of the heavenly spheres, thus facilitating sailing by the stars. He was the first to find out how to count from one to ten, from ten to fifty, and from fifty to a thousand.[55]

Palamedes described his greatest contribution, the invention of writing, this way:

"I invented remedies for forgetfulness. These remedies include syllables that are without speech and, yet, speak. I invented writing for men's knowledge, so a man absent across the ocean might have reliable knowledge of all matters back there in his house. The dying man might write down the size of his wealth when bequeathing it to his sons, and the sons know it. Finally, the troubles that afflict men when they start quarrelling—a written tablet does away with such quarrel and prevents the telling of lies."[56]

Gorgias, a rhetorician and philosopher born around 483 BCE in Leontini, Sicily, confirms the virtues Sophocles and Euripides assigned to Palamedes. Gorgias wrote a speech defending Palamedes against the accusations of Odysseus that during the Trojan War, Palamedes committed treason. Gorgias has Palamedes say the following to the jury about himself, which is remarkably similar to what Aeschylus said of what Prometheus did to help the early Greeks.

---

[53] Apollodoros, *Library* Epitome 3.6–8, tr. Michael Simpson (Amherst: University of Massachusetts Press, 1976); Hyginus, Fabulae 105, ed. P. K. Marshall (Munich/Leipzig 2002), in *Anthology of Classical Myth*, ed. Stephen M. Trzaskoma et al. (Indianapolis/Cambridge: Hackett Publishing Company, 2004) 247–248. A medieval Greek scholar of the twelfth century, John Tzetzes, also summarizes the history of the hatred of Odysseus against Palamedes: *Allegories of the Iliad* Prolegomena 871–874, 927–928, 961–996, 1030–1166, trs. Adam J. Goldwyn and Dimitra Kokkini (Cambridge, MA: Harvard University Press, 2015).

[54] Virgil, *The Aeneid* 2.122.

[55] Sophocles, *Fragments* Nauplius Fragment 432, ed. and tr. Hugh Lloyd-Jones (Loeb, 1996).

[56] Euripides, *Fragments* Palamedes 578, ed. and tr. Christopher Collard and Martin Cropp (Loeb, 2008) 46–59.

Palamedes tells the soldiers he advanced the cause of humanity and civilization with his discovery of military tactics. In addition, he takes credit for "written laws, the guarantees of justice; writing, the instrument of memory; weights and measures, the convenient means of commercial exchange; number, the guardian of goods; powerful beacons and very swift messenger services—and, last but not least, droughts, a harmless way of passing the time."[57]

Plato also gives credit to Palamedes for inventing numbers and counting while contributing to the organization of the army in battle formation.[58]

Plato mentions Palamedes in several of his works and gives him a preeminent role in the recovery of human fortunes after the last destructive flood before his time. Plato says that floods and plagues pushed humans close to extinction several times, each era after a global catastrophe resorting to reinventing civilization. So, about a thousand years before his time, Plato credits Daidalos, Orpheus, and Palamedes for life-saving discoveries that helped the Greeks rebuild their society and culture. Plato says that Marsyas and Olympus made pioneering contributions to the art of music; and Amphion invented the lyre.[59]

We know practically nothing about Olympus. Marsyas and Amphion were extraordinary musicians. Marsyas was also a satyr and a follower of the god Dionysos. He made the fatal mistake of challenging Apollo, god of music, to a musical contest.

Marsyas lost and a Scythian referee flayed him alive. A tributary of the River Meander in Asia Minor bears the name of Marsyas.

Xenophon has Socrates mentioning the unjust death of Palamedes.[60] And Plato has Socrates expecting to meet Palamedes after death.[61]

Blessed Is the Man Who Has Learned History

Sophocles was particularly cautious in matters of technology and power. Man, he said, was ingenious beyond all measure, the conqueror and master of wild animals, fish and birds, horses and mountain bulls. With his ceaseless

[57] Gorgias of Leontini, *Defense of Palamedes* 28–30 (tr. John Dillon and Tania Gergel, *The Greek Sophists*, London: Penguin Books, 2003).

[58] Plato, *Republic* 7.522d.

[59] Plato, *Laws* 3.677a–d.

[60] Xenophon, *Apology* 26.

[61] Plato, *Apology* 41b.

cultivation of the land, man is even wearing down the oldest of the gods, Earth. Traditions, the laws of the land, and divine justice temper man's technical ingenuity and, often, reckless daring and conquests. For this reason, Sophocles concluded, there were many terrible, powerful, and wondrous things in the world, but none more terrible, more powerful, and more wondrous than man. He used his cunning and inventive mind for evil or for good.[62]

Like Sophocles, Euripides had penetrating insights into human nature. He praised those who asked questions, using their minds in searching and learning from the natural world. He admired history when history meant both narrative of events and investigation for the sake of knowledge, the basis of science. He described his enthusiasm about history/science as follows:

"Blessed is the man who has learned history—gaining knowledge of the world through science. Such a man has no intention of either harming his fellow citizens or acting unjustly. Instead, he observes and studies the order of immortal and ageless nature, its origins and how it was formed. Men with this knowledge are never tempted to do shameful things."[63]

This is Euripides at his best: profound and straightforward, catching the rising scientific spirit of Greek civilization.

Hippocrates probably shared Euripides's affection for humans immersed in nature. Hippocrates was also a defender of human health and science. Science makes healing possible, and healing is necessary for a viable society. He urged his physician colleagues to serve the ill, often offering their services for nothing. He was convinced that love for people and science and technology go together.[64]

Hippocrates also decided that, of all things affecting human survival and health, good food was the most important. Food was medicine and medicine was food. Wheat and barley were two of those essential medicinal foods. The Greeks made bread with wheat and cakes with barley. Hippocrates put good food at the center of his healing art or medicine, *ietriken*. This was the application of reason and science in the service of human wellbeing and welfare. The result was the discovery of good food, which replaced a bad diet that caused pain, disease, and death.[65]

---

[62] Sophocles, *Antigone* 332–375, ed. Mark Griffith (Cambridge Greek and Latin Classics, 2003).

[63] Euripides, *Fragments* Antiope, Uncertain Fragment 910 (ὄλβιος ὅστις τῆς ἱστορίας ἔσχε μάθησιν...) tr. mine.

[64] Hippocrates, *Precepts* 6.1–7.

[65] Hippocrates, *Ancient Medicine* 3.31–54.

The Water Aqueduct of Samos

Good food and human survival depend on water. Herodotus praised the citizens of the island of Samos in the Aegean for digging through a mountain and bringing water to their polis. He also said the citizens of Samos dug a meter-wide and nine-meter-deep channel along the length of the tunnel. This carried piped water from a large spring to the town. The master builder of this tunnel was an engineer from Megara called Eupalinos. Another achievement of Samos was a mole in the sea. This enclosed the harbor in deep water. The third thing they built, Herodotus says, is the "largest temple ever seen." A local architect by the name of Rhoesos designed the temple.[66]

Eupalinos of Megara, 550–500 BCE, was an architect and engineer who built the water aqueduct of Samos sometime before 530 BCE.

Wealth, Trade, Crafts, and Science

A junior contemporary of Herodotus, the comedian Aristophanes, c. 445–c. 385 BCE, had something profound to say about the origins of technological inventions and crafts. He put money at the root of all power and, therefore, technology. Money explains why one man cobbles shoes, another works in bronze and someone else in wood; and another washes sheepskin or sells onions. According to Aristophanes: "If wealth regained his sight and spread around equally, men would abandon trade and science. If trade and science disappeared from the world, who would be interested in working metal, building a ship, sewing, making a wheel, working with leather, making bricks, doing the wash, tanning hides, or cultivating the land with the plough in order to reap the fruits of Demeter? Yes, why would anyone work if it were possible to live in idleness without a thought for any of these things?"[67]

The Athenian rhetorician Isocrates, 436–338, did not share the critical ideas of Sophocles or Aristophanes on the uses of science and technology. He was a booster of the technological advantages of civilization one could expect and enjoy, especially in Athens. He explained why: Athenians are so much better off than other Greeks in thought and speech that they become the teachers of others. It's not surprising, he said, that Athens has made the name "Greek"

[66] Herodotos, *The Histories* 3.60.

[67] Aristophanes, *Wealth* 160–164, 166–167, 510–516.

more appropriate to a way of life. Greeks are those who share in our education and culture (*paideusis*) rather than in our birth.[68]

Isocrates gave a dazzling definition of being Greek, which echoed the achievement of Greek civilization for the next two millennia. His thought followed the evolutionary model of Hesiod, the champion of rural Greece. This was not accidental, as advances in technology found their most useful application in rural Greece.

Wheat and barley and olives and the grapevine saved Greeks from the deadly effects of bad diet. Milling grains with millstones improved the quality of the flour. By the third century BCE, Greeks used pairs of rounded stones to grind their wheat and other grains. The stones turned with the employment of a crankshaft and pilot and handle. Eventually, perhaps in the early first century BCE, the Greeks invented and used water mills for grinding their grains.

Antipater of Thessalonike wrote a poem in which he celebrated the use of waterpower in making flour. He said that slave girls could sleep late now that Demeter had ordered the Nymphs to run the water mill for grinding grains.[69] Similar techniques improved the extraction of the precious oil out of the stone fruit of olives and wine out of the grapes. Olive presses evolved from flat beds with stonerollers to moving in a circular manner.

Equally important in the life of the farmers, that is, the lives of most Greeks, was stock breeding for healthier and stronger domesticated animals and the selection of seeds for better, disease-resistant, and higher-yielding food crops.

Theophrastos, student of Aristotle, a great botanist and agronomist, explained the place of food crops in nature, including the role insects played in farming. He suggested what farmers could do to get a good harvest. He explained the diseases that the weather brought to trees and described the diseases of pulses and cereals. Like Hippocrates, he clearly thought that the influence of the climate and soil on farming was decisive. He also reported how vegetables fight insects and weeds.[70]

These extraordinary scientific insights, following on the original biological contributions of his teacher, Aristotle, were some of the first ecological principles and biological pest management observations in history.

---

[68] Isokrates, *Panegyrikos* 32–33, 40–42, 47–50.

[69] *The Greek Anthology* 9. 418.

[70] Theophrastos, *History of Plants* 8.7.6–7; 2.2.7–12; 4.14.2–14; 7.5.4–6, ed. Kaktos Philological Team (Athens: Kaktos, 1998).

Theophrastos had a lasting influence among farmers, botanists, and students of the natural world for several centuries. Farmers increased their technical skills. The Roman writer Pliny the Elder reported in the late first century that farmers "have tried every possibility" in improving their trees. Grafting, Pliny said, has been perfected.[71]

To this very ancient tradition of technical innovation, polytheism, which blended nature and the gods, was an asset. It favored technics and science/philosophy.

[71] Pliny, *Natural History* XV.57, tr. John F. Healy (London: Penguin Books, 1991).

Chapter 2

A Sense of Wonder

As for many ancient people, astronomy was seminal to the culture and perspective of the ancient Greeks. Much of their relationship to the natural world derived from their study of the stars and constellations. This chapter delves into the great pioneers of astronomy, the natural world, and the Cosmos.

Anaximander and Pythagoras, for example, started the Greeks thinking about a geometrical Cosmos of order and harmony. They started using and advocating mathematics as a language to read the natural world and the Cosmos. Other early thinkers made additional contributions to natural philosophy, including the discovery and articulation of the atomic theory of Democritus and the heliocentric cosmological theory of Aristarchos of Samos. This great tradition gave rise to the invention of the Antikythera Mechanism, which I will discuss in chapter 5.

Mythic Language for Reading the Heavens

Plato spoke about "a sense of wonder [*to thaumazein*, τό θαυμάζειν], being the proper passion of a philosopher. In fact, philosophy has no other foundation."[72] Aristotle, too, thought the origins of philosophy and science lay in our wondering, especially our interest in explaining the celestial phenomena of Helios (Sun) and Selene (Moon), and the stars and the Cosmos. He said

---

[72] Plato, *Theaetetus* 155d (μαλα γαρ φιλοσόφου τούτο το πάθος, το θαυμάζειν, ου γαρ άλλη αρχή φιλοσοφίας η αυτή).

the man who wonders considers himself ignorant. The philosopher seeks truth "in order to escape from ignorance," and pursues "science in order to know," not for any utilitarian purpose.[73]

Like the people of so many ancient cultures, the early Greeks had a rich tradition of ascribing mythological explanations to the origins and powers of the heavenly bodies. These cosmic entities took the form of their gods who influenced the culture of humankind through their actions. However, instead of worshipping the gods blindly, Greek philosophers and scientists used the complex myths surrounding the gods to create new theories of the Cosmos. Myths gave them a language with which they could begin to explain the natural world around them.

Greek myth began with the pre-Olympian, elemental gods of the universe they called Titans. Signifying the fundamental portions of the Cosmos, the great gods Ouranos (Sky) and Gaia (Earth) gave birth to twelve children, one of whom was Hyperion. With another Titaness, Thea, he fathered the two most important heavenly bodies: Helios and Selene. Helios succeeded his father in the heavens and his sister Selene succeeded her aunt, Phoebe, the first Moon goddess. Helios bathes the world with light by driving his four-horse "golden-yoked" chariot across the sky from east to west. Selene soothes the darkness of night with her gentle rays.

The Greeks knew that neither humans nor nature could survive without the warmth and light of the Sun. The Sun ripened their crops. The rising and setting of the Sun ordered their days. The light of the Sun allowed them to see, to work, and to create. Plato traced the meaning of the name "Helios" to the Doric word *halios*, meaning "the gathering of people observing the rising of the Sun and its ceaseless rolling while moving around the Earth." Plato also explained that the name of Selene came from *selas* (bright light) and *phos* (light).[74] The third- or fourth-century novelist Heliodoros of Emesa, Syria, described Helios and Selene as "the purest and brightest of the gods."[75]

---

[73] Aristotle, *Metaphysics* 982b12–27.
[74] Plato, *Kratylos* 409a–c.
[75] Heliodoros, *Ethiopian Story* 10.4, 41, tr. Walter Lamb and ed. J. R. Morgan (Everyman, 1997).

Helios: Copper coin, Malaga, Spain, late 3rd century BCE.
Courtesy Numismatic Collection, Alpha Bank, Athens.

Perhaps equally important to the origins of Greek cosmogony, cosmology, and astronomy was another Titan, Atlas (one who endures or carries). He was the son of the Titans Iapetus (the piercer or craftsman) and Clymene (daughter of Oceanus, the ocean). He was the guardian of the pillars that held up the sky. He is often pictured as holding its great burden on his shoulders.

Atlas was much more than a mighty god. He supported the entire understanding of the Cosmos on his shoulders. According to the Roman architect and engineer of the first century BCE, Vitruvius Pollio, Atlas taught humankind "the Sun's course and the revolutions of the Moon and all the stars."[76] Indeed, according to the Greek historian Diodoros, who was a contemporary to Vitruvius, Atlas was thought to be the founder of science itself. He is said to have perfected astrology and established astronomical science. He invented the movement of the heavens and taught that the stars and the heavens have the shape of spheres. He is said to have written the first astronomical textbook as well.

When Atlas met demi-god Herakles, he educated him in the science of the heavens, passing on to him all that he knew about Cosmos symbolically

[76] Vitruvius, *On Architecture* 6.7.6, tr. Frank Granger (Loeb, 1985).

by getting him to accept the weight he bore. Herakles freed himself and then brought that knowledge to Hellas.[77] Thus, Herakles won the affection of the Greeks not only as their greatest hero, but as one of the first philosophers and scientists to open the nature of the heavens to them.

The creation of the constellations continued when Atlas married Pleione, an Oceanid nymph. She gave birth to seven daughters at Cyllene in Arcadia of Peloponnesos. These girls took their mother's name and became the Pleiades: *Alcyone, Merope, Kalaino, Elektra, Sterope, Taygete,* and *Maia*.[78]

After Zeus and the Olympian gods overthrew the Titans, Zeus placed the seven daughters of Atlas in the constellation Pleiades, from the verb *pleo*, to sail. Pleiades rose at the beginning of sailing season in the Mediterranean. In the *Odyssey*, they guide Odysseus home to Ithaca.

The entire pantheon of the stars we have come to know derives from the Olympian Greek gods. The myths of the creation of each constellation represent powerful stories of the population of the heavens. In due course, I will argue that those constellation myths were rooted in practical and metaphysical concerns about the Earth and the Cosmos.

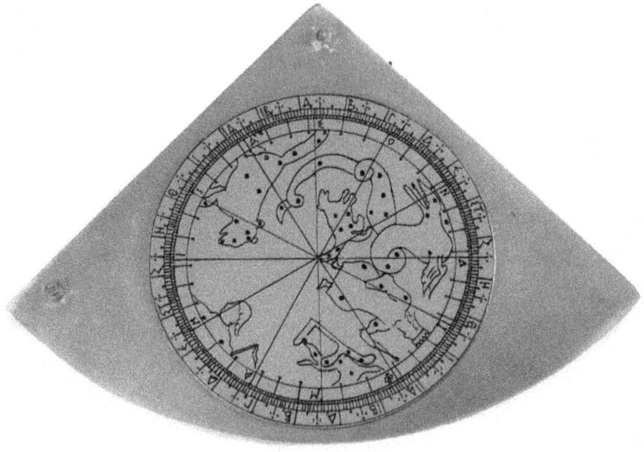

Constellations in the back side of Hipparchos' quadrant (Plinthis). We see Arcturus (bear) and Draco (snake) between the great and little bear. Herakles and his club stand ominously over Draco. Courtesy Dionysios Kriaris.

[77] Diodoros, The *Library of History* 3.60.1–2; 4.27.4–5.

[78] *Gods and Heroes of the Greeks: The Library of Apollodoros* 7.3.10, tr. Michael Simpson (University of Massachusetts Press, 1976); Diodoros, *The Library of History* 3.60.4–5.

The names of the constellations, for example, give us clues of natural philosophy that preoccupied the Greeks from time immemorial. Constellations represent the earliest Greek speculation that eventually crystallized into astronomy.

Not merely Homer but Hesiod mentions the constellation Pleiades. The god of metallurgy, Hephaistos, puts the constellation of the seven sisters on the Shield of Achilles. The moment Kalypso allows Odysseus to leave her island, Odysseus keeps looking at the Pleiades as his guide to finding his home island, Ithaca.

Seeing the Pleiades means the start of the harvest, a real achievement in scientific observation. The Pleiades attracted the attention of Orion, a giant hunter from Boeotia. Several gods came to the assistance of the Pleiades. Dionysos blinded Orion. Then Artemis or Gaia killed Orion, who unwisely either bragged he could defeat Artemis in a contest of discus throwing or he boasted he would kill all wild animals, in which case, Gaia sent a scorpion that stung him to death.

The constellation Hyades may have also been daughters of Atlas. The name Hyades comes from the verb *yo*, which means sending rain; so, the constellation Hyades is related to the rainy season. Greek poets thought of the Pleiades as doves and the Hyades as swine.

The other major constellation, Arcturus, the Bear or Ursa Major, is also known as Bootes, the plowman or farmer as he stands behind a bullock, cow, or ox to cultivate the land. Dionysos, son of Zeus and god of wine, honored his earliest follower Ikarios, his daughter, Erigone, and their dog, Maira, by putting them in their own constellations next to Bootes. Erigone became Virgo and Maira, Canis Minor, the Lesser Dog.

Castor was mortal, the son of Tyndareos, king of Sparta. His mother, Leda, was daughter of the Aetolian king Thestios. Polydeukes was an immortal demigod, the son of Zeus and Leda. The brothers were inseparable. When Castor was dying from a lethal wound, Polydeukes begged his father, Zeus, to let him share his immortality with Castor. Zeus agreed and immortalized Castor and Polydeukes, the Heavenly Twins, in the stars of the constellation Gemini.

The constellation Aries includes the ram that carried Phrixos, son of the cloud goddess *Nephele*, from Orchomenos in Boeotia to Aia, the capital of Kolchis in the far east of the Black Sea. Kolchis was the kingdom of Aietes, son of the Sun god Helios. Phrixos sacrificed the ram to honor Zeus. Its skin was the Golden Fleece that *Iason* (Jason) and the Argonauts brought to Greece.

Sailing with the Argonauts, Herakles killed the snake *Ladon* guarding the apples of Hesperides. Zeus immortalized *Ladon* in the constellation Draco, between the Great and Little Bears. Herakles, his club ominously hanging over the head of Draco, is in his own constellation next to the snake. Another constellation related to Zeus was that of Taurus, the bull-like form Zeus took to abduct Europa, daughter of Agenor, son of Poseidon and Libya. Zeus, as a handsome bull, brought Europa to Crete where he made love to her and she bore him three sons: Minos, Rhadamanthys, and Sarpedon.

In chapter 1, I recounted the story of Minos and Pasiphae getting married. Pasiphae, however, cheated him. She mated with a bull and gave birth to the monster Minotaur: having the body of a man but the head of a bull. Ariadne, daughter of Minos, helped the Athenian prince Theseus to kill the Minotaur. Yet Theseus left Crete for Athens without Ariadne, who had fallen in love with him. Dionysos took Ariadne to Olympus, where he made her immortal, giving her as a wedding present a golden crown made by Hephaistos. Dionysos then set this crown as the constellation Corona Borealis.

One of the constellations is associated with Troy, the land of fine horses, according to Homer. One of the Trojan kings, Tros, married *Kallirhoe*, daughter of the river god *Skamandros*. One of their sons, *Ganymedes*, was so handsome that the gods brought him to Olympus as a cupbearer to Zeus. In time, Zeus immortalized *Ganymedes* as the constellation Aquarius, the Water-carrier.[79]

Isaac Newton tried to date the events of ancient history with astronomical calculations. He studied Greek mythology and he convinced himself that most of the constellations represented persons and events before the Argonaut expedition. He assigned the invention of constellations and the first heavenly sphere among the Greeks to Cheiron the Centaur, teacher of the greatest Greek heroes, including Herakles and Achilles.

Newton studied the astronomers Eudoxos, Aratos, Hipparchos, and Ptolemaios. He said Cheiron built a "primitive" sphere in 939 BCE for the Argonauts' perilous journey to Colchis for the recovery of the Golden Fleece. Newton also says that one of the Argonauts, Mousaios, constructed a heavenly sphere on which he painted the constellations of Cheiron to guide the Argonauts to "sail by the stars."[80]

---

[79] Jenny March, *The Penguin Book of Classical Myths* (London: Allen Lane, 2008) 121–297.

[80] Frank E. Manuel, *Isaac Newton Historian* (Cambridge, MA: Harvard University Press, 1963) 78–87.

The Greeks invented the constellations to understand the heavens and the natural world on Earth. They associated the constellations with the seasons of winter, spring, summer, and fall. But before these astronomical and agrarian divisions of the year became accepted as distinct periods linked to weather and harvest, the seasons meant something much broader.

Out of the chaos of time, the gods began to put order into the Cosmos. The seasons were ideas for the establishment of civilized communities on Earth. That original concept of the Greeks' view of the seasons went all the way to the beginnings of their mythology.

Cosmic Justice and Equality

The second wife of Zeus was the Titaness Themis, goddess of law and order. Themis also means customs established in relation to the unchangeable order of the universe. The children of Zeus and Themis were the three *Horai*, Hours: *Eurynome* (Lawfulness); *Dike* (Justice); and *Eirene* (Peace).[81] Eventually, the *Horai* became associated with more concrete divisions of time: the year, the month, the day, and the hours.

Theokritos of Syracuse, the poet who lived in the early third century BCE in Alexandria, called the *Horai* "soft-footed" and "beloved," the "slowest of the immortals" bringing gifts to human beings.[82] Indeed, time, Chronos, was a god with interest and powers over the fate of human beings.[83]

Perhaps, it was "radiant" Themis, mother of the seasons, who inspired Greek philosophers to add justice and equality to enrich their scientific understanding of the workings of the heavens. This happened perhaps because Cosmos in Greek signifies a beautiful and just world order defended by the father of the gods, Zeus.

Zeus accorded his fellow divinities respect and equanimity. This immutable world order made equality inevitable and necessary for the rule of justice in nature and the heavens.

---

[81] Hesiod, *Theogony* 901–903.

[82] Theokritos, *Syracusan Women at the Festival of Adonis [in Alexandria]: Idyll* 15.101–105, ed. and tr. A. S. F. Gow (Cambridge: Cambridge University Press, 1950).

[83] *The Greek Anthology* 7.245.

This insight inspired philosophers to add the concept of cosmic equality to their understanding of the universe. Cosmic equality underlined the vast differences between the immortals of the heavens and humans. Even among gods, the supreme ruler, Zeus, had to step closer to the equality of the rest of the divinities. An early ancient Greek natural philosopher, Anaximander, spoke about cosmic equality and justice. I will say more about him shortly. Gregory Vlastos, a twentieth-century scholar of ancient Greek philosophy, described this cosmic equality as the "naturalization of justice" and "new commonwealth of nature," which became a model for Greek democracy.[84]

This natural philosophy had its roots in the earliest period of Greek culture. It emerged as one of the core interests of the thinkers of the sixth and fifth centuries BCE. Natural philosophy also became part of the theory of health. According to Alkmaion of Kroton, Italy, a Greek physician who flourished around 470 BCE: "What preserves health is the equality of environmental factors and food—moist and dry, cold and hot, bitter and sweet. The supremacy of any one of them causes disease ... The cause of disease is an excess of heat or cold causing overindulgence or deficiency of food ... Disease may come about from environmental causes, from the quality of the water, other environmental conditions or hard work or torture. Health, on the other hand, is a harmonious blending of moist, dry, cold and hot weather."[85]

These ideas, including equality in politics, were part of the Greeks' understanding of the Cosmos. The gods, after all, took the form of the stars and phenomena in the natural world. The Greeks made the gods anthropomorphic, the better to become familiar with them. The Earth herself was a goddess. So, the early Greek natural philosophers were stargazers and students of the Earth. They tried to figure out the structure and purpose of a divine universe and how it related to their lives on Earth.

[84] Gregory Vlastos, "Equality and Justice in Early Greek Cosmologies," Classical Philology, Vol. 42, No. 3 (July 1947), 156–178.

[85] Aetios, "On the Opinions of the Philosophers" V.30.1 in *Greek Medicine from the Heroic to the Hellenistic Age: A Source Book*, ed. James Longrigg (New York, Routledge, 1998) 31.

Thales: First in Astronomical Wisdom

Thales was born in Miletus (c. 624–545 BCE). Some Greeks thought he was the "first in wisdom among astronomers."[86] Thales linked the natural world with the Cosmos.

The great contribution of Thales was his theory that water was the basic stuff of life, matter, and the universe. He postulated that it was such a fundamental element of the Cosmos that the Earth floated on it, keeping it in place.

Plato thought of Thales as a practical philosopher and inventor skilled in craftsmanship.[87] Aristotle theorized that Thales got this idea about water from seeing that the nourishment of all things is moist and the seeds of all things have a moist nature, water being the source of moisture.[88]

Anaximander: The Boundless Power of Nature

Anaximander of Miletus (c. 610–545 BCE) was a student of Thales. But unlike Thales, he did not think a concrete substance like water could give rise to the Cosmos. He envisioned instead an invisible, everlasting, divine force he named *apeiron* (boundless). This indefinite, boundless, indestructible, and immortal *apeiron* is the first principle and mother of all things. It is in motion, a condition conducive to the separation of something from it.

That "something" gave birth to natural forces such as hot and cold. The hot became fire, light, the Sun and the stars. The cold became dark mist, air and The Earth. Hot and cold were moist but the heat of fire made them dry. These opposite forces, acting on each other, gave rise to change and life.

Anaximander was convinced justice and equality mingled with hot, cold, dry, and moist. These were eternal cosmic forces and the heavens, no less than the polis, would be inconceivable without them. The equity, symmetry, and reciprocity of these forces safeguarded the order of the Cosmos and natural conditions that governed it: heat and cold, moist and dry, light and dark, substance and void. No one of these forces could dominate the rest because they

---

[86] *The Greek Anthology* 7.83.

[87] Plato, *The Republic* 10.600a.

[88] Aristotle, *On the Heavens* 294a28–33; *Metaphysics* 983b6–27.

pay a steep price. Necessity and justice see to it that changes on Earth and the Cosmos are never capricious.[89]

After all, a perfect and divine power like the *apeiron* had to live up to its reputation. So, Anaximander incorporated it *into* nature, making nature the almighty queen of the universe: boundless; eternal; present in everything, and hidden. Boundless nature was nature with boundless power.[90]

Anaximander's almighty nature set order in the heavens. The stars formed a circle of fire separated from the fire in the Cosmos by dark mist. In addition, circles and spheres supported the spinning stars in the heavens. In other words, Anaximander thought of the stars like rings of fire, very much like chariot wheels bathed in mist or air. These chariot wheels had holes for the movement of fire and the appearance of the stars. The Sun has its ring of fire. Another ring accommodated the Moon and another the fixed stars.[91]

Anaximander put the Earth at the center of these three concentric rings: that of the Sun, the Moon, and the fixed stars. This means the Earth at rest was in the middle and center of the celestial sphere because it was equidistant from all points of the circumference of the celestial sphere. That way, being at rest, in equilibrium, kept the Earth free of domination.

The Earth was at the center of the universe. It could not move up or down or sideways. Neither could the Earth move simultaneously in opposite directions. The Earth, Anaximander says, is at rest because of its similarity to the Cosmos and because of necessity.[92]

Next to these cosmological contributions, Anaximander was equally ingenious and original in explaining the origins of humans. He invented the theory of biological evolution, saying in effect humans evolved from other animals: nearly two and a half millennia before Charles Darwin. He explained the origins of life in the primeval seas: "The first animals were produced in moisture, enclosed in thorny barks." He postulated that evolution of humanity as well:

"In the beginning humans were born from other kinds of animals, since other animals quickly manage on their own, and humans alone require lengthy

---

[89] Simplicius, *Commentary on Aristotle's Physics* 24.13–21; Aristotle, *Physics* 3.4 293b10–15 in *A Presocratics Reader*, ed. Patricia Curd and tr. Richard D. McKirahan (Indianapolis: Hackett, 1995) 12.

[90] Radim Kocandrle and Dirk L. Couprie, *Apeiron: Anaximander on Generation and Destruction* (Cham, Switzerland: Springer Nature, 2017) 99–100.

[91] Hippolytus, *Refutation* 1.6.3–5 in A *Presocratics Reader* 13.

[92] Aristotle, *On the Heavens* 295b10–16.

nursing ... There arose from heated water and Earth either fish or animals very like fish. In these [fish] humans grew and were kept inside as embryos up to puberty. Then finally they burst out and men and women came forth already able to nourish themselves."[93]

The Agora Shone Like a Star in the Sky

Anaximander and other early natural philosophers brought their political insights of justice and equality to their understanding and explanation of the workings of the Cosmos. That was inevitable. It's impossible to separate experience from thinking about the universe. In addition, these thinkers were experimenting with a new political system.

Long before the sixth century BCE, democracy was gaining ground all over the Greek world. For example, Euripides has Theseus, the greatest hero of Athens, defending democracy: Athens, Theseus says, is free. People, not the rich, reign.[94] Theseus lived sometimes in the thirteenth century BCE, a generation before the Trojan War.

In the island of Samos around 510 BCE, the tyrant Polycrates died and willed the power of the state to Maiandrios. However, Maiandrios did not follow in the footsteps of Polycrates. He did not want anything to do with tyranny. He immediately erected an altar to Zeus *Eleutherios* (Zeus the Liberator) and told the citizens of Samos that he placed state power in the middle and declared *isonomia* (equality before the laws) for them.[95]

That experience of abandoning the arbitrariness, violence, and dominance of monarchy, tyranny, and oligarchy for democracy was vital for freedom and self-respect. It marked them, shaping their political and, therefore, scientific thought. And, interestingly enough, democracy found most of its adherents in the agora.

Agora (assembly of the people, gathering place) comes from the Greek verb *ageiro* (to gather together). Agora was a very old political concept and institution. We come across agora in the Homeric epics, which dates it all the way to the second millennium BCE.

[93] A *Presocratics Reader* 11–14.

[94] Euripides, *The Suppliant Women* 4005–408.

[95] Herodotos, *The Histories* 3.142.

For example, the second book of the *Odyssey* of Homer gives us a glimpse of the beginnings of agora. The story opens with the son of Odysseus, prince Telemachus of Ithaca, "handsome as a god," ordering the heralds to summon an assembly. When "flowing-haired Achaeans" take their place in a circle outside the palace of Odysseus, Telemachus rises up and denounces the suitors for plaguing his mother.[96]

The roots of the Athenian agora lead us to tyrants and democrats alike. Tyrant Peisistratus grasped the political significance of agora and in the 560s BCE built two shrines dedicated to Zeus Agoraios (Zeus of the Agora) and Apollo Patroos (Apollo of our Fathers).

Plato calls the agora a sacred ground.[97] It was that and much more. After the reforms of Solon, in the early sixth century BCE, and Cleisthenes in the late sixth century BCE, Athens secured the ascent of democracy. Agora then became the very center of public life.

Athenians went to the agora for several reasons. Agora had the advantages of what we call the marketplace. So, according to Plato, people visited the agora for myriad reasons: searching for things to buy and sell, including foreign currency.

Agora officials checked the quality and prices people paid for grain and food and for other goods.[98] Aristotle said agora should also be a place for reflection and studies.[99] In addition, agora was the center for religious festivals, athletic competitions, and politics. This was the choice ground where the Athenians constructed government buildings for administration, the courts, and the Assembly.[100]

Like the Achaeans of the Trojan War era, male Athenian citizens would go to the agora to talk to each other about the polis. They would form a circle for debate. Any one of the citizens could address the rest from within the circle. He spoke as an equal. And he felt perfectly safe.

---

[96] Homer, *The Odyssey* 2.1–260; *The Homeric Hymns* 296–297.

[97] Plato, *Laws* 6.778c–d.

[98] Plato, *Laws* 8.849, 11.917–918.

[99] Aristotle, *The Politics* 7.12.1331a19–1331b18.

[100] John Mck. Camp II and Diana Buitron-Oliver, *The Birth of Democracy: A Catalogue of an Exhibit* (Athens: Genadeios Library and the American School of Classical Studies, 1993) 16 (in Greek).

So, Anaximander and other natural philosophers put the Earth at the center (middle) of the universe in order to give it stability, the stability they experienced or expected in the agora of the polis.

The center or middle avoided the political extremes of the monarchy-tyranny-oligarchy and dire poverty. The middle was the center. The center was the middle. Such symmetry in the heavens and polis traced its common origins in the geometry of political stability the Greeks had or wished for their communities. Greeks put political power at the center-middle of the common and public place in the agora of the polis. That middle of the agora shone like a star in the sky.

Plato and Aristotle put the middle at the center of their political philosophies: advocating moderation in everything. The "middle" class was born from this early conviction the polis and the heavens moved around the order and harmony of the middle.

Aristophanes, the great fifth-century BCE Athenian comic playwright, also toyed with the cosmological image enveloping the middle of the agora in his hilarious cloud-cuckoo-land of "Birds." He shows Athenians might have entertained the notion their agora had potential cosmic connections and consequences.

He chose the fifth-century BCE astronomer Meton to make his point. He probably appreciated Meton had made the heavens less mysterious and more predictable. How did he do that? Meton, I will be arguing later on in this book, favored the nineteen-year calendar for bringing order and stability to measuring time. His nineteen-year calendar accurately checked the picture of the heavens: showing the positions of the stars in the sky as being exactly the same as those of nineteen years ago. It turned out the nineteen-year calendar was also useful and reliable for events in the polis.

No wonder Aristophanes teased Meton in order to illustrate the connections of the agora to the Cosmos. He has Meton planning the agora of Athens:

"METON: I will measure the land of the Agora with my level ruler here. I will also square the circle of the Agora. Your Agora will then stand at the center. Straight roads will lead to its hub—just like a star whose center forms a circle but whose rays shine in straight lines."

"PEISETAIROS: I think he's really Thales!"[101]

---

[101] Aristophanes, *Birds* 1002–1009.

The real Thales of Miletus was also caught with the symbolism of the political importance of the agora. And being an Asian Greek, Thales was also sensitive to the political power of Persia. The great historian Herodotus, also an Asian Greek, has Thales urging the leaders of the Panionians (Greeks living in poleis in Asia Minor, Ionia) to move their political fortunes to the agora of the island of Teos because Teos was in the middle of Ionia.

Thales proposed the unification of Ionia around 547 BCE, sometime after Cyrus, king of Persia, defeated Croesus, king of Lydia. The plan of Thales would have established a single Ionian government council in Teos and the member poleis would be downgraded to equal demes.[102]

Just like the Ionians of Asia feared Persia, the Athenians feared Sparta. In fact, while Aristophanes was writing his plays and Plato was growing up, Athens and Sparta were fighting the bitter and lengthy Peloponnesian War. Nevertheless, Athenians had the delusion of invisibility, though they had a master. That master was the law. Obeying the law made them equals. Citizens could not take over the state. However, they commanded and obeyed themselves and everyone else at the same time.

They became free from all domination. The Earth at rest, like a democratic polis, was stable and independent of external influences. Symmetry connected it to the Cosmos.[103]

Beginnings of Understanding of How the World Works

Anaximander absorbed some of those political ideas. Yes, we know him primarily because of his scientific work. He broke new ground in the sciences and philosophy. His cosmology includes an infinite number of worlds in the universe. Nature is eternal, self-regulating, and the mightiest power in the Cosmos.

The ancient geographer Strabo cites Eratosthenes, a third-century BCE natural philosopher, mathematician, mapmaker, geographer, and librarian in the Library of Alexandria, giving Anaximander credit for having constructed the first map of the world.[104] A modern French scholar, Jean-Pierre Vernant,

---

[102] Herodotos, *The Histories* 1.170.

[103] Jean-Pierre Vernant, *Myth and Thought Among the Greeks* (New York: Zone Books, 2006) 213–233.

[104] Strabo, *Geography* 1.1.11.

claims Anaximander was the first natural philosopher who used geometry for studying and understanding the Cosmos.[105]

Other scholars, G. S. Kirk, J. E. Raven, and M. Schonfield, are convinced Anaximander "made a comprehensive and detailed attempt to explain all aspects of the world of man's experience. ... His account of nature, though among the earliest, was one of the broadest in scope and most imaginative of all."[106]

No doubt, Anaximander was the first natural philosopher who gave us valuable scientific insights of how the world works. He narrowed the gap between heavens and Earth. His theory of evolution was a stroke of genius.

Pythagoras: Founder and Father of Divine Philosophy

Pythagoras (569–c. 475 BCE) was another original natural philosopher. He probed the Cosmos through mathematics and observations of the natural world. His teacher, Pherekydes of Syros, considered him the most outstanding philosopher in all of Greece.[107] Late antiquity literature praises Pythagoras to the heavens: he was the founder of mathematics, astronomy, music, and philosophy.[108]

About a millennium after Pythagoras, a Platonic philosopher, Iamblichos, in late third to early fourth centuries of our era, describes Pythagoras as "the founder and father of divine philosophy."[109] Proklos, historian of mathematics and Platonic philosopher from the fifth century of our era, said Pythagoras "transformed mathematical philosophy into a scheme of liberal education."[110]

Pythagoras was the grandson of Ankaios. Ankaios was descendant of Zeus and founder of the polis Same in the island of Cephalonia. The oracle of Apollo at Delphi asked Ankaios to colonize an island in the Aegean, which Ankaios named Samos. Pythagoras was born in Samos.

After his early education in Samos, Pythagoras lived in Egypt for about twenty years and Phoenicia for some ten years. He studied with priests and visited those countries' sacred monuments. He returned to Samos and visited

---

[105] Vernant, *Myth and Thought Among the Greeks*, 213.

[106] Kirk et al., eds., *The Presocratic Philosophers*, 101, 142.

[107] *The Greek Anthology* 7.93.

[108] Charles H. Kahn, *Pythagoras and the Pythagoreans: A Brief History* (Indianapolis: Hackett, 2001) 1.

[109] Iamblichus, *On the Pythagorean Life* 1.2, ed. and tr. Gillian Clark (Liverpool: Liverpool University Press, 1989).

[110] Proklos, A *Commentary on the First Book of Euclid's Elements* 65.

Miletus, where he spent time and studied under Thales and Anaximander. When he returned to Samos, he found himself in a hostile political environment. He packed his things and left his country. He moved to Croton in Magna Graecia (Greater Greece in Southern Italy). He established a community and a school, where he taught for about forty years.

Pythagoras expanded the tradition of Thales and Anaximander in using geometry for measuring and understanding the natural world and the Cosmos. The ancients credit him for perfecting geometry.[111] The Pythagorean theorem—the square of the hypotenuse of a right triangle equals the sum of the squares of the other sides of the triangle—mirrors that tradition.

Pythagoras believed that the Earth and the other heavenly bodies were spheres moving around an invisible fire at the center of the universe. He blended music, mathematics, and astronomy in his search and discovery of a perfect and eternal harmony of natural laws, order, and beauty in the Cosmos.

Pythagoras was certain the stars made music. His theory of how that happens was a paean to the hidden beauty and order of the universe. He connected the mathematical ratios of the sounds he heard from his monochord musical instrument to the diversity of sounds of motion and speeds, sizes, and positions of the stars. He was entirely absorbed with the "celestial harmonies of the cosmos." He was probably alone in hearing and understanding the universal harmony and music of the spheres. The song was also "more complete and satisfying than any human melody ... of subtle and exceptional beauty."[112]

Once Pythagoras visited a Peloponnesian polis named Phlios. In his discussions with Leon, the ruler of Phlios, Pythagoras introduced himself as a philosopher. Leon asked for an explanation. Pythagoras said that a philosopher was like a spectator at the Olympics. The spectator was at the games not for financial profit or fame but for the satisfaction of a sense of wonder. This meant seeing clearly, observing, gazing, and searching for an understanding of the nature and the causes of the competing athletes from the athletes themselves or, in other instances, from wondering about society or the natural world. A philosopher did all that for the pleasure of discovering the truth.[113]

The greatest beneficiaries of the philosophy of Pythagoras were the Greeks. Iamblichos may be exaggerating the influence of Pythagoras, but not by much.

[111] Laertios, *Lives of Eminent Philosophers* 8.11.
[112] Iamblichos, *On the Pythagorean Life* 15.65.
[113] Laertios, *Lives of Eminent Philosophers* 8.8.

He says that Pythagoras "triggered all the exact sciences and branches of knowledge, everything that gives the soul true vision and clears the mind blinded by other practices, so that they [the Greeks] may see the real principles and causes of all there is."[114]

This was especially true with mathematics and the numbers as eternal first principles. Pythagoras borrowed his theory of numbers from Orpheus, a second millennium BCE natural philosopher and musician of enormous originality and influence. The Orphic Hymns are full of allusions shedding light on natural phenomena and mathematical knowledge.[115]

Plato absorbed the thought of Pythagoras, thus assuring his legacy.

Herakleitos: Nature Loves to Hide

Pythagoras had a critic in Herakleitos of Ephesos, who was born sometimes around 540 BCE. Herakleitos was like a prophet: enigmatic, provocative, honest, arrogant, and insulting. He is supposed to have said he was the sole man who discovered wisdom.[116] He also said knowing lots of things does not necessarily teach understanding or wisdom; otherwise, Homer, Hesiod, and Pythagoras would have known better. He was equally offensive towards other Greek thinkers.

Despite his objections to expansive knowledge, Herakleitos had the world and the Cosmos in his thought. His ambition was to guide all the people of the world everywhere. He wrote a book about nature and donated it to the temple of Artemis in Ephesos. The book addressed the Cosmos, politics, and the gods. It's possible he made his writing obscure intentionally to avoid the contempt of the masses.

He was convinced the best of people care for immortal fame, not for wealth or popularity. But most people, he said, "graze like cattle." He hoped only a select few would read his book.[117] However, his book did not make it to our times. What we have instead are fragments of his thought that other

[114] Iamblichos, *On the Pythagorean Life* 6.31.

[115] Maro K. Papathanasiou, *Cosmological and Cosmogonic Thinking in Greece During the Second Millennium BCE* (Doctoral Dissertation, second edition, School of Physics and Mathematics, University of Athens, 2016) (in Greek).

[116] *The Greek Anthology* 7.79.

[117] Laertios, *Lives of Eminent Philosophers* 9.1–6.

thinkers quoted in their works. These fragments, accurate or not, are extremely hard to decode. Even Plato had difficulty understanding Herakleitos.[118] From these fragments, the picture of Herakleitos is incomplete, probably biased, and certainly not clear.

Nevertheless, Herakleitos was important in clarifying the human relation to the natural world and the Cosmos. He proposed that the Cosmos is run by divine *logos*. Now logos in Greek means word, speech, things said, account, and explanation. Herakleitos, however, thought of logos as a common and unifying objective knowledge and truth. And while logos was unchanging, it sparked and supervised a river of change in the world. Anyone understanding logos, Herakleitos assures us, has a key to the understanding of the natural world and the Cosmos.[119]

The material manifestation of logos was fire, the defining element of change and creation. Fire is perpetually changing but, in reality, remains the same. Fire is perpetually creating and consuming things and the world forever, destiny being the arbiter of this cosmic cycle of creation and destruction.

Natural change came from the merging of opposites like hot and cold and dry and wet. Opposites, Herakleitos said, cooperate. The most beautiful harmonies come from opposition. We know health by disease, good by evil, wellbeing and satisfaction by hunger, and leisure by hard work. Herakleitos admitted it is difficult to notice how opposing forces agree. He said to look carefully at the bow and the lyre. Without these opposing forces, the world and the Cosmos would be a pipe dream. But the continuous clash of opposite forces creates harmony underlying the sustainability of the world. In addition, Herakleitos said existing things in this world are full of souls and gods. But Herakleitos went further. A world full of gods did not mean a peaceful world. He saw strife and war as necessary for birth and creation and concord and peace, fueling destruction by fire.[120]

War, Herakleitos said, is our father and king. It brings to light who among us is god-like and who is merely a person and who is a slave and who is free.

---

[118] Plato, *Symposium* 187a.

[119] G. S. Kirk et al, eds., *The Presocratic Philosophers* 181–212 (second edition, Cambridge: Cambridge University Press, 1999); *A Presocratics Reader*, 30–41.

[120] *A Presocratics Reader* 30–41; Laertios, *Lives of Eminent Philosophers* 9.1–17.

As for justice, he said, it is primarily competition, a powerful force responsible for all things coming into being.[121]

This is no war talk by Herakleitos. War is another way of saying the struggle of opposites is real, and the life force of harmony coming out of such conflicts keeps the world in existence. What we don't see in the design of things is more harmonious than in the stuff we see. Just like Anaximander, Herakleitos said that nature loves to hide.

The utterances of Herakleitos are insightful. The man was strange but had penetrating views on humans and the Cosmos: everything does flow. One cannot step in the same river twice. No man or god made the world: it has always existed, being an everlasting fire that eventually dies and comes up again. Every day we have a new Sun. Its movement around the Earth will always remain the same. However, should the Sun change its motion, justice will dispatch its agents, the Erinyes, to bring the Sun to trial.[122]

These observations, like the oracular sayings of Apollo at Delphi, enriched early Greek thought. Herakleitos obviously borrowed from his predecessors, especially Anaximander and Pythagoras.

Parmenides: What Is Leads to Truth

Another important influence in the understanding of the Greeks and their exploration of the natural world and the Cosmos comes from Parmenides of Elea in Southern Italy (c. 520–after 450 BCE). Parmenides demonstrated the unreliability and impossibility of the constructs of the future and the past, what he termed "the coming to be and the passing away." He said that only the present—what is—could be deemed reliable. The existing reality of the Cosmos is complete, whole, unchanging, perfect, unique, un-generated, and imperishable. Studying the Cosmos as is leads to truth.

Thinking and speaking the truth is necessarily thinking and expressing the reality of life or being. So, a human or animal being is the ultimate truth in the world. "**Nothing comes out of nothing**," he said, and "**the Cosmos is eternal, having neither a beginning nor an end**." He studied the Cosmos and, in his mathematics, gave us accurate measurements of the time planets

[121] Kirk et al., eds., *The Presocratic Philosophers* 193.

[122] Kirk et al., eds., *The Presocratic Philosophers* 194–212.

take to go around the Sun and reappear in the same spot in the night sky. Those numbers were critical in the 2021 theoretical reconstruction of the Antikythera computer as it might have been in the second century BCE (chapter 7).

Parmenides wrote a poem in Homeric hexameter in which he claimed that the daughters of the Sun-Helios, driving a chariot pulled by mares, led him to the goddess Aphrodite, who told him he had to learn all things. He relied on reason and objectivity alone. He advanced natural science and called for people to diminish the biases they bring to their endeavors. He rejected what he called "beliefs of mortals" because those beliefs come from the senses, which are influenced by an ever-changing world.[123]

Empedocles: Earth, Fire, Air, and Water

While Parmenides insisted on reason and here and now as the arbiters on what is real and true, another important statesman, healer, and natural philosopher, Empedocles (495–435 BCE) from Akragas, Sicily, turned to the senses for understanding the world and grasping the truth. One of his landmark contributions was the establishment of the four basic building blocks of the universe: Earth, fire, air, and water.

He postulated a Cosmos of equal but competing forces, none of which had the primary influence on the universe. Love and Strife, which Empedocles considers immortal and main powers of the Cosmos, balance each other. They control the four eternal elements. Love combines with the elements, forming a massive divine organism he called Sphairos. But this super god does not last for long since Strife separates the elements into their independent state.[124]

Like Anaximander, Empedocles thought in terms of evolution. He is reported to have said he had already been "a boy, a girl, a shrub, a bird and a warm-blooded fish of the sea." He also informed his friends in Akragas that "he was not a mortal man but an immortal god."[125]

---

[123] *A Presocratics Reader*, 44–51; See also: Arnold Hermann, *The Illustrated To Think Like God: Pythagoras and Parmenides: The Origins of Philosophy* (Parmenides Publishing, 2004).

[124] *A Presocratics Reader* 61–71; Paul T. Keyser and Georgia L. Irby-Massie, eds., *The Encyclopedia of Ancient Natural Scientists: The Greek Tradition and its Many Heirs* (Routledge, 2008) 283–284.

[125] *The Greek Anthology* 9.569.

Anaxagoras: Scaling the Heavens Searching for Truth

Anaxagoras also thought of the evolution of the species. He said animals evolved from one another, the driver being the changes taking place in the environmental conditions of moisture and heat, and those of the Earth. He was a physicist and cosmologist who looked at the sky as his "fatherland." He firmly believed that he was born "to study the Sun and the Moon and the heavens."[126] In fact, he became an insightful natural philosopher in the great tradition of the thinkers of the seventh and sixth centuries BCE. The Muse of astronomy kept him constant company. He said *Selene* (the Moon) had no light of her own, but the Sun-Helios always bathes her with his brightness.[127]

Imagine the audacity and genius of this man who, in addition to his scientific work, advised Athens' premier politician, Pericles. Anaxagoras was not Athenian. He was born around 500 BCE in Ionia (Asia Minor) in a polis known as Klazomenai. But he spent most of his life in Athens. Athenians admired him so much they called him "mind." His cosmology replaced chance and necessity with mind as the reason for the order and harmony of the Cosmos.[128] Upon his death, the citizens of Klazomenai wrote an inscription on his grave in which they said Anaxagoras scaled the heavens searching for truth.[129]

The Origins of Atomic Theory

Out of the ideas of Thales, Anaximander, Pythagoras, Parmenides, Empedocles, Anaxagoras, and other natural philosophers, there emerged in the second half of the fifth century BCE the atomic theory. This was one of the most spectacular scientific developments of all time.

Leucippus and Democritus of Abdera, a polis in Thrace in Northern Greece, developed the first scientific model of the fundamental unit of the universe: the atom. Only three hundred fragments survive from the work of Democritus and two from that of Leucippus. Most of what we know of their theories was summarized largely by Aristotle and other ancient commentators on Greek science.

---

[126] Laertios, *Lives of Eminent Philosophers* 2.7; 2.10.
[127] Kirk et al., eds., *The Presocratic Philosophers*, 380–381.
[128] Ploutarchos, *Pericles* 4.4.
[129] Laertios, *Lives of Eminent Philosophers* 2.15.

According to Aristotle, Democritus and Leucippus penetrated the very fabric of existence unlike any natural philosophers before them. They discovered matter is made up of atoms, invisible particles that cannot be cut. They explained that the atoms come to be and pass away by changes in their groupings and positions. Since "the truth lay in the appearance, and the appearances are conflicting and infinite, Democritus and Leucippus made the atoms infinite in number."[130] Aristotle admires Leucippus and Democritus for the daring of their speculation, in essence that nothing exists but atoms and space.

They implied the key to the Cosmos and the structure of matter had two basic ingredients: atoms and the void, both eternal; an infinite number of invisible, impenetrable, and indivisible atoms move in infinite void, creating and destroying countless worlds. These atoms are different in size and shape with geometrical characteristics. When they come together, they form matter and trigger change.

Ancient Greeks thought highly of Democritus. He was the "glory" of his homeland, Abdera. He was a philosopher and a scientist who dug deep into the secrets and laws of the natural world and the Cosmos. And despite his achievements and wisdom, he laughed at the frailties of life. Time engulfed everything.[131]

For another glimpse of the atomic world of Democritus, one goes to Epicurus. He arrived in Athens in 306 BCE from Samos. He developed his own "apolitical" philosophy of making the individual invisible and unnoticed to the authorities. He did not find any satisfaction with the many Greek gods but he did not preach against them.[132]

Rather, he accepted the atomic theory of Democritus and used it to explain the world. Lucretius, a Roman poet of the first century BCE, was so spellbound with Epicurus and his ideas, especially his core teaching of the atoms and the void, he wrote an entire book, *On the Nature of the Universe*, immortalizing the work of Democritus and Epicurus. But Lucretius's hero was Epicurus. He said Epicurus's genius "outshone the race of men and dimmed them all, as the stars are dimmed by the rising of the fiery sun."[133]

---

[130] Aristotle, *On Generation and Corruption* 315b6–11.

[131] *The Greek Anthology* 2.131–135.

[132] Tim Whitmarsh, *Battling the Gods: Atheism in the Ancient World* (New York: Alfred A. Knopf, 2015), 197.

[133] Lucretius, *On the Nature of the Universe* 3.1044–1045, tr. R. E. Latham (Penguin Books, 1994).

Diogenes Laertius also extolled the atomic theory of Democritus. He did his writing in the mid-third century, 700 years after Aristotle and 800 years after Democritus. Laertius had access to the writings of both Democritus and Aristotle. He summarized the ideas of Democritus as follows:

The first principles of the universe are atoms and empty space—everything else is merely thought to exist. Worlds are unlimited; they come into being then, in time, perish. **Nothing can come into being from nothing**. The atoms are unlimited in size and number. They exist in the whole universe in a vortex, giving rise to all composite things—fire, water, air, and the Earth. These composite things are conglomerations of atoms. And it is because of their permanent nature that these atoms are unalterable. The Sun and the Moon have been made of such smooth and spherical atomic masses. The same thing can be said of the soul, which is identical with reason.

According to Laertius, Democritus believed we see by virtue of the impact of images upon our eyes. All things happen by virtue of action or *necessity*, the vortex being the cause of the creation of all things. The end of all action is *tranquility*, a state in which the soul continues calm and strong, undisturbed by any fear or superstition or any other emotion. "The qualities of things exist merely by convention. **There is nothing but atoms and void space in nature**."[134]

Atomism remains a singular contribution of Greek thought, serving as the stimulus for the development of modern atomic theory and physics. Atomism was the ultimate "crown of Greek philosophical achievement" before Plato.[135] Without the benefits of complicated instruments, the atomists looked into matter and the Cosmos, revealing the innermost secrets of how the universe works.[136]

A sense of wonder permeates the work of the great natural philosophers of ancient Greece. Gazing at the skies, they drew upon their observations of the forces of the Cosmos. Through the interpretation of their mythic traditions and logical and scientific observation, they sought answers to the fundamental questions of the universe.

The next chapter explains how this sense of wonder, philosophy, found expression in the building of the Parthenon and, at the same time, blossomed in the thought of Plato, who taught Aristotle, who invented science.

---

[134] Diogenes Laertios, *Lives of Eminent Philosophers* 9.44–45, tr. R. D. Hicks (Loeb, 1995).

[135] The *Presocratic Philosophers*, 433.

[136] On August 19, 2015, PBS presented a three-part series, "The Mystery of Matter: Search for the Elements," in which it ignored the Greek contributions to the atomic theory.

Chapter 3
Untouched by Time

The Athenians built the Parthenon to honor their patron goddess of wisdom and craftsmanship, Athena.

A near-flawless monument to architecture, science, and engineering, the Parthenon is a timeless symbol of the contributions of the ancient Greeks to a new model for the understanding of the natural world through reason, measurement, and scientific technology. Built in the fifth century BCE, this iconic temple is the creative progenitor of the Antikythera Mechanism.

As if to confirm the perfection of the Parthenon, there came into being Plato, a moral philosopher who taught Aristotle, who invented science, as we understand it. This chapter discusses the construction of the Parthenon as well as the thought of Plato and Aristotle. Between the Parthenon, Plato, and Aristotle, there's a tradition of technology, philosophy, and scientific theory that led the Greeks to the Alexandrian Era and the Antikythera computer.

The Parthenon

Alexander Tzonis, professor of architectural theory at the University of Technology, Delft, Holland, connects the building of the Parthenon to the advancement of science. Tzonis reports that the change that took place in Greece between the eighth and fifth centuries BCE was epoch-making: "Greece produced a new way of constructing and construing the world, which was unprecedented in its systematic rigor and embedded with new disciplinary institutions. There was no place for falsehood or accident in this system," he wrote.[137]

[137] Alexander Tzonis and Phoebe Giannisi, *Classical Greek Architecture: The Construction of the Modern* (Paris: Flammarion, 2004) 184.

Polytheism, natural philosophy, and advanced craftsmanship culminated in an extremely sophisticated form of sculpture, architecture, and city planning of the classical age. Poleis filled with thousands of statues and dozens of great temples, including the Parthenon in Athens. How the Greeks constructed the Parthenon has yet to be understood, much less surpassed.

Building the Parthenon was, first of all, a massive public works project. Ploutarchos has something to say about the Parthenon. He lived from about 46 to 120 in our era. He was a Greek polymath, philosopher, prolific writer, and priest of Apollo. He wrote almost six centuries after the Athenians built the Parthenon. He left a few clues of what and who made the Parthenon possible.

He reported that the materials used for the construction of the Parthenon included marble, bronze, ivory, gold, ebony, and cypress wood. The technicians who shaped these materials to form the Parthenon included carpenters, molders, bronzesmiths, stonecutters, dyers, gold and ivory experts, painters, embroiderers, and embossers. Add to these rope makers, weavers, leather workers, road builders, and miners. Then there were sailors and pilots who carried the marble by land and sea and trainers and drivers of yoked animals that did other indispensable work. All in all, some 200 craftsmen and 50 sculptors did the lion's share in the building of the Parthenon. We still use the tools they used.

Ploutarchos also says that the builders of the Parthenon tried to outdo themselves in the beauty of their handiwork, which was inimitable in its perfection and grandeur. Ploutarchos was equally effusive in his praises of Pericles, under whose leadership and administration the Parthenon came into being. The works of Pericles, he said, were done "in a short time for all time."

Each one of those works was so fresh, vigorous, and beautiful that, according to Ploutarchos, it was at once ancient; in fact, each work like the Parthenon looked as if "untouched by time. The ever-green breath of an ageless soul was infused into the works of Pericles."[138]

William Martin Leake, a British traveler and philhellene, visited Athens in the 1810s. In 1821, he published his study of "The Topography of Athens" where he praised the "magnificent" Parthenon, "which, by its united excellences of materials, design, and decorations, was the most perfect ever executed."[139]

---

[138] Ploutarchos, *Pericles* 12.6–7–13.1–3, tr. Bernadotte Perrin (Loeb, 1967).

[139] William Martin Leake, *The Topography of Athens*, second edition, 2 vols (London: J. Rodwell, 1841) I, 334.

Another philhellene, the French philosopher Ernest Renan, visited the Acropolis in 1865 and, like Ploutarchos, fell in love with the beauty and sacredness of the Parthenon. He saw the ideal crystallized in Pentelic marble. He admitted that "Greece had created science, art, philosophy and civilization; but the scale failed me. When I saw the Acropolis, I have had the revelation of the divine." In addition, Renan equated the beauty of the Parthenon with "absolute honesty," reason, and the respect Greeks had towards their gods. He said the hours he spent on the Acropolis were "hours of prayer" to Athena Pallas.[140]

A twentieth-century student of ancient Athens, R. E. Wycherley, noted that the Parthenon was "The culmination of Greek architecture.... The temple dominated the Acropolis and was the crowning glory of the city as a whole. It was elaborately designed, and no care or expense was spared in building and decoration to make it a worthy offering to Athena and a splendid symbol of the power and achievement of Athens."[141]

E. Guhl and W. Koner, eminent German scholars, also of the twentieth century, are certain that works of architecture "produce the grandest and most powerful impression and give the most distinguishable character to the life of a nation." This was particularly true with the Greeks who "were enabled and gifted more than any other nation" in rendering "the innermost nature of their genius in external works of art."

Just like they designed the Antikythera Mechanism as a window to the heavens and Earth, they brought together the celestial and temporal in the Parthenon. According to Guhl and Koner, "The temple became the rallying-point of everything good, noble, and beautiful, which we still consider as the glory of Greek culture and refinement."[142]

The American classics scholar, Jon D. Mikalson, agrees. He found divine origins in Greek architecture. He spoke about the "inclination" of the Greeks "to give to their gods only what was beautiful." The result of such a relationship, he concluded, "filled their cities and villages, the richer ones more than the poorer ones, with temples, statues, and dedications of unsurpassed beauty."

---

[140] Ernest Renan, *Prayer on the Acropolis*, tr. from the French by Xenophon Dantis (Athens: Dionysios Petsalis, 1962) 15–26. See also the French text of Renan and its translation into Greek by Rigos Kappatos, *Prayer on the Acropolis* (Athens: Ekate, 2015).

[141] R. E. Wycherley, *The Stones of Athens* (Princeton University Press, 1978) 105.

[142] E. Guhl and W. Koner, *The Greeks: Their Life and Customs* (London: Senate, 1994) 1–3.

Then Mikalson added: "Most of what we think of as characteristically Greek in architecture, sculpture, mythology, lyric poetry, tragedy, and comedy owed its origins and, especially in the Classical period, its development, to the religious institutions and practices of the Greek people. The cultural environment in which the Greek individual lived, whether in Athens or Sparta or Thebes, was significantly determined by his religion and that of his ancestors."[143]

The Parthenon was an embodiment of Greek religion. Like an ageless celestial mirror, the Parthenon also has been reflecting the dazzling power, artistic achievements, and technical knowledge of Athens, the premier Greek polis in the fifth century BCE.

Cosmos Within a Cosmos

Careful observation of the Parthenon shows that there are no straight lines or right angles in the building. Instead, its architecture incorporates and takes advantage of deviations from the perfectly vertical or horizontal. The columns are a striking example of these virtues. The Parthenon looks like it responds under its own weight: The steps curve upward, the columns swell slightly in the middle and tilt inward while the metopes, panels adorned with sculpture, tilt outward. The corner columns are a little bit thicker than the rest of the columns. The overall impression in looking at the Parthenon is one of beauty and harmony.

Jeffrey Hurwit, professor of history at the University of Oregon, thinks the harmony of the Parthenon "is dependent on a certain mathematical system of proportions." But the harmony of the Parthenon comes from values complementing mathematics. The Parthenon is more than a building.

The Parthenon is a sculpture resembling the human body, which explains all those deviations from the perfectly mathematical. Hurwit puts it like this: "These deviations from the straight, from the perpendicular, from the perfectly vertical, from the perfectly horizontal are analogous to the curvatures and the swellings and the irregularities of the human body. And in that sense, the Parthenon strikes me as being a sculptural as well as an architectural achievement."[144]

---

[143] Jon D. Mikalson, *Ancient Greek Religion* (Blackwell Publishing, 2005) 196.

[144] November 2007 interview of Jeffrey Hurwit by Gary Glassman, producer of "Secrets of the Parthenon," WGBH, NOVA, PBS.

Lothar Haselberger, professor of Roman architecture at the University of Pennsylvania, calls the refinements of curvature "delicate deviations from strict regularity." These refinements went into the building of nearly all Greek temples after the Persian Wars. Their purpose was making the building shape "its own life and truthful presence. This truth could lie in a bent structure, yet never in deceptively bending the truth."

Haselberger explains the technique of curvature as follows: "Once the curve, with a maximum rise of no more than a few fingers' breadth, was created in the building's substructure, it had to be faithfully preserved through the entire structure as it rose—course by course—up to its entablature, pediment, and roof. Seemingly straight lines and surfaces became subtle curves and spherical planes, leaving not a stone, column, or angle unaffected. Within the entire high-precision structure, not a single horizontal element remained unbent, so to speak."[145]

Alexander Tzonis sees the "intentional deviations from regularity" as a striking refinement in the architecture of the Parthenon. In addition, he sees the triumph of reason in its construction:

"The orderly arrangement—the spacing and shaping—of the columns [of the Parthenon] reinforced the visual discontinuity, thereby generating a vibrant 'illusory-contour' that set the edifice, as an individual entity, apart from its surroundings.... It created a *temenos* [sacred precinct for a temple]—no longer in a purificatory, but rather in a cognitive sense. It made a '*kosmos* within *kosmos*,' or 'world within the world,' free of pollution and spatial anomalies. More significantly, by erecting a colonnade and arranging the interior components of the building in a coherent way, this was a major step in the pursuit towards the formation of a coherent system of architectural thinking and spatial reasoning," he wrote.[146]

Edith Hamilton, a British classical scholar of the early twentieth century, would have congratulated Tzonis. She, too, saw the perfection of Greek civilization in the Parthenon. She wrote:

"In the Parthenon straight columns rise to plain capitals; a pediment is sculptured in bold relief; there is nothing more. And yet—here is the Greek

[145] Lothar Haselberger, "Bending the Truth: Curvature and Other Refinements of the Parthenon" in *The Parthenon: From Antiquity to the Present*, ed. Jenifer Neils (New York: Cambridge University Press, 2005) 101–157.

[146] Tzonis and Giannisi, *Classical Greek Architecture*, 111, 190.

miracle—this absolute simplicity of structure is alone in majesty of beauty among all the temples and cathedrals and palaces of the world. Majestic but human, truly Greek. No superhuman force as in Egypt; no strange supernatural shapes as in India; the Parthenon is the home of humanity at ease, calm, ordered, sure of itself and the world."[147]

The Dialogues and Mathematics of Plato

The Peloponnesian War temporarily put the brakes on enlightenment in Greece. But the Parthenon was built before the war, a permanent reminder of Greek achievement. Then came Plato and Aristotle. Their works illustrate the continuation of the theoretical possibilities and rigor of Greek civilization.

Most of Plato's life and all of Aristotle's were in the fourth century BCE. This was also the age of Alexander the Great and the spread of Hellenic culture and scientific institutions in the Mediterranean, especially in the Middle East (chapter 6). Both Plato and Aristotle benefited from the life of Socrates, 469–399 BCE, the moral philosopher of Athens and teacher of Plato. Socrates spent his life trying to teach Athenians virtue, who put him to death for that presumption. Plato immortalized Socrates in his dialogues where Socrates is usually the main character.

In 399 BCE, Meletos, a young Athenian, took Socrates to court, charging him with corrupting the young and making up gods. Socrates walked to the court at the Basileus Stoa, the Porch of the King, for a hearing and while waiting for the court to open, he met Euthyphron, a young man who was suing his father for the death of one of his workers.

Socrates and Euthyphron talked about the meaning of piety, and Socrates quickly showed Euthyphron he was committing impiety by suing his father. Euthyphron was simply confusing piety with what pleases the gods. He could not understand that piety is piety because it is pious and infused with justice, a condition that pleases the gods.

Socrates explained to Euthyphron he was related to Daidalos, the great Athenian sculptor, architect, and engineer who worked for Minos in Crete during the second millennium BCE. Daidalos made statues that were so perfect

[147] Edith Hamilton, *The Greek Way to Western Civilization* (New York: The New American Library, 1957) 45–46.

and life-like, that they looked as if they moved (chapter 1). Socrates admitted that, without even trying, he outdid Daidalos because he constructed words that made others move.[148]

Socrates was a stonemason but, in fact, he was an architect of words and methods, a philosopher that, like a physician aiding a woman give birth to a baby, brought out the knowledge and wisdom inherent in all people. His method, taken up by Plato and Aristotle, tolerated no mistakes or falsehoods in the study of nature and society: how people ought to live their lives or how a polis ought to organize its culture and political freedom.

Socrates asked Euthyphron to teach him what piety is, and Euthyphron's answers showed him to be a confused young man who thought he knew more about theological issues than he actually did. Socrates dissected the words of Euthyphron just like a doctor would dissect a body to discover the right place and direction of the vessels out and to the heart. The attention of Socrates was clarity, meaning, and consequences of what he and Euthyphron said to each other.

Just like the "Poetics" of Aristotle was a guide to clear thought, the dialogues of Plato and the works of Aristotle constitute the poetics of Hellenic civilization. Here's where knowledge—in science and technology, politics and literature, tragedy and comedy, philosophy, and all other fields of ideas—went through its first peer review for originality and usefulness. In addition, and this is very fundamental, both Plato and Aristotle followed Pythagoras in using mathematics to decipher and understand the Cosmos, becoming the forerunners of the scientific revolution.

The dialogues of Plato reveal wisdom about culture and the universe. In fact, each of the dialogues is a model of the Cosmos because each probes and uncovers order, beauty, and truth. And according to Jay Kennedy, a historian and philosopher of science at Manchester University, the dialogues of Plato are embedded in a Pythagorean universe that is musical and mathematical.

Plato organized his dialogues around a certain number of lines and around the number twelve, which is important in music theory and is attributed to Pythagoras: the total number of lines of each dialogue almost always divide by twelve to give us a whole number. For example, the "Apology" has 1,200 lines,

[148] Plato, *Euthyphron* 11b–d.

giving 100 lines per twelfth; the *Republic*, with 12,000 lines, gives 1,000 lines per twelfth.

Plato inserted important speeches or ideas in the middle of the dialogue; bad or negative ideas in the ten- to eleven-twelfths, and good things like courage and the gods in the eight-twelfths of the lines of dialogue.

In 2010 Jay Kennedy reported breaking the "code" Plato used in organizing his work according to stichometry, or measurement of the number of the lines, "stichoi," of the text. Kennedy drew fascinating connections between the ideas in the dialogue, where Plato put those ideas in the text, and their relationship with the number twelve. In other words, Kennedy zeroed in on the line organization of the dialogues. He concluded that behind such a structure there lies a musical mathematical structure very much like that of the Cosmos.

"There is an analogy between, on the one hand, an imperceptible music reflecting proportions in the constitution of the Cosmos and, on the other, a musical scale which organizes a dialogue and yet is submerged beneath its surface—an unheard melody finally accessible to reason and measurement," he wrote.

"Just as the individual in the *Republic* mirrors the order of the polis, and the individual in the *Timaeus* mirrors the order of the Cosmos, each dialogue is another microcosm mirroring macrocosmic principles of order: 'the dialogue is a cosmos.' "[149]

Kennedy's insight takes us another step closer to Plato, who breathed mathematics, cosmology, and political idealism in everything he did. Another scholar, Proklos, who lived in the fifth century, from 410 to 485, about 900 years after Plato and a millennium and a half before Kennedy, brings us even closer to the spirit of Plato and his contribution to science.

Proklos was a Platonist and director of the Platonic Academy in Athens. He was a polymath with a special love and understanding of philosophy and mathematics. In fact, he wrote extensively about the history of mathematics. His voluminous commentaries on Plato and Euclid are insightful and extremely valuable. They preserve the originality of ancient Greek thought. What is also of interest is that Proklos could be so comprehensive and original at a time—the fifth century of our era—when the Christians were making

[149] J. B. Kennedy, "Plato's Forms, Pythagorean Mathematics, and Stichometry," *Apeiron*, 2010, 1–31.

Hellas a palimpsest: violently scrapping the Hellenic culture of the country and writing the legends of Jesus in its place.

Perhaps it was this dangerous historical time, and the courage one had to have to defend Hellenism, that explains the admiration of a student of Proklos, Marinos of Neapolis, for his teacher. Marinos said Proklos was "the prince of all the sciences."[150] But Marinos, just like his master, Proklos, said practically nothing about the Christians eating Hellas alive in front of their own eyes. Only fear explains such silence.

According to Proklos, Plato used mathematics for the construction, purification, and elevation of the soul. Mathematics dissolved the mist of ignorance from the mind of men. Plato used numbers for dividing the soul while binding it together "with proportions and harmonious ratios." Proklos also says that Plato deposited in the soul the "primal principles of figures, the straight line and the circle."[151] The circle for Plato and the Greeks was special. Proklos captured the place of the circle in Greek thought when he said the circle was the "first and simplest and most perfect of the figures."[152]

Proklos was right. Plato lived mathematics. In book 7 of the *Republic*, he raises arithmetic and geometry to the heavens, arguing that science draws the soul towards truth, in a sense, sciences guiding the eyes of the soul to see the truth. Plato also sprinkled his dialogues with mathematical terms and tried to convince students of philosophy of the vital role of mathematics in the study and understanding of nature and society.

Mathematics had consequences for science. According to Raphael Demos, a distinguished twentieth-century Greek Platonic scholar, the embrace of the mathematical method by modern science was Plato's "cherished doctrine."[153]

Proklos also described Euclid as a mathematician who belonged to the "persuasion" of Plato, one who felt comfortable in his philosophy: so that he structured his mathematical masterpiece, the *Elements*, like an introduction and application of Platonic philosophy to geometry.

---

[150] Marinos of Neapolis, *The Life of Proclus* 6.26–27, tr. Kenneth Sylvan Guthrie, ed. Al. N. Oikonomides (Chicago: Ares Publishers, 1977).

[151] Proklos, *A Commentary on the First Book of Euclid's Elements* 1.16.

[152] Proklos, *A Commentary on the First Book of Euclid's Elements* 147.1–2.

[153] Plato, *Selections*, ed. Raphael Demos (New York: Charles Scribner's Sons, 1927) v.

Both philosophy and the *Elements* are based on axioms; the former metaphysical and the latter mathematical. Proklos saw the goal of geometry as necessary for the "construction of the so-called Platonic figures."[154]

These figures were regular polyhedra, solids with surfaces made up by several polygonal faces. There are five polyhedra, each known by the number of its faces: (1) the tetrahedron or pyramid with 4 triangular faces; it is constructed with 24 triangles and represents the element fire; (2) the hexahedron or cube with 6 square faces; it is made up of 24 triangles and represents the element Earth; (3) the octahedron, made up of 48 triangles, has 8 triangular faces; it represents the element air; (4) the dodecahedron with 12 pentagonal faces; and (5) the icosahedron, built with 120 triangles. It has 20 triangular faces; it is the element water.

Johannes Kepler, 1571–1630, a distinguished German astronomer, used the polyhedra of Plato for a mathematical universe in which the distances of the planets from the Sun were proportional to spheres inscribing and circumscribing the five regular solids: tetrahedron, cube, octahedron, dodecahedron, and icosahedron. The planetarium of Kepler mirrors this cosmic view.

The space physicist Xenophon Moussas says Plato's polyhedra are equivalent to the elementary particles and atoms of modern physics.[155]

In his cosmological masterpiece, *Timaios*, Plato connected the tetrahedron with fire, the hexahedron with Earth, the octahedron with air, the icosahedron with water, and the dodecahedron with the Cosmos.[156]

I discuss *Timaios* later on in this chapter.

What Do We Know of Aristotle's Life?

I have cited Plato's student Aristotle repeatedly. But before we examine his original contributions to biology, we will sketch his life. Where did he come from? Who was he? Answering these questions makes it easier to grasp the extraordinary importance of this philosopher, and especially appreciate his passion for the natural world.

---

[154] Proklos, *A Commentary on the First Book of Euclid's Elements* 1.10.29–30; 2.4.66–69.

[155] Xenophon Moussas, *The Antikythera Mechanism: The Tablet: The First Mechanical Universe: The Most Ancient Computer and Astronomical Instrument* (in Greek, Athens: Union of Greek Physicists, 2012) 180.

[156] Plato, *Timaios* 55c–56c.

Aristotle was born in 384 BCE in Stageira, Chalcidice, Northern Greece. His father, Nikomachos, was a relative to Asklepios, who was son of Koronis and Apollo. Asklepios was the god of medicine. Nikomachos was also the court physician to Amyntas II, king of Macedonia. Amyntas was the father of Philip II, who was father of Alexander the Great.

At the age of seventeen, in 367 BCE, Aristotle left Stageira for Athens, where he enrolled at Plato's Academy.

Aristotle studied under Plato for twenty years until Plato died in 347 BCE. With the death of his teacher, Aristotle left Athens and went to Assos near Troy in northwest Asia Minor. This small but strategically placed kingdom was ruled by a friend of Aristotle named Hermeias. Aristotle married Pythias, who was possibly the niece, daughter, or sister of Hermeias. Aristotle had a daughter with Pythias.

Aristotle stayed in Assos from 347 to 345 BCE. Fresh from the Academy of Plato, he started a school in Assos, where he taught ethics, philosophy, and science. He also started his biological research. Place names and species unique to the Troad ecosystem appear in Aristotle's works.

However, this was a time of radical transformation in the geopolitics of Macedonia, Greece, and Persia. Hermeias built a strong but secret alliance with King Philip II of Macedonia. He was willing to have Assos become the bridge for Philip's invasion of Persia. Persia reacted with violence to Hermeias's pro-Macedonian policies. In 345, under deception, a Greek general working for the Persian king convinced Hermeias to negotiate with him. The Persians arrested Hermeias and took him to Susa, where they tortured and crucified him. With the death of Hermeias, Aristotle left for Lesbos, where he continued his teaching and biological studies from 345 to 342 BCE.

The murder of Hermeias made Aristotle furious. He already had a low opinion of non-Greeks or barbarians. Now he would probably have loved to avenge the death of his friend. He praised Hermeias for having been a model of Homeric courage, comparing him to Herakles and Achilles. He wished that the Muses, daughters of goddess Memory and Zeus, would make Hermeias immortal.[157] Indeed, he and Hermeias agreed that Persia had always been a destabilizing force and threat to Greece.

[157] Laertios, *Lives of Eminent Philosophers* 5.7–8.

The Persian danger to Greece was acute in the early forth century BCE. This was especially obvious after the defeat of Athens by Sparta. Persia took advantage of a divided Greece and its agents and gold exacerbated the deep political crisis of the country.

Indeed, Isocrates was convinced that Persian influence was so widespread in Greece that Persia, in effect, controlled the country. The Persian king, whom Greek politicians called "the Great King," could probably station his troops in Greek poleis (city-states). He taxed those city-states and listened to their complaints against each other.[158]

So, the role of Persia in Greek affairs was a hot topic of conversation and politics for decades. Greek patriots resented Persia terribly. For example, Gorgias, c. 483–c. 385 BCE, a distinguished teacher of oratory from Leontini, Sicily, addressed the Greeks at the Olympics in 392 BCE[159] and urged them to resolve their political differences and attack the Persians.

Isocrates did something similar in his magnificent *Panegyrikos logos* (panegyric speech) in 380 BCE during one of the Panhellenic athletic and religious festivals. In a sense, he told the Greeks that fighting and defeating the barbarians, by which he meant the Persians, was a matter of life and death. Having a secure peace among Greek states demanded the Greeks "unite to make war against the barbarians." The risk from such a war would cement Greek unity.[160]

The discussion of Aristotle and Hermeias about Greek nationalism, Greek identity, and the Persians was typical of what Greeks thought about the security of their country for a long time.

Gorgias and Isocrates spoke to crowds in the Panhellenic Games because the sanctuaries in Olympia, Delphi, Isthmus of Corinth, Nemea, and Athens were, primarily, centers for Greek unity and nationalism. Manolis Andronikos, Greek archaeologist who unearthed the tombs of Philip II and his family in Aegae (Vergina), Macedonia, northern Greece, agreed. He said that during the Panhellenic festivals athletics and music delighted the Greeks and brought back memories of common origins and of belonging to one country:

"Their wise men came to win glory and praise and to speak to them of their past and their future. Here the Greeks forgot their own particular city and all that divided them; full of elation, they lived a common life, albeit ephemeral,

[158] Isocrates, *Panegyrikos* 120–121.
[159] Philostratos, *Lives of the Sophists* 1.492–494.
[160] Isocrates, *Panegyrikos* 173–187.

and began to speak of a broad motherland without borders but bounded only by the hearts of men. At Olympia, Delphi, the Isthmus, Epiros, Rhodes, and Nemea, the small countries became fused into one, which the Hellenes ... called by their common name, Hellas."[161] Panos Valavanes, professor of Greek archaeology at the University of Athens, also connects the sanctuaries to the shaping of Greek identity.[162]

Aristotle and Hermeias were living in that emerging Hellas. They agreed Macedonia was capable of defending Greece. But Greece had to be a unified country, and Aristotle had no doubt that Macedonia could and would unite Greece. A united Greece could easily rule the world. Indeed, Aristotle was certain of that. In his *Politics*, he thought the Europeans living in the cold climate of northern Europe were courageous people but without the necessary intelligence to have civilized states or rule others. In other words, Europeans lived in rogue states. The Asians, however, were intelligent but deficient in courage with the result of living lives of subject people and slaves. Only the Greeks, according to Aristotle, combined intelligence and courage and lived in a temperate climate between the Europeans and the Asians. This enabled them to maintain their political independence and to be the best governed people in the Mediterranean. So, Aristotle had no doubt a united Greece could become the master of the world.[163]

It was in the context of this political discussion that Hermeias recommended Aristotle to King Philip.[164] Philip, of course, had been a friend of Aristotle's father.

In 342, Philip invited Aristotle to tutor his son Alexander, then thirteen years old. Ploutarchos says Philip gave Aristotle and Alexander a school in the beautiful precinct of the Nymphs near Mieza, a mountain village close to a town known as Beroea (Verria). The Nymphaeum had stone benches and shaded walkways. Aristotle taught Alexander, including his closest friends: Hyphaestion, Kassander, son of Antipater, and Ptolemy, son of Lagos. In time, Antipater and Ptolemy became kings.

---

[161] *The Eternal Olympics: The Art and History of Sport*, ed. Nicolaos Yalouris (New Rochelle, NY: 1979), 73.

[162] Panos Valavanes, *Sanctuaries and Games in Ancient Greece: Olympia, Delphi, Isthmia, Nemea, and Athens* (In Greek, Athens: Kapon, 2004), 126.

[163] Aristotle, *Politics* 1327b18–37.

[164] Jaeger, *Aristotle*, 110–121.

Aristotle instructed these young men in philosophy, science, history, and international politics. He probably highlighted the political differences between the Persians and the Greeks, stressing the necessity of Macedonia leading a united Greece against Persia. In 338 BCE, while Aristotle was still tutoring Alexander, Philip crushed Athenians and Thebans in Chaeronea.

For the first time in their history, most of the mainland Greeks were now united, though they did not appreciate seeing Macedonian forces in their poleis. Athens resented Philip and the Macedonians. Aristotle, however, most likely welcomed Philip's victory.

Aristotle also taught his students literature and epic poetry. He edited the *Iliad* of Homer for his students. The *Iliad* became Alexander's guidebook to military excellence. Achilles, the greatest Greek hero in the Trojan War, captured Alexander's soul. He modeled his life after that of Achilles. Alexander kept his dagger and the *Iliad* always under his pillow. Ploutarchos says Alexander loved Aristotle more than his father. Alexander learned medicine from Aristotle and was grateful to him for teaching him how to lead a noble life.[165] But more than anything else, Aristotle was fundamental in the formation of Alexander's Greek consciousness, courage, and intellect. Aristotle also prepared Alexander for his conquest and rule of Persia and the world. Knowledge has always been power.

In 336 BCE, a young soldier named Pausanias assassinated Philip. Alexander, now twenty years old, became King Alexander III. He immediately began preparing Macedonia and Greece to invade Persia.

At that moment, Aristotle left the Macedonian court for Athens. He started teaching at a school dedicated to Apollo Lykeios (of the wolf) or Lyceum. This school was part of a sacred grove near Athens. Aristotle used to teach while walking through a covered court in the grove of the school. That's why his philosophy came to be known as Peripatetic—from *peripatos* (walking).

Alexander appointed Antipater to govern Greece and Macedonia. Antipater was a friend and protector of Aristotle. But when in 323 BCE Alexander the Great died, Aristotle decided the time had come to leave Athens. He sensed the rising anti-Macedonian sentiment in the city and, as he probably said, he did not want Athens to repeat its first crime against philosophy when Athens, in 399 BCE, put Socrates to death for "impiety." In fact, an Athenian accused Aristotle for "impiety." But Aristotle did not bother to appear in the Athenian

[165] Ploutarchos, *Alexander* 7.1–4; 8.2–4.

court or refute the accusations against him. He appointed Theophrastos, an outstanding student and natural philosopher he had met in Lesbos, to direct his school and gave him its library, including his books and manuscripts. He left Athens for Chalkis, Euboea, home of his mother. He died in Chalkis in 322 BCE at the age of sixty-two.

Diogenes Laertios saved Aristotle's will.[166] Antipater was the executor of Aristotle's will. In this personal document Aristotle shows himself to be affluent, liberal, full of love for his family, and compassionate for his slaves. He gave freedom to some of his slaves immediately and others were promised freedom on the marriage of his daughter. His wife, Pythias, who died before him, asked her bones be placed in the same grave with the bones of Aristotle. In his will, Aristotle confirms the wish of Pythias, that his bones be buried in the same grave with the bones of his deceased wife.

After the death of Pythias, Aristotle lived with another woman, Herpyllis, who gave him a son, Nikomachos. Aristotle speaks of Herpyllis with affection and gratitude. He asked that Herpyllis be given one of his two houses with enough money and servants for a comfortable life.

These biographical facts help us understand that Aristotle was more than logic and science. He was a decent man and a Greek patriot. He knew Greece had to become one unified country to maintain its freedom and independence.

Aristotle and the Natural World

Aristotle grew up at a time when careful observation and study of the natural world were established traditions. Hippocrates, Asklepios, and Greek physicians had set the foundations of medical science. It's quitc possible Aristotle's love for animals (studying, dissecting, cataloguing, and learning from them) had its roots in the widespread tradition of medical science.

Aristotle, like Plato, studied all nature, indeed the entire universe inhabited by immortal heavenly beings and mortal humans, plants, and animals. His preference was the natural world, especially animals because, as he put it, "we are living in their midst."[167]

---

[166] Laertios, *Lives of Eminent Philosophers* 5.11–16.

[167] Aristotle, *Parts of Animals* 644b.28.

Aristotle urged us to be curious and affectionate towards all animals, including the lowest among them. Animals explain the why and causes of things. Animals give us extraordinary pleasure that cannot be measured. It would be strange and absurd, he said, if we enjoyed looking at paintings and statues of animals but failed to delight much more in the animals themselves.

Aristotle, painting by Evi Sarantea. Courtesy Sarantea.

Aristotle kept reminding us that animals are works of nature, which does nothing in vain. He hoped people did not act like children and fail to examine even the humblest and meanest of animals. "There is something marvelous and beautiful in all-natural beings," Aristotle wrote. He mentioned that once strangers visited the philosopher Herakleitos. They found him warming himself at his kitchen stove so they hesitated to go into his house. Herakleitos, however, invited them to enter, saying, "There are gods even here." In like manner, Aristotle was convinced that "we should study all animals because all of them are natural and beautiful. Nature is full of purposefulness: There is nothing accidental in animals and other works of nature. Rather, the presence of natural virtues dominates all animals and other works of nature." Furthermore, Aristotle insisted that the function and purpose of the works of nature are inherently beautiful. If someone says that the study of animals is an unworthy pursuit, Aristotle advises that person he ought to say the same thing about the study of himself. Consider, Aristotle says: "There's plenty of repulsion in looking at the blood, flesh, bones, blood vessels and similar parts making up the human body."[168]

There's both perfection and extraordinary insight in the work of Aristotle. No other scholar or scientist so far has approached his genius. And it is highly unlikely that any scholar or scientist of the present or the future will ever reach the heights of his wisdom.

Aristotle invented zoology as a science. He made masterful contributions to poetry, politics, rhetoric, ethics, and philosophy. Aristotle shone as a pioneering great biologist and, as a result, his influence in the sciences has been comprehensive and lasting.

Aristotle did field studies for several years all over the Greek world, including the island of Lesbos, the home of his best science pupil, Theophrastos. Aristotle worked with meticulous care, defining, organizing, describing, and evaluating animals. For example: he observed sacks a little larger than grapes coming down the Hypanis River in the Cimmerian Bosporus (Southern Bug River, Kerch Strait, Ukrainian-Russian border) at the time of the summer solstice. He noticed a winged animal with four legs coming out of the sacks. The insect, Aristotle says, "lives and flies all over until the evening. But with the setting of the Sun its life dries out and it dies, living but one day. For this reason, we call it ephemeron." This is the mayfly.

Aristotle studied and reported on the *ichneumon* wasps, which are smaller than the typical wasps. These tiny wasps cut the heads off spiders and then carry their bodies to a hole in a wall that they smear with mud after they lay their grubs in it. The developing young eat the dead spider bodies and become hunter-wasps.[169]

The biological studies of Aristotle include dozens of such scientific observations that probably took hundreds of hours of very careful and expert study. He reported that female animals are more cunning and care more for the young than male animals, though they are less courageous. For example, if the female

[168] Aristotle, *Parts of Animals* 645a7–30.
[169] Aristotle, *Historia Animalium* 5.552b17–553a2.

cuttlefish gets into trouble, the male stands by to help, but if danger hits the male cuttlefish, the female runs away.[170]

Aristotle studied the life of honeybees, their remarkable organization as societies of extremely useful insects, each honeybee devoted to a specialized function: most gathering nectar and pollen from flowering plants and trees for making honey, others caring for the grubs, and others guarding the hive. Aristotle was impressed by the cleanliness and the "political" purpose of honeybees in creating cohesive societies coexisting with humans.

He admired dolphins for being the fastest animals on Earth and in the sea. He described the dolphins as mammals with a gentle disposition and nature, and a passionate attachment to their young.[171]

In addition, Aristotle studied the sea urchin. He left us a remarkably accurate and fascinating account of the anatomy and physiology of this secretive animal, using its spines like feet. Aristotle likened the sea urchin to a "lantern" because its surrounding skin is missing.[172] The mouth parts of the sea urchin are still known as Aristotle's lantern.

Aristotle sought zoological information from people who made a living from animals: hunters; fishers; hog, eel, and horse breeders; cattlemen; shepherds; farmers; and midwives.

In Lesbos with Aristotle

In early June 2018, I visited Lesbos, hoping I would get slightly closer to Aristotle. My friend Ioannes Hatzopoulos, emeritus professor of environmental studies at the University of the Aegean, became my guide. Ioannes and I started seeing Lesbos with a visit to his home on the outskirts of Mytilene, the beautiful capital of the island. Rural Lesbos is olive tree and grapevine country. The goddess Athena and god Dionysos would have loved it. The countryside is crisscrossed by narrow roads and small traditional homes buried behind trees and gardens. Here one hears birds singing and smells the aromas of fruits and wildflowers. Rural Lesbos is ancient Greece.

[170] Aristotle, *Historia Animalium* 9.608b15–20.
[171] Aristotle, *Historia Animalium* 9.623b5–627b22; 631a9–631b3.
[172] Aristotle, *Historia Animalium* 4.530a32–531a7; *Parts of Animals* 4.679b28–681a9.

The moment we reached his secluded green grove, we walked straight to the fruit trees. I ate plums, peaches, and apricots. Biting the peaches was eating sweetness and aroma. Now, in rural Lesbos, I started dream-walking to my youth; running to fig trees and through vineyards without shoes. I would crawl under the vineyard and eat bunches of grapes slowly, enjoying each little sphere of pleasure.

After dinner, Ioannes and I drove to the seashore of Pyrgon Thermis, where we walked near the water. It was getting dark and the natural world was fading. We drove to downtown Mytilene and walked the streets in the harbor. I admired several houses, including the city hall, built with the grace and beauty of classical architecture.

The Sanctuary at the Center of Lesbos

Next day we drove to the Messon sanctuary, a fourth-century BCE large classical Greek temple built in the strategic center of Lesbos—in the sanctuary in the middle. This temple was dedicated to the worship and celebration of Zeus, Hera, and Dionysos. But the Messon sanctuary had an additional quality and practical usefulness. It served like the Olympics in Greece. It was an inducement to reconciliation and peace between the city-states of Lesbos, especially the powerful and competing poleis of Mytilene, Methymna, and Eressos. The Messon temple is thirty-five kilometers from Mytilene.

I was astonished by the attractiveness of the natural world embracing the Messon temple—a natural world that almost certainly attracted the passion and scientific curiosity of Aristotle. It is a valley surrounded by foothills of mountains with names like Lepetymnos and Pyrraion. Pine trees and olive trees are everywhere. In addition to the mountains, the valley is graced by the Kalloni gulf and lagoon and the nearby ancient polis of Pyttha and the colorful villages dotting the seaside of Kalloni. This is where young Aristotle did most of his biological research during the years 345 to 342 BCE.

Aristotle's scientific research sparked the golden age of Greek science. He taught that "**Nature does nothing in vain. Every living thing in the natural world has a purpose.**"

The Christians, however, destroyed Aristotle's world in Lesbos. They nearly wiped out even the stones of the Messon temple. Some of the columns and stones of the Messon sanctuary they used for the Christian basilica they built over the temple. But the stones and columns they did not use, they burned or smashed to very small pieces. This dreadful fact comes from a relatively recent Greek archaeological study of the Messon sanctuary.[173]

My friend and I ate Kalloni fish like *sparos*, *kephalos*, and *sardines*, which Aristotle studied. We then visited the wetlands in the lagoon of Kalloni and admired the grace and beauty of grebes, ducks, herons, and stilts. Those birds also spoke to Aristotle as much as they spoke to me. In the absence of human intervention, there's a wonderful eternity in the natural world.

History of Animals

Aristotle did not do his research for the sake of research alone. He often used his data to support his theoretical vision. He admitted that when evidence was not sufficient, he withheld judgment until more observations created the necessary knowledge for the support of theory. He stressed the seminal importance of facts in science, not allowing theories to obscure research unless those theories agreed with the facts.[174]

He was the first student of nature who approached nature with a thorough program of empirical research, fully expecting data for an explanation of the phenomena he studied.

The zoological works of Aristotle include nine books on the *History of Animals*, five books on the *Generation of Animals*, and four books on the *Parts of Animals*. He also wrote two shorter studies on the *Motion of Animals* and on the *Progression of Animals*.

The *History of Animals* is about the facts Aristotle discovered in his comparative zoological studies. He explains these facts in the *Parts of Animals*. He studied something like 500 animals. His keen observations and insights on the embryology, physiology, anatomy, and behavior of animals are original and largely accurate. They astonish me to no end. I studied zoology at the University of Illinois, but my classes on animals barely touched the philosophy in Aristotle's biology. Aristotle studied animals deeply. He loved them and

[173] Greek Ministry of Culture, *ἐν τῶ ἴρω τῶ ἐμ Μέccω* (*In the Sanctuary in the Middle*) Mytilene, 2004, pp. 22–25, 76–81.

[174] Aristotle, *Generation of Animals* 760b27–33.

thought they were beautiful. He understood them as living beings relating to each other and to the perfect natural world that brought them into life.

Aristotle's biological observations mirror so much knowledge and wisdom that, like the Antikythera computer, were more than a millennium ahead of their time. They constitute the greatest contribution any human being has ever made to science.

Aristotle's biology, says the French scholar Pierre Pellegrin, "constitutes such an important enterprise that it is hard to find anything comparable in the history of science."[175] About a third of the surviving Aristotelian texts deal with animals and the natural world. The interests of Aristotle are "rooted in living things."[176] Aristotle had a large footprint in Greek and Western culture, especially science.

E. R. Dodds, a British classical scholar, dates the flowering of Greek science with the establishment of the Lyceum by Aristotle in 335 BCE. He says that from 335 BCE until the end of the third century BCE, Greek rationalism "appeared to be on the verge of final triumph." This was also "the great age of intellectual discovery" during which Greek science changed "from an untidy jumble of isolated observations mixed with a priori guesses into a system of methodical disciplines." In mathematics and astronomy, the Greeks were supreme down to the sixteenth century.[177]

Dodds forgot Hippocrates, whose life speaks of rationalism and disciplinary science, not of an untidy jumble of guesses. Archimedes and the third century BCE were another clear expression of rationalism, which lasted well beyond the end of the first century BCE. But Dodds is right that Aristotle and the establishment of the Lyceum in 335 BCE in Athens made a difference in Greek science. For about thirteen years, 336 to 323 BCE, Aristotle taught probably hundreds of students on the rigors and beauty of science and philosophy. He created a library and a Mouseion in his school. The Mouseion was the house of the goddesses of learning, which for Aristotle, meant doing advanced studies in science, philosophy, and the natural world. In chapter 6, I examine how these two great Aristotelian traditions, a library and the Mouseion for the

[175] Pierre Pellegrin, "Aristotle," in *Greek Thought: A Guide to Classical Knowledge*, ed. Jacques Brunschwig and Geoffrey E. R. Lloyd, tr. Catherine Porter (Cambridge, MA: Harvard University Press, 2000), 567.

[176] John Herman Randall, Jr., *Aristotle* (New York: Columbia University Press, 1960) 220.

[177] E. R. Dodds, *The Greeks and the Irrational* (Berkeley: University of California Press, 1951) 236–237.

encouragement of scientific studies, enriched the Alexandrian Age and made possible the scientific revolution of the Alexandrian Era, especially the discoveries of the third and second century BCE.[178]

Reading the biological works of Aristotle, one has the impression Aristotle was the precursor of modern biologists, especially Charles Darwin, 1809–1882. When William Ogle sent Darwin his Oxford translation of Aristotle's *Parts of Animals*, Darwin thanked him profusely for the pleasure he felt from reading Aristotle. "I have rarely read anything which interested me more," Darwin wrote. "From quotations which I had seen I had a high notion of Aristotle's merits, but I had not the most remote notion what a wonderful man he was. [Carl] Linnaeus and [Georges] Cuvier have been my two gods, though in very different ways, but they were mere schoolboys to old Aristotle.... I never realized, before reading your book, to what an enormous summation of labour we owe even our common knowledge."[179]

Linnaeus was a Swedish botanist who flourished in the eighteenth century, 1707–1778; and Cuvier, the French father of modern comparative anatomy and paleontology, lived in both the eighteenth and nineteenth centuries, 1769–1832. They were Darwin's "gods." They were, in fact, two of the greatest scientists of Europe who had studied Aristotle.

Georges Cuvier praised Aristotle to heaven, making him a benefactor of humankind and one of the immortals. He erroneously thought that philosophy before Aristotle was entirely speculative, full of abstractions. In fact, Cuvier argued that science did not exist before Aristotle. "Science," he said, "came entirely out of the head of Aristotle just like Athena, fully armed, came out of the head of Zeus." Cuvier had no doubt Aristotle, "the pupil of Plato, discovered and demonstrated more truths and did more scientific work, in one life of 62 years than 20 centuries after him." Cuvier had tremendous respect for Aristotle's "History of Animals." He considered it the "only study of comparative anatomy" down to the seventeenth century. The "History of Animals," he said, is more than a series of descriptions of different animals. It is closer to general anatomy where Aristotle explains the differences and the resemblances

[178] Orjan Wikander, "Gadgets and Scientific Instruments," in *The Oxford Handbook of Engineering and Technology in the Classical World*, ed. John Peter Oleson (New York: Oxford University Press, 2008) 785–799.

[179] Darwin to Ogle, February 22, 1882, in Francis Darwin, ed., *The Life and Letters of Charles Darwin*, 2 Vols. (New York: Basic Books, 1959) II, 427.

of animals, stressing the comparative examination of their organs. It is in the *History of Animals* that Aristotle lays the foundations of "the great classifications with the most perfect exactitude." The *History of Animals* of Aristotle is a work of the "most remarkable and astonishing perfection," qualities that characterized the contributions Aristotle made to most branches of the science of zoology.

Cuvier summarized his views and respect for Aristotle as follows:

"We should consider Aristotle one of the greatest observers who ever lived. Without any doubt his genius for classifying was the most extraordinary ever produced by nature.... The great divisions and subdivisions of the animal kingdom by Aristotle are astonishingly precise. They have almost all resisted additions by science."[180]

A twenty-first-century student of Aristotle, Armand Marie Leroi, professor of biology at the Imperial College, London, suggested that, "Darwin's works are infused with Aristotle."[181] They are. Indeed, Aristotelian influence is all over Darwin's thought not because Darwin read much Aristotle but primarily because Darwin learned from his two gods, Linnaeus and Cuvier, who were serious students of Aristotle.

Did Aristotle have a theory of evolution? Yes, he did. But he and Darwin (and Darwin's followers, including Leroi) understood and spoke of evolution in slightly different terms. Aristotle lived at a time the Greeks thought of the Earth and the natural world as alive and divine. Darwin lived at a time his fellow Europeans thought of the Earth as a pile of "resources" waiting for exploitation by the mighty industrializing (machine-like) human. This included animals, which Aristotle saw as members of his divine Cosmos and paths to science, truth, and beauty.

Critics of Aristotle stumble over Greek natural philosophy. They cannot understand why Aristotle would consider men and animals potentially perfect and divine beings. So, naturally, the critics find Aristotle's teleology (the ultimate potential perfect purpose of the natural world, society, and the Cosmos) problematic. That is, they ask, why is Aristotle so concerned with purpose? Perhaps, the critics fail to understand that teleology comes from two words: *telos* and *teleios*, that mean fulfillment, purpose, and perfection. In addition,

[180] Georges Cuvier, *Histoire des sciences naturelles*, 5 vols (Paris: Fortin, Masson, 1841) I, 130–149, 432.
[181] Armand Marie Leroi, *The Lagoon: How Aristotle Invented Science* (New York: Viking, 2014), 275.

they put themselves at a higher moral plane than Aristotle. They accuse him of favoring males over females; and of suggesting that some humans are incapable of using reason, a condition that makes these humans "natural slaves." This misunderstanding, and a pathological urge to appear fashionable at a time of the raging race priorities in America, probably explain the offensive epithet of calling Aristotle a "jerk" by Agnes Callard of the University of Chicago.[182]

Darwin, too, was incapable of understanding Aristotelian teleology, the idea that the world was perfect and that nature does nothing in vain. Darwin was rather cautious, not given to large theories. Unlike Aristotle, he could not possibly think of perfection in the natural world at a time of so much human-induced destruction. The fossil record told him of lost worlds of larger animals. But Darwin's fellow humans had become the living dinosaurs wrecking the natural world. Who would now protect the surviving species?

Like the undemocratic, tough, and class-ridden society of his country in the nineteenth century, Darwin thought of nature dressed in tooth and claws. The strongest survive. Darwin called this ceaseless struggle natural selection. Out of this perpetual selection, the killing of the weak by the strong, there emerged Darwin's "Origin of Species," his explanation of how the world of plants and animals made it to his time.

Darwin's education and knowledge were limited by Christian metaphysical assumptions, all contrary to science. But Darwin persisted and his biological explorations opened his mind to the deep wisdom of the natural world. Like Aristotle, he detected connections between animals, plants, and inanimate things. He would have probably agreed with Aristotle's explanation of the workings of nature:

"Nature moves very slowly and in tiny steps from inanimate things to animals. In such constant evolution it is difficult to see the middle or boundaries of change. Plants stand next to lifeless things. Plants, however, move towards life. Some of them are more alive than others. The entire genus of plants compared to other bodies looks full of life. But plants compared to animals look rather lifeless ... There's a continuous evolution of plants towards animals."[183]

Next to Darwin, another British scientist explained Aristotle to the world. This was the naturalist and translator of Aristotle, D'Arcy Wentworth

[182] Agnes Callard, "Aristotle is a Jerk, but We Shouldn't Cancel Him," *The New York Times*, July 23, 2020.

[183] Aristotle, *Historia Animalium* 588b4–11.

Thompson, who did his work in the early twentieth century. Thompson reported that, for all intents and purposes, the oldest European universities, Bologna, Paris, and Oxford, really came into being to study Aristotle.

Thompson loved Aristotle. He had no doubt Aristotle was the "great biologist" of antiquity and a "very great naturalist." In fact, in natural history, Aristotle is entirely like us. His language, method, and thinking became our language, method, and thinking. "He had familiar knowledge of a thousand varied forms of life, of bird and beast, and plant and creeping thing. He was careful to note their least details of outward structure, and curious to probe by dissection into their parts within. He studied the metamorphoses of gnat and butterfly and opened the bird's egg to find the mystery of incipient life in the embryo's chick. He recognized great problems of biology that are still ours today, problems of heredity, sex, of nutrition and growth. Of adaptation, of the struggle for existence, of the orderly sequence of Nature's plans. Above all, he was a student of Life itself," Thompson wrote in 1913.

Thompson was right. Aristotle was primarily a great biologist. "We know," Thompson said, "that the history of biology harks back to Aristotle by a road that is straight and clear, but that beyond him the road is broken and the lights are dim. And we have seen that biology was no mere by-play of Aristotle's learned leisure, but was a large intricate part of the vast equipment of his mind."[184]

In addition, Thompson summarized the influence of Aristotle as follows:

"So, it comes to pass that for two thousand years and throughout all lands men have come to Aristotle. Arab and Moor and Syrian and Jew treasured his books while the western world sat in darkness; the great centuries of Scholasticism hung upon his words; the oldest of our Universities, Bologna, Paris, Oxford, were based upon his teaching, yea, all but established for his study. Where he has been, seen or unseen, his influence remains; even the Moor and the Arab find in him, to this day, a teacher after their own hearts; a teacher of eternal verities, telling of sleep and dreams, of youth and age, of life and death, of generation and corruption, of growth and of decay: a guide to the book of Nature, a revealer of the Spirit, a prophet of the works of God."[185]

---

[184] D'Arcy Wentworth Thompson, *On Aristotle as a Biologist* (Oxford: Clarendon Press, 1913) 11–30.

[185] Thompson, "Natural Science," in *The Legacy of Greece*, ed. Richard Livingston (New York: Oxford University Press, 1969) 159–160.

Werner Jaeger, a German classicist also writing in the early twentieth century, described the influence and genius of Aristotle like this: "The name Aristotle," he said, "suggests impersonality, timelessness, intellectual sovereignty over the whole world of abstract thought throughout long stretches of history, and scholastic idolatry.... In his works everything is the most perfect, polished, logical art.... [Aristotle was] the compelling organizer of reality and of science."[186]

Another scholar, George Sarton, professor of the history of science at Harvard, wrote in the mid-twentieth century that the achievements of Aristotle "are astounding and many of the results attained by him kept their validity for two thousand years."[187]

In the late 1950s, Philip P. Wiener and Aaron Noland, editors of the *Journal of the History of Ideas*, called Aristotle "the encyclopedist of ancient Greece." They traced his influence on Euclid and praised his authority in logic, biology, and "all the other natural sciences." They said Aristotle had "profound influence on cosmology, ethics, political theory, and esthetics. To most scholars of the Middle Ages he was 'the master of all those who know.'"[188]

In 1975, the American philosopher Anthony Preus concluded that the biological work of Aristotle was "crucial" for the advancement of sciences in antiquity. In fact, the biology of Aristotle became the model for biological and scientific research in the Mouseion of Alexandria. The Mouseion expanded the Aristotelian work in the life sciences and became influential in biology and medicine for centuries.

Aristotle "had developed a comprehensive theoretical system which strongly encouraged biological investigation," Preus wrote. The metaphysical assumptions of Aristotle for animals were interesting and compelling. Preus said Aristotle was "the first philosopher we know to make biological individuals metaphysically primary." And just like the Pythagoreans and the Platonists expanded the importance of mathematics by "giving metaphysical priority to

[186] Werner Jaeger, *Aristotle: Fundamentals of the History of His Development* (London: Oxford University Press, 1967) 368–406.

[187] George Sarton, *Ancient Science and Modern Civilization* (Lincoln, NE: University of Nebraska Press, 1954) 10.

[188] Philip P. Wiener and Aaron Noland, "The Classical Heritage," in Wiener and Noland, eds., *Roots of Scientific Thought* (New York: Basic Books, 1957) 7–8.

the objects of that science," Aristotle made biology interesting by developing the philosophical foundations of biology.

"Aristotle is one of the few people who investigated biological phenomena as an essential and central part of his philosophical investigations," wrote Preus. "The close relationship between metaphysical theory and scientific curiosity is largely responsible ... for the many unique characteristics of Aristotle's biological works, and for much of their influence and charm."[189]

In the late 1990s, Christos Evangeliou, professor of Hellenic philosophy at Towson University in Maryland, credited Aristotle for "developing a comprehensive system of rational thought" that included the "intelligible and divine realm of Nous [mind]."[190]

In our times, as I already mentioned, it is Armand Marie Leroi who has brought Aristotle back to the central stage of science. His 2014 book, *The Lagoon: How Aristotle Invented Science*, is a valuable account of Aristotle's original work in biology. Aristotle, says Leroi, wrote "beautiful scientific works" about the "living world in all its beauty." Indeed, he did. Aristotle did more than invent zoology. He "invented science itself." His books educated countless men and women for the past two millennia. But, in our times, Aristotelian ideas made a difference in the education of men like Carl Linnaeus and Georges Cuvier, and, through them, Charles Darwin.

Leroi is a student of Darwin. Like Darwin, his admiration of Aristotle came as much from studying Aristotle's biology as it did from studying Linnaeus's and Cuvier's studying of Aristotle. Those who made modern science, says Leroi, "absorbed the very structure of his [Aristotle's] thought. And so, his thought became our thought, even when we do not know it is. His ideas flow like a subterranean river through the history of our science, surfacing now and then as spring, with ideas that are apparently new but are, in fact, very old."[191]

In general, Aristotle was the apotheosis of the scientific and philosophical spirit of the Greeks. His biology is comprehensive, scientific, and philosophical. It covers all life. Even Leroi becomes lyrical in his appreciation of Aristotle's

---

[189] Anthony Preus, *Science and Philosophy in Aristotle's Biological Works* (New York: Georg Olms, 1975), 259–260.

[190] Christos C. Evangeliou, *The Hellenic Philosophy: Between Europe, Asia, and Africa* (Binghamton University, 1997) 70.

[191] Leroi, *The Lagoon* 7–9, 91–93. Leroi blundered badly in his assessment of Plato, whom he accused of being a totalitarian, an enemy of science, and other such nonsense (88).

achievement. Aristotle explains the metamorphosis of mater into living beings. His logic, says Leroi, captures how nature paints the self-creation of life as a work of art.[192]

The eleventh-century Arab scholar Said al-Andalusi went beyond lyricism in exalting Aristotle. He thought of Aristotle as a divine being. Al-Andalusi loved Aristotle's scientific and philosophical choices. "Aristotle became the leader of the philosophers and one who united all the virtues of the scholars. No one can object if Allah assembled the world in one individual [Aristotle]," he wrote.[193]

The Greek gods, too, had assembled the world in Aristotle. The golden age of Greek science, starting in the late fourth century BCE and going all the way to the second century of our era, would have been inconceivable without Aristotle. It is possible that the *Elements* of Euclid was also a model for the application of Aristotelian science.

Simple, Uniform, and Orderly Movement of the Stars

During the century of Aristotle, the Greeks advanced mathematical astronomy with their use of geometry in modeling the movement of the stars. Eudoxos of Knidos, a mathematical genius second only to Archimedes, started the Greeks on this path. Thomas Heath, the pioneer modern British scholar on the history of Greek science, is right in suggesting that Eudoxos "invented a geometrical hypothesis for explaining the movements of the planets which for ingenuity and elegance yields to none."[194]

Eudoxos earned the praise of scholars from ancient times. Diogenes Laertios, for example, says that Eudoxos "discovered the properties of curves."[195] And Proklos suggests that Euclid systematized "many of the theorems" of Eudoxos in his *Elements*.[196] Eudoxos studied under Plato who probably learned some of his mathematics from his pupil. Eudoxos was behind the solid geometrical figures or *polyhedra*, which Plato put at the center of his cosmological system.

[192] Leroi, *The Lagoon*, 202.

[193] Said al-Andalusi, *Book of the Categories of Nations*, tr. and ed. Semaan I. Salem and Alok Kumar (Austin: University of Texas Press, 1991) 21–30.

[194] Thomas Heath, *Aristarchus of Samos* (First publishes 1913, Elibron Classics, 2007) 191.

[195] Diogenes Laertios, *Lives of Eminent Philosophers* 8.90.

[196] Proklos, *A Commentary on the First Book of Euclid's Elements* 68.

Eudoxos argued that the movement of the stars is simple, uniform, and orderly. He said that each star moved on one or more spheres, which rotated around the Earth as their common or homocentric center.[197] For example, according to ancient commentators, Eudoxos arranged the stars in a circular movement that ends up looking like the figure eight: each planet is located on the equator of a sphere that moves uniformly inside another sphere; each of the two spheres has different poles but rotates in the opposite direction; the combined motion of the planet looks like a hippopede, a horse fetter shaped very much like the number eight.[198]

Eudoxos was making his astronomical proposals in the fourth century BCE, a time when the Greek vision of the heavens was on the same page with the sixth-century BCE cosmology of Anaximander: the Earth was still at the very center of the universe, the Earth being a very large sphere but a sphere that was motionless. There was another sphere, which accommodated the fixed stars. That sphere rotated every day around the Earth. The Sun, the Moon, and the planets existed between the sphere of the fixed stars and the Earth. They, too, spun around the Earth.

Plato and the Cosmos

Now we return to Plato. He thought the Greeks lived by the sea like ants or frogs in a marsh. They occupied but a tiny portion of a very large Earth. Their country extended from the River Phasis to the Pillars of Herakles.[199] The River Phasis is now the River Rioni in the Asian country of Georgia, bordering Russia and the Black Sea. The ancients thought of the River Phasis as the dividing line between Asia and Europe. And the Pillars of Herakles are the Straits of Gibraltar, the narrow passage between Europe and Africa.

Plato saw the Earth inundated by hollows, which are full of water, mist, and air. These hollows are also the homes of men and animals, all connected with underground channels and rivers. The largest surface river circling the Earth is Ocean—*Okeanos*. There is another large river running in the opposite direction of *Okeanos* and the Greeks called it Acheron. This river flows through

---

[197] Aristotle, *Metaphysics* XII.8.1073b17–1074a14.

[198] Heath, *Aristarchus of Samos*, 190–212.

[199] Plato, *Phaidon* 109b, ed. C. J. Rowe (Cambridge Greek and Latin Classics, 2001).

deserts and under the ground where it forms the Acherusian Lake. This lake, Plato assures us, is a transition station for the souls of the majority of the dead. After a certain appointed time at the *Acherusian* Lake, the souls return to the Earth as new beings.

The headwaters of a third river spring somewhere between the first two. Not far from its source, this river drops into a region burning with fire, forming a lake larger than the Mediterranean Sea. The river, boiling water and mud, winds its way past the *Acherusian* Lake and drops deep into the Earth and runs into *Tartaros* or the Underworld, a region under the Earth for the punishment of heinous criminals. The lava of the burning river (*Pyriphlegethon*) ends up at several areas of the world.

The fourth river (*Stygion*) is opposite to the *Pyriphlegethon* River. The *Stygion* River flows into a wild blue-gray region forming lake Styx. Once the waters of the *Stygion* drop into lake Styx, they become extremely powerful. They rush in the opposite direction of the *Pyriphlegethon* and the *Acherusian* Lake but don't mix with their waters.

Plato puts the Earth, like other stars, in the pure sky or ether. He assures us that if one were to look at the Earth from space, the Earth would be like a spherical ball surrounded by twelve strips of leather: dazzling in color. One region of the Earth is green and of extraordinary beauty; another is golden; and yet another is white, even whiter than chalk or snow. Plants, trees, and flowers fill the surface of the Earth with fruit, color, and beauty. In addition, the Earth has precious stones, gold, silver, and other metals. Seeing the Earth for Plato is the ultimate pleasure—a sight for the blessed.[200]

I have already discussed the mathematical interests of Plato, his learning from the Pythagorean tradition in number and science. That tradition also gave him the idea that souls are immortal, moving from one being to another.

Plato used his moral philosophy and considerable fluency in mathematics and astronomy, which he thought as a "blessed science."[201] His dialogue *Timaios* was about the making of the Cosmos from stuff in chaos. That was also the time when time itself did not exist and the Earth was without form. *Timaios* has the gods using geometry, the triangle being the key to the atomic structure of the elements of life: Earth, air, water, and fire.

[200] Plato, *Phaidon* 109a–113c (111a2–3: ώστε αυτήν ιδειν [την γην] είναι θέαμα ευδαιμονων θεατων).
[201] Plato, *Epinomis* 992d.

Thus, geometry was a model of order and measurement and symmetry, regularity and harmony. Geometry opened the divinely ordered Cosmos to inspection and measurement, including understanding of the laws of motion of the star gods. In addition, Plato considered geometry, mathematics, and the sciences essential to elevating the soul to truth. From there, men would be able to grasp the beautiful and the good.[202]

With such understanding, humans could model their behavior from the lessons of universal order. Plato urges us to approximate the troubling motions within us to the unerring motions of the stars.

First of all, humans must coordinate the motions of their bodies with those of their souls, and then approximate those motions with the motion of the Cosmos. Athletics take care of the motions of the body in the same way the use of reason and doing good bring the motions of the soul in unison with the motions of the universe. Humans could then become more beautiful and better, fulfilling the main purpose of philosophy.

Order thus must prevail in the body, soul, family, and the polis. Only then, Plato says, does the perfection of the soul, body, family, and polis shine: "Communion, friendship, orderliness and temperance and justice bind together heavens and Earth," he wrote. "This universe is called Cosmos or order, not disorder or misrule ... geometrical equality or proportion is mighty both among gods and men."[203]

Order—*taksis*—and Cosmos for Plato mirrored the eternal universe and the beautiful. But order for Plato was also the necessary antidote to the political disorder and decay in Athens of his time. This also explains his inspiration from the world of being and eternal forms and models, a world of perfection. This was a world of "unchanging and harmonious order, where nothing can do or suffer injustice, where all is in order according to reason."[204]

Next to this perfect universe, there was another Cosmos of number, harmony, proportion of musical strings, and the heavens of the stars. This Cosmos mirrored the perfect being and the eternal forms. But Plato looked at Cosmos as the structure of the soul and of the polis, geometry linking them, bringing heavens and Earth together.

---

[202] Plato, *Republic* 527b–531c.

[203] Plato, *Gorgias* 503e, 507e–508a.

[204] Plato, *Republic* 500c.

The connection Plato made between the soul, the state, and the universe is probably from Pythagoras. No wonder he made the Pythagorean "Timaios" the protagonist of his story. *Timaios* is an extraordinary cosmological document that had a profound influence in antiquity, and not an insignificant effect in the Middle Ages all the way to the Renaissance. One has to wait until Isaac Newton in the eighteenth century for another cosmological statement comparable to the *Timaios* of Plato.

Plato employed theology, philosophy, and science in the construction of a perfect Cosmos, a Republic of gods and citizens. This ideal polis was modeled on a prototype in the mind of Zeus, a super father god he called *demiourgos* or handicraftsman, the name Greeks gave to those skilled in making useful things with their hands. According to "Timaios," this god did what he did out of pure goodness: "The divine craftsman of the Cosmos was good, that is, he had no envy," Plato wrote. "Because he had no jealousy, he wanted everything to become as much like himself as was possible.... The divine craftsman wanted all things to be good and nothing imperfect. He found the visible universe in chaos and replaced disorder with order because he was convinced order was so much better than disorder.... He constructed the Cosmos with a soul in which he implanted reason. He wanted to create a world that would be excellent and supreme. Thus, by the providence of the divine craftsman, the world, truly alive with a soul and intelligence, came into being."[205]

Plato saw the mathematical hand of *demiourgos* in the fabrication of the universe. He mixed his *polyhedra* with his elementary particles—Earth, fire, air, and water—in order to show the geometrical structure of the particles. Differences among these particles trigger motion as sameness demands rest. And just as the Cosmos is a perfect sphere, its motion is a perfectly circular one.[206]

But one has to see this Cosmos to admire its perfection. So, Plato put special emphasis on vision, which he explained as pure fire inside the eye. That fire becomes one with the fire outside the eye forming a ray of vision. When this ray of vision strikes an object, it generates motions that the ray sends to the whole body before they reach the soul, at which point we can see.[207]

That way we see the motions of the heavens, a gift of the gods. Heavenly motions give us time and, therefore, the opportunity to study and learn from

[205] Plato, *Timaios* 29e–30c.
[206] Plato, *Timaios* 33b.
[207] Plato, *Timaios* 47ab, 61c–68d.

nature. The result of such study is philosophy, by far the greatest gift the gods bestowed on us.

This world, the Cosmos of Plato, is made up of stars. The movements of these stars are circular; reason governs the star gods; and all rational movement is circular. This astronomical model, which advanced astronomy, assumes regular circular motions for the stars and may also explain irregular movements in the heavens.

The astronomy of Plato was testimony to his understanding of the Earth. He thought the Earth vast and circling around a pole extending through the universe. Plato also accepted the idea of the geocentric Cosmos, endowed with reason, an eternal soul, beauty, life, divinity, and permanence. The patterns in the sky, he said, are very precise, the heavens being like embroidery. The star gods are rational and beyond interference in the workings of nature on Earth or in the vast universe.[208]

In fact, Plato saw the Cosmos like a community of gods whose existence was as clear as daylight. In his *Laws*, one of the persons of the dialogue, Kleinias, says: "Just look at the Earth, and the Sun, and the stars and the universe; look at the fair order of the seasons, and the division of them into years and months. They certainly furnish proofs of the existence of the gods. Moreover, all Hellenes and barbarians believe in them."[209]

In the seventh book of the *Republic*, Plato describes a practical need for astronomy because it helps people understand the phenomena of seasons, months, and years, thus facilitating farming, sailing, and the military.[210]

The Heavens of Aristotle

The insightful and original Platonic cosmology must have loomed large in the mind of Plato's greatest student: Aristotle. Yet, Aristotle did not see the world with Plato's eyes, though he was a "Platonic philosopher."[211]

Aristotle starts his cosmology with the shape of the Earth. He was convinced the Earth was like a sphere. He brought even common sense in support

208 Plato, *Phaidon* 97–99, 108–110; *Republic* books 7 and 10; and *Timaios*.

209 Plato, *Laws* 886.

210 Plato, *Republic* 7.527d.

211 Christos C. Evangeliou, *Essays and Poems on Aristotle* (2 vols., Sioux City, IA: Parnassos Press, 2016) I, v.

of that contention. He asked: "How else would eclipses of the Moon show segments shaped as we see them? The Moon appears to us each month in a variety of shapes, including straight, gibbous, and concave. However, in eclipses the outline of the Moon is always curved. Considering that the Earth causes the eclipses of the Moon because it passes between the Sun and the Moon, the form of the eclipse mirrors the shape of the Earth. Here's then why the Earth is spherical."[212]

Aristotle also considered the circular Earth a relatively small star that had no motion. He placed the center of the immovable Earth right at the center of the Cosmos. The inspiration for these observations, or hypotheses, came to Aristotle from his philosophical conviction that the Earth and the entire universe were special: they were immortal.

He wrote his cosmological *On the Heavens* to explain the eternal nature of the geocentric universe about which he reached the following conclusions: only one in existence; more than one universe is impossible; built by no super god; has no beginning or end, being exempt from decay and generation; and, therefore, the universe has always existed and will always exist, being immortal like the gods.

"All men have some understanding of the nature of the gods. In addition, all who believe in the existence of the gods, be they barbarians or Hellenes, agree in giving the highest honor to them.... In all of time past, and according to our inherited records and how far back they reach, no change appears to have taken place either in the structure of outermost heaven or in any of its proper parts. Also, the common name for the eternity of the universe, which we inherited from our distant ancestors, speaks to us to this day that our ancestors shared our cosmological ideas. It's clear that the same ideas come to the minds of men not once or twice but again and again. Thus, our distant ancestors decided that the material of the universe, going beyond air, Earth, fire, and water, deserved a name of its own: *aither*, ether, something that runs forever, an eternity of time," Aristotle wrote.[213]

Aristotle explained that the heavenly spheres, themselves star gods, move in a circular motion forever because of *eros*-love or the mighty attraction-gravitation

[212] Aristotle, *On the Heavens* 2.14.297b14–31.

[213] Aristotle, *On the Heavens* 1.3.270b6–25; on the immortality of the one universe: 1.9.277b26–2.14.296a34.

existing between them and an unmoved mover god or gods who are immortal and non-material.[214]

The Brightness of the Stars

Adding the cosmologies of Plato and Aristotle, we have an impressive and evolving explanation of how the Cosmos works. This geocentric astronomical system, however, was certainly elegant and mathematical, and, according to Plato and Aristotle, immortal. The gods made it up and ran it forever. Yet, Plato and Aristotle failed to explain the universe fully.

Autolykos of Pitane, born between 360 and 290 BCE, explained the workings of the celestial sphere and the risings and settings of the fixed stars. He tried to account for the different brightness of the stars on the basis of their unequal distances from the Earth. His two surviving works, "On the Rotating [Celestial] Sphere" and "On the Risings and Settings [of the Stars]," are the oldest complete astronomical texts.[215] Autolykos and most Greek scientists stayed with the geocentric Cosmos of their making.

The planets that Eudoxos named—Kronos, Zeus, Ares, Aphrodite, and Hermes—were the planets people knew in the ancient world. Eudoxos also put the fixed stars visible in Greece into the constellations we know today. Eudoxos borrowed the constellations Orion, Pleiades, Hyades, and the Bear/Bootes from Homer and Hesiod.

I continue the story of Greek cosmology in chapters 5 to 7. However, we first need to take a look at Archimedes, who was also an astronomer. The next chapter is about him.

---

[214] Aristotle, *Metaphysics* 12; *Physics* 8.

[215] Autolykos, *On the Rotating [Celestial] Sphere* and *On the Risings and Settings [of the Stars]*, 2 books, in *The Complete Surviving Works of Autolykos of Pitane*, ed. Evaggelos Spandagos, in Greek (Athens: Aithra, 2002).

Chapter 4

Archimedes: Our Model and Ideal Scientist

Archimedes was a genius. His mathematical and engineering inventions became some of the pillars of the golden age of Greek science. He is credited with the earliest mechanical model of the Cosmos. He borrowed the technology of the mechanical gears from an early contemporary engineer named Ktesibios. The gears drove the Antikythera Mechanism.

The Science and Engineering of Archimedes

In one of his books, *Psammites* (*The Sand-Reckoner*), Archimedes attempted to measure the size of the universe by calculating the number of the grains of sand necessary to fill the sphere of the fixed stars. That number turned out to be a huge one: something of the order of ten to the sixty-third power—10^{63}. He correctly measured the angle of seeing the Sun in the sky: thirty-two to twenty-seven sixtieths of a degree. "The diameter of the Sun," he said, "is about thirty times greater than the diameter of the Moon and not greater ... the diameter of the Sun is greater than the side of the chiliagon [a thousand-sided polygon] inscribed in the greatest circle in the (sphere of the) universe."[216]

Archimedes was the greatest Greek mathematician of the ancient world and, next to Aristotle, the greatest scientist who ever lived. He was born in Syracuse in 287 BCE. He was also a philosopher, an astronomer, a physicist,

[216] Archimedes, *The Sand-Reckoner* 136.22–137.13.

an engineer, and an inventor. In fact, like Aristotle before him, he set the foundations of Greek and Western science. In a metaphorical sense, all Western science is a series of comments to Archimedes.

Marshall Clagett, an American historian who studied the influence of Archimedes on the thought of the Middle Ages, said one can barely exaggerate the importance of Archimedes in the history of science. He was an admired model in his own times and later ages. We see his name in the works of great scientists. Galileo, for example, paid his debt to Archimedes by mentioning him "over one hundred times and in almost Homeric hyperbole, using such expressions as *suprahumanus Archimedes, inimitabilis Archimedes, divinissimus Archimede.*" Galileo praised Archimedes because Archimedes employed mathematics "in the treatment of physical problems as well as in the originality and fertility of his mathematical techniques."[217]

Archimedes invented a variety of machines and fields of science like statics, hydrostatics, and combinatorics. Out of this creativity he set the foundations of calculus and mathematical physics. He was a mathematician of "astounding originality," says N. M. Swerdlow, astronomer and professor of the history of science at Caltech. There is nothing in the work of Archimedes that he did not discover. We continue to study his works.[218] His writings were essential for the rebirth and evolution of science. The scientists of the Renaissance, and not merely Galileo, "venerated" Archimedes "for his bold inventions," adopting him as "a strict model."[219] Since the Renaissance, scientists have been looking up to Archimedes. He is the "model and ideal" they try to "follow and attain."[220]

Archimedes contributed a lot to the scientific revolution of the third century. His father, Pheidias, was an astronomer who gave him a good education. He probably did his advanced studies in Alexandria. However, Archimedes spent most of his life in Syracuse. He put his scientific discoveries and reflections

[217] Marshall Clagett, *Archimedes in the Middle Ages*, Volume I: *The Arabo-Latin Tradition* (Madison: The University of Wisconsin Press, 1964) 1.

[218] N. M. Swerdlow, "The Recovery of the Exact Sciences of Antiquity: Mathematics, Astronomy, Geography," in *Rome Reborn: The Vatican Library and Renaissance Culture*, ed. Anthony Grafton (Washington, DC: Library of Congress, 1993) 131.

[219] Giuseppe Boscarino, "The Mystery of Archimedes: Archimedes, Physicist and Mathematician, Anti-Platonic and Anti-Aristotelian Philosopher," in *The Genius of Archimedes*, 313–322.

[220] Rene Dugas, *Mechanics in the Seventeenth Century* (Neuchatel, Switzerland: Griffon, 1958) 18.

in letters he wrote to Greek scientists, including his friend Eratosthenes working in Alexandria.

Eratosthenes was the director of the great Library of Alexandria. Among his several achievements, he measured the circumference of the Earth. Kleomedes, an astronomer who lived about 400 years after Archimedes and Eratosthenes, described how Eratosthenes did that astounding feat of science.

Eratosthenes assumed that Alexandria and Syene were in the same meridian. Syene, modern Aswan, Egypt, is on the banks of the River Nile 5,000 stades (= 900 kilometers) south of Alexandria. Eratosthenes knew that during the summer solstice at noon at Syene, the gnomon stood naked without a shadow, as the Sun was straight overhead. A gnomon was part of a sundial; a vertical stick or metal for measuring the shadow of the Sun.

Two hundred years after Eratosthenes, the geographer Strabo confirmed that during summer solstice, at noon, at Syene, the Sun cast no shadow on sundials.[221]

I have already mentioned Heliodoros, the Greek novelist of late antiquity. Writing about 600 years after Eratosthenes, he says that "in the region of Syene the rays of the Sun stand directly overhead at the summer solstice ... and so the water also at the bottom of the cistern receives a clear illumination for the same reason."[222]

Eratosthenes was the first scientific cartographer to make the first map of Egypt and the first map of the world. He chose Syene and Alexandria for his experiment to find the circumference of the Earth. He completed his calculation of the circumference of the Earth by examining the sundial at noon during summer solstice in Alexandria. If the Earth were flat, a gnomon at noon in Alexandria, like in Syene, would leave no shadow. But Eratosthenes knew the Earth was round. So, the Alexandria sundial marked the shadow of the Sun. Eratosthenes measured the angle the Sun's rays made with the vertical gnomon of the sundial during noon at summer solstice. That angle formed by the Sun's rays and a line from the center of the Earth was about seven degrees, or 1/50 of the meridian circle. He knew the distance between Syene and Alexandria was 5,000 stadia. One stadium is about 180 meters. He divided that distance by the angle and estimated the distance equivalent of one degree of the meridian was 700 stadia. Thus, the circumference of the Earth was 360 × 700 = 252,000 stadia.

---

[221] Strabo, *Geography* 2.5.7.

[222] Heliodoros, *Ethiopian Story* 9.22.

The number 360 represents the total degrees of the great circle or meridian of the Earth. The result of Eratosthenes's measurement of the circumference of the Earth, 252,000 stadia, is 24,662 miles,[223] a figure that is extremely accurate, the error being less than 1 percent of the modern estimation.

Eratosthenes's success, his science in measuring the meridian/circumference of the Earth, illustrates his command of optics, geometry, and geodesy. He practiced science, as we know it. His theoretical model was his knowledge of the Earth, which he depicted on a drawing. Russo says this was a "brilliant example of the power of scientific method: by going back and forth between the real world and the model, he gained information about the unknown regions of the earth, which no Ancient had ever seen." In fact, Russo reports that even scientists at the time of Galileo could not "improve on Eratosthenes's findings." Russo argues that the Greek kings of Egypt (Ptolemies) funded the work of Eratosthenes, being fundamental in drawing accurate maps. Eratosthenes was "the highest-ranking person in charge of the government's scientific policy."[224] No wonder that Archimedes corresponded with Eratosthenes.

The science letters of Archimedes, that is, his works, are models of brevity, conciseness, and clarity. Scholars like Eratosthenes copied and spread them throughout the Greek world. The surviving works of Archimedes came to us through medieval Greece or Byzantium, the Eastern Graeco-Roman Empire headquartered in Constantinople and founded by Emperor Constantine in 330.

The Archimedean works include *On the Equilibrium of Planes*, two books; *Quadrature of the Parabola*; *On the Sphere and Cylinder*, two books; *On Spiral Lines*; *On Conoids and Spheroids*; *On Floating Bodies*, two books; *Measurement of a Circle*; *Sand-Reckoner*; *The Method of Mechanical Theorems*; *The Cattle Problem*; *Stomachion*; and *Construction of the Regular Heptagon*. This last work came from the Arabic corpus of Archimedes.

In chapter 6, I will discuss the policy of the Macedonian kings ruling Egypt and funding advanced scientific research. They reorganized learning and research to fit the proposals of Aristotle, chief of which was about the investigator's respect for reality in either the natural world or the world of humans.

[223] Kleomedes, *On the Circular Motion of the Heavenly Bodies* I. 10.52, tr. Stylianos V. Spyridakis and Bradley P. Nystrom, *Ancient Greece: Documentary Perspectives* (Dubuque, Iowa: Kendall/Hunt, 1985) 268–270.

[224] Russo, *The Forgotten Revolution*, 68–69, 276.

Understanding of that reality, which Parmenides described as what is, would then become the heart of science, giving birth to fields of knowledge with their own rules; understanding of reality would also become a source of speculation and theoretical elaboration in the definition of a problem and its solution. Science would become a systematic way for searching for knowledge and truth.

Archimedes led the way in mathematics and engineering. Ancient and modern students of Archimedes admire his mathematical insight, his technological innovations, and his patriotism. He used his theoretical and engineering work to advance the wellbeing of his society. He put at the disposal of Syracuse all his knowledge and machines, sacrificing his life defending his country.

Marco Ceccarelli, Laboratory of Robotics and Mechatronics, University of Cassino, Italy, reports that the work of Archimedes exudes the virtues of "enthusiasm and optimism in mechanism design in developing technology for enhancing society and quality of life."[225] Geymonat is also convinced the work of Archimedes had social significance. After all, Archimedes did much more than drawing diagrams in the sand. According to Geymonat: "Archimedes—the inventor of the pump screw, ship-destroying machines of war, hydraulic winches, planetariums, and so forth—served as the classic example of one who could employ mathematics for use in everyday life."[226]

Archimedes combined geometry and mechanics in the interpretation and solution of problems. Wilbur Knorr said that Archimedes left a legacy of "practical inventions" and "masterful insights" from the "applications of geometry in mechanics and hydrostatics," which are exemplary of the "fruitful liaison of theory and practice that underlies great achievement in the physical sciences."[227] He accomplished the quadrature of the parabola, which resembles the squaring of the circle. He measured a segment of the parabola being four-thirds times the triangle it encloses.

Reviel Netz, professor of classics at Stanford University and translator of the works of Archimedes, summarized the importance of Archimedes as follows:

"Archimedes's methods of measurement formed the basis for reflections leading on to the calculus, while Archimedes's statics and hydrostatics formed

---

[225] Marco Ceccarelli, "The Mechanics of Archimedes: Towards Modern Mechanism Design," in *The Genius of Archimedes*, 177–187.

[226] Geymonat, *The Great Archimedes*, 89.

[227] Wilbur Knorr, "Archimedes," in *Greek Thought*, 553.

the basis for reflections leading on to mathematical physics. In this sense, the text of Archimedes is with us, simply, as modern science."[228] For example, an object in water loses weight equal to the volume of the water it displaces. In other words, the object is buoyed by a force equal to the weight of the water the object displaces. The combined weight of the water and the immersed object must equal that of the water of the same volume. Weight is the gravity pulling the object downwards and buoyancy is a force pulling the object upwards. These principles make up the science of hydrostatics, which Archimedes explained in his work "On Floating Bodies." The weight of the object is measured on a scale and the buoyancy is its volume.

In an international conference on the legacy and influence of Archimedes, June 8–10, 2010, in Syracuse, two distinguished scientists, Stephanos Paipetis from Greece and Marco Ceccarelli from Italy, summed up the general consensus of the sixty contributors as follows:

"Archimedes' works are still of interest everywhere and, indeed, an in-depth knowledge of this glorious [Archimedean] past can be a great source of inspiration in developing the present and in shaping the future with new ideas in teaching, research, and technological applications."[229] Archimedes also remains the paradigm for the science and technology that went into the Antikythera machine. A few scholars even connect Archimedes to the science of the twenty-first century. "Today the ancient scientist [Archimedes] is considered both a great theoretician and a model of scientific ingenuity," Geymonat wrote. "It cannot be overemphasized that, as an inventor, Archimedes was truly inimitable, inasmuch as he obtained results that were both innovative and original. He presented these results simply and without grand eloquence, maintaining a tone of professionalism that transcends mere self-confidence, revealing the enormous potential of human intelligence."[230]

Vincent De Sapio, Sandia National Laboratory, and Robin De Sapio, Orinda Union School District, California, also highlight the modernity and continuous usefulness of Archimedes merging geometry with mechanics. They speak of the "application of mechanical reasoning to geometric problem

[228] Reviel Netz, tr. and ed., *The Works of Archimedes*, Volume I: *On the Sphere and the Cylinder* (Cambridge: Cambridge University Press, 2004) 17.

[229] Stephanos Paipetis and Marco Ceccarelli, eds., *The Genius of Archimedes—23 Centuries of Influence on Mathematics, Science and Engineering* (New York: Springer, 2010) vi.

[230] Geymonat, *The Great Archimedes*, 89, 91.

solving" by Archimedes. In other words, Archimedes used mechanical metaphors in his interpretation of abstract mathematics.

According to Vincent and Robin De Sapio, Archimedes's "approach to thinking about problems may in fact be his greatest contribution [to science]." Archimedes explained his thinking in his "Mechanical Theorems," which is full of insights with "vast utility in addressing modern problems over two millennia after Archimedes' time."[231]

Geymonat also argues that, for the first time, science and technology "melded" in the life and works of Archimedes. Such a union benefited both. Archimedes did not create a school but he "cultivated science with the spirit of an engineer rather than of a professor."[232] Geymonat should know. He is a professor.

The Roman Assassination of Archimedes

Archimedes was related to the king of Syracuse, Hieron II. During the second Punic War, 218–201 BCE, between Carthage and Rome, Syracuse sided with Carthage. Roman troops encircled Syracuse, demanding its capitulation. Archimedes rushed to the defense of his country, employing his engineering and scientific skills for the construction of powerful weapons in order to defend Syracuse. Thanks to his ingenuity, Syracuse was invulnerable. The Romans, however, were waging a life and death war against the Carthaginians. They concluded that Syracuse had to be captured.

They also knew that Archimedes was the brain behind the defenses of Syracuse. They wanted him, alive or dead. A traitor made that possible. In 212 BCE, the Romans captured Syracuse and slaughtered Archimedes.

The Greek historian Polybios, c. 200–118 BCE, described how Archimedes's military weapons "frustrated" and defeated superior Roman forces besieging Syracuse from land and sea. Marcellus, the Roman commander, admitted that Archimedes used his ships to "ladle sea-water into his wine-cups" while whipping out of the wine party the Roman soldiers attacking Syracuse with siege engines.

231 Vincent and Robin De Sapio, "Mechanical Advantage: The Archimedean Tradition of Acquiring Geometric Insight from Mechanical Metaphor," in *The Genius of Archimedes*, 493–506.

232 Geymonat, "Archimedes in Ancient Roman World," in *The Genius of Archimedes*, 361–368.

The artillery of Archimedes was "extraordinarily effective," spreading death and destruction between the besieging troops and crushing the engines of the enemy. Archimedes's war machines used iron-grappling hooks, lifting ships and men, and killing them by dropping them from a height. Polybios admired the "genius" of Archimedes, becoming "an immense, almost a miraculous asset."[233]

Other ancient sources also mention that Archimedes used cannons against the Romans. Leonardo da Vinci, 1452–1519, the Renaissance genius of art and technology, also reported that Archimedes invented a steam cannon. He called it "Architronito," which was "a machine of pure copper" throwing "iron balls with great noise and fury." Cesare Rossi, University of Naples, agrees that, in fact, Archimedes had steam cannons that "hit the Roman ships with incendiary projectiles."[234]

The Greek historian Diodoros described how Archimedes smashed the Roman fleet with the machines that Polybios described. He probably had read Polybios. But in contrast to Polybios who said nothing about Archimedes's burning mirrors, Diodoros did. He was from Sicily and did his work about a century after Archimedes. He reported that when Marcellus moved his fleet away from the walls of Syracuse, Archimedes continued to destroy Roman ships by hurling against them stones the size of wagons. Diodoros also reports that Archimedes used mirrors as fiery weapons against the fleet of Marcellus:

"Archimedes employed a hexagonal mirror, and, at an appropriate distance from this mirror, he set up similar but small quadrangular mirrors. Archimedes used metal plates and small hinges for adjusting the mirrors. He clearly designed these mirrors to catch the full energy of the Sun at noon, both during the summer and winter. The mirrors reflected the Sun's rays onto the ships with the result a fearsome heat and fire engulfed the barges. The burning ships were at a distance of an arrow's flight from the mirrors. Archimedes' burning mirrors reduced the Roman ships to ashes. Thus, did the old man, Archimedes, and his mirrors, vanquish Marcellus,"[235] Diodoros wrote.

---

[233] Polybios, *The Rise of the Roman Empire* 8.3–7, tr. Ian Scott-Kilvert and ed. Frank W. Walbank (New York: Penguin Books, 1979).

[234] Cesare Rossi, "Archimedes' Cannons Against the Romans," in *The Genius of Archimedes*, 113–131.

[235] Diodorus Siculus, *The Library of History* 26.18.

Other Greek intellectuals repeated the story of Archimedes's burning mirrors. The great physician Galen[236] and the brilliant satirist Lucian[237] of the second century mentioned Archimedes burning the Roman fleet with his mirrors.

A medieval Greek scholar of the twelfth century, John Tzetzes, repeated the story Diodoros related of Archimedes and the mirrors. He reports Archimedes used solar weapons against the Romans. He constructed special mirrors that used the rays of the Sun for burning Roman ships at the distance of a flying bow.[238]

Silius Italicus, c. 26–101, a Roman politician who wrote the epic of the Second Punic War, 218–201 BCE, described the fate of Roman ships in the harbor of Syracuse, facing "the cunning device of an unexpected weapon." Silius said this weapon was "an iron-clawed grapnel," which used its "iron claws" and "unyielding jaws" for killing Roman soldiers and warships. Silius explained that the "ingenuity" and "cunning" of Archimedes were more powerful than the military force of Marcellus. Archimedes kept the troops of Marcellus in disarray.

Silius praised Archimedes: "[Archimedes brought] immortal glory on his city ... [This was] a man whose genius far surpassed that of other sons of Earth. He was poor in this world's goods, but to him the secrets of heaven and Earth were revealed. He knew how the rising Sun portended rain when its rays were dull and gloomy; he knew whether the Earth is fixed where it hangs in space or shifts its position: he knew the unalterable law by which Ocean surrounds the world with the girdle of its waters; he understood the contest between the Moon and the tides, and the ordinance that governs the flow of Father Ocean. Not without reason men believed that Archimedes had counted the sands of this great globe: they say too that he had moved ships and carried great buildings of stone, though drawn by women only, up a height."[239]

However, Silius Italicus was not entirely honest in his reporting on Archimedes. He made Marcellus look pretty good, almost a peacemaker, even

---

236 Galen, *Mixtures* 3.657–658, in *Selected Works*, tr. P. N. Singer (New York: Oxford University Press, 1997).

237 Lucian, *Ippias* 2, in *Luciani Opera*, ed. M. D. Macleod, 2 Vols. (Oxford: Oxford University Press, 1972).

238 Ioannis Tzetzae, *Historiae* 2.106–159, 12.964–987, ed. Petrus Aloisius M. Leone (Naples: Instituto di Filologia Classica, University of Naples, 1968); *Selections Illustrating the History of Greek Mathematics* XVIIa, tr. Ivor Thomas, 2 vols. (Loeb, 1941).

239 Silius Italicus, *Punica* 14. 320–352, tr. J. D. Duff (Loeb, 1950).

though he admitted Marcellus "poured forth all the horrors of war" around Syracuse. He also explained the death of Archimedes as an awful accident: "Archimedes was calmly poring over a figure traced in the sand, when the great disaster came down upon him." Silius Italicus includes this fiction in the same breath he called Archimedes a "defender of his country."[240]

Ploutarchos, a younger contemporary to Silius Italicus, had probably read the same sources about Archimedes as Silius did. He repeated Silius's fiction but his report on Archimedes is more detailed. Ploutarchos, who lived about 300 years after Archimedes, read the histories of Polybios and Diodoros. He was also aware of what Roman writers had said about Archimedes. But, like his Greek colleagues, Polybios and Diodoros, he loved Archimedes.

He said that Archimedes had so much scientific knowledge and invented so many useful machines that his reputation was almost divine. He was the "soul and mind" of the defense of Syracuse against the Romans from 214 to 212 BCE. Like Polybios and Diodoros, Ploutarchos emphasized the military superiority of the engines of Archimedes in defending Syracuse against Marcellus's army. He painted a vivid if incomplete and inaccurate picture of the science of Archimedes, especially his engineering achievement.

"Archimedes possessed such a lofty spirit, so profound a soul, and such a wealth of scientific theory, that although his inventions had won for him a name and fame for superhuman sagacity, he would not consent to leave behind him any treatise on this subject of machines. He regarded the work of an engineer and every technical skill that ministers to the needs of life as ignoble and vulgar," Ploutarchos wrote. "Archimedes devoted his earnest efforts to those [geometrical] studies the subtlety and charm of which are not affected by the claims of necessity. These studies, he thought, are not to be compared with any others; in them the subject matter vies with the demonstration, the former supplying grandeur and beauty, the latter precision and surpassing power. For it is not possible to find in the geometry of Archimedes more profound and difficult questions treated in simpler and purer terms." Ploutarchos was convinced that Archimedes was consumed with a passion for science. He continued:

"Some attribute the success of Archimedes to his natural endowments ... For no one could by his own efforts discover the proof for the solution of a geometrical problem, and yet as soon as he learns it from Archimedes, he thinks

[240] Silius Italicus, *Punica* 14. 676–678.

he might have discovered it himself; so smooth and rapid is the path by which he leads one to the desired conclusion. The stories about him are credible. For example, Archimedes was under the influence of some familiar and domestic gods. His obsession with geometry was such that he would forget even his food and neglected the care of his person. He would trace geometrical forms in the ashes. And when he took a bath, he used olive oil in drawing geometrical figures on his body.

"He was possessed by a great delight, and in truth, he was a captive of the Muses. And although he made many excellent discoveries, he is said to have asked his kinsmen and friends to place over the grave where he would be buried a cylinder enclosing a sphere, with an inscription giving the proportion by which the containing solid exceeds the contained."[241] Archimedes proved a sphere is two-thirds the volume and surface area of its circumscribing cylinder.

Ploutarchos certainly was right in saying Archimedes was obsessed with science, especially geometry. But he was dead wrong in suggesting that, somehow, Archimedes denigrated all things technical and mechanical as being marks of brutish upbringing and labor.[242] We don't know why, in this instance, Ploutarchos misunderstood Archimedes so badly. After all, Archimedes loved both geometry and mechanics, using them interchangeably for a better understanding and solution of problems.[243]

Archimedes and the Greeks did not consider science and technology different disciplines but siblings, one dependent on the other, being different faces of the same coin. A Greek scholar, T. P. Tassios of the National Technical University of Athens, called this habit of thought "cross-fertilization" of science and technology.[244]

Three centuries separated Ploutarchos from Archimedes. He served the Romans, who could have cared less about the subtleties of science. Like the story of the death of Archimedes, which I am examining shortly, Ploutarchos did not really understand Plato, supposedly set against technology.

[241] Ploutarchos, *Marcellus* 17.3–7; Archimedes's defense of Syracuse: 14–19, tr. Bernadotte Perrin (Loeb, 1990). As usual, I changed slightly some parts of this translation.

[242] Ploutarchos, *Marcellus* 17.3–4.

[243] Archimedes, *Method* in *The Works of Archimedes*, tr. and ed. Thomas Heath (First published 1897, 1912, New York: Dover, 2002), Supplement, 13.

[244] T. P. Tassios, "Cross-Fertilization of Science and Technology in the Time of Archimedes," in *The Genius of Archimedes*, 347–359.

Ploutarchos shared the Roman bias of his predecessor historians, Roman and Greek. These men lived in a superpower named Rome that wiped out Carthage and Corinth and crushed ruthlessly the political independence of Greece. They probably had no doubt the Roman Empire would last forever. Whether Greek or Roman, they saw the world through Roman eyes. They had either heard or witnessed how the Romans had dismembered and looted Greece. They had no illusions about Roman kindness towards Greece. Besides, at the time of the death of Archimedes, in the late third century BCE, Greek kingdoms were still obstacles to the Roman ambition to dominate the world.

Thus Greek (and Roman) writers reported that Archimedes died "accidentally" during the terrible upheaval of the invading forces and despite the orders of the Roman general Marcellus to spare his life. Ploutarchos says Marcellus felt bad that the "great and beautiful Syracuse" was sacked. His greatest affliction, supposedly, was the death of Archimedes.

Like Silius Italicus, Ploutarchos has Archimedes absorbed in the solution of some geometrical problem when Roman troops were all over the city slaughtering and plundering. In such an unlikely situation, Archimedes met his death at the hands of an unknown Roman soldier who killed him because Archimedes refused to follow his order.

Another unlikely story Ploutarchos repeats is that Archimedes wanted to please Marcellus so he put in a box his mathematical instruments like sundials, spheres, and quadrants. Archimedes then tried to reach Marcellus but a soldier, thinking the box was full of gold, killed him.[245]

This story of the death of Archimedes is bizarre, making Ploutarchos an apologist for Marcellus. Archimedes was the mortal enemy of the Romans and Marcellus in particular. He had built his war engines very close to the city walls. Stones, missiles, and other weapons rained on the invading Romans from all directions. These invisible weapons caused so much chaos and death among Roman soldiers that, according to Ploutarchos, "they seemed to be fighting gods" rather than men.

Ploutarchos quotes Marcellus describing Archimedes as a "geometrical Briareos," surpassing in strength and violence one of the three-hundred-handed monsters, sons of Ouranos (Sky) and Gaia (Earth). Archimedes, Ploutarchos

[245] Ploutarchos, *Marcellus* 19.

says, was like the commander-in-chief of Syracuse, all fighting Greeks integrated into his war machines devastating the Roman attackers who dreaded anything on the walls of Syracuse that looked like an engine of war.[246]

Under these circumstances of invasion and war when Archimedes was constantly employing his weapons against the Romans, it is extremely unlikely he would isolate himself in his home or any place else to solve a geometrical problem—at least not during the Roman attacks against Syracuse. I agree with the Italian scholar Cettina Voza that the killing of Archimedes has all the ingredients of a state-sponsored murder that silenced a great enemy. After all, the Roman command must have been mortified with the prospect that Archimedes could have joined Hannibal, their Carthaginian nemesis.

Voza did an assessment of the literary sources on the death of Archimedes and concluded, as I did, the traditional version of why Archimedes died was unreliable. She says the death of Archimedes was "a state-sponsored assassination." That way Rome got rid of Archimedes, its "indomitable" enemy. Moreover, Rome destroyed Syracuse and inflicted a blow against science. Under the "great Syracusan," Archimedes, science "had reached one of the highest peaks in its development."[247]

The relatives of Archimedes buried him in Syracuse and on his tomb, they fulfilled his wish, engraving a cylinder enclosing a sphere. Archimedes proved that the surrounded sphere was two-thirds of the cylinder that enclosed it.

A hundred and thirty-seven years later, in 75 BCE, Cicero served in Sicily as an official. He liked Greek culture and knew of Archimedes. He "discovered" his tomb and the diagram of the cylinder around a sphere, an immortal symbol of the great truths of Archimedes, mathematical truths that made him famous in his time and all time after that.

Cicero says the tomb of Archimedes was "overgrown with thorn bushes and thickets." He lamented that Syracuse, "the most famous city of the Greek world, once even the most learned, would have been ignorant of the memorial of its most keen-witted citizen," Archimedes. Conveniently enough, Cicero forgot that it was the Romans who devastated Syracuse. Then he asked: "Who is there who has anything to do with the Muses, that is with humanity and learning, who would not prefer to be this mathematician?" Cicero was certain

---

[246] Ploutarchos, *Marcellus* 16–17.

[247] Cettina Voza, "The Death of Archimedes: A Reassessment," in *The Genius of Archimedes*, 507–513.

that the mind of Archimedes was "nourished by seeking out and pondering theories, accompanied by the delight he gave his own cleverness, which is the sweetest sustenance of souls."[248]

Archimedes and the Laws of the Heavens

More than 300 years later, in the fourth century, the Greek mathematician Pappos of Alexandria praised Archimedes as a great scientist and engineer who put mechanics to work for human wellbeing.

"Archimedes of Syracuse is acknowledged by some to have understood the cause and reason of all these [mechanical] arts [and crafts]," Pappos wrote; "for Archimedes alone applied his versatile mind and inventive genius to all the purposes of ordinary life ... that remarkable man is universally honored and held in esteem, so that his praises are still loudly sung by all men."[249]

Another Greek of the fourth century praised Archimedes as nearly divine. This was a poet with the Latin name of Claudian. This man retained his many-gods Hellenic religion and made a living at the Roman imperial court in the West, eulogizing the emperor and court officials in Latin. Claudian knew Archimedes had built a planetarium, a kind of mechanical universe. But he imagined Archimedes's celestial mechanism was a glass sphere. He described that glass universe as follows:

"When Jupiter saw the heavens in a small glass sphere, he laughed and addressed the gods in such words as these: "Has the power of mankind's careful striving come this far? Is my handiwork now mocked in a fragile sphere? Look! By his skill an old man of Syracuse has copied the laws of the heavens, nature's reliability, and the ordinances of the gods. An animate force shut within attends the different stars and moves along the living mechanism with its regulated motions. A pretend Zodiac [circle] runs through its own private year, and the simulated Moon waxes with the new month. Bold diligence now rejoices to set its own heaven turning and regulates the stars through human intellect."[250]

---

[248] Cicero, *Tusculan Disputations* 5.64–65, tr. J. E. King (Loeb, 1945, 1971).

[249] Pappos, *Mathematical Collection* 8.1–3 in *Greek Mathematics*, tr. Ivor Thomas, 2 vols. (Loeb, 2002) II, 619.

[250] Claudian, *Shorter Poems* 51 in *Greek and Roman Technology: A Sourcebook*, ed. John W. Humphrey et al. (London: Routledge, 2002) 58.

Claudian mentioned the Zodiac. This is an imaginary great circle around the celestial sphere. It is almost identical to the ecliptic, the Sun's path among the stars. Another way of seeing the Zodiac is like a belt in the sky, six to twelve degrees on both sides of the ecliptic. The Zodiac is divided into twelve equal segments enclosing twelve signs of mythical animals and heroes decorated by stars. The Greeks called these signs katasterisms or constellations. The Zodiac signs correspond to the twelve months of the year. These signs/constellations represent stars connected to important human activities like harvesting and sowing of crops. Chapter 7 includes a more extensive discussion of the constellations.

Back to Archimedes: About 1,500 years after Claudian, Thomas Heath, the nineteenth-century British editor of Archimedes, had no doubt that Archimedes was "perhaps the greatest mathematical genius that the world has ever seen."[251]

Archimedes used the diagrams of geometry in measuring curvilinear objects. He brought together mathematics and physics, setting the foundations of calculus and, therefore, helping us to understand the meaning of infinity. This knowledge of measuring curves and, in general, using mathematics as a language of nature also explains the universe. Archimedes filled a "tool kit" for the reinvention of the sciences of the physical world, the kind of science that Galileo Galilei and Isaac Newton developed.

Archimedes and the Greeks knew all about the actual infinity. Euclid certainly did and, possibly, the philosopher Zeno of Elea, c. 490–after 445 BCE, was well acquainted with the concept of infinity. Sometimes Greeks used infinity but, purposefully, they avoided it. The Greeks "were ahead of the infinity game ... Archimedes was capable of producing the kind of science of physics that Galileo and Newton produced. He made the decision not to—other things occupied his mind."[252]

I already cited the influence of Archimedes on Galileo. Galileo had a tremendous respect for Archimedes, whose mathematical physics became the core and spark of his own science. He started studying Archimedes early on in his life. He admitted that he found "infinite astonishment" in the works of Archimedes.[253]

---

[251] Heath, ed., *The Works of Archimedes*, v.

[252] Netz and Noel, *The Archimedes Codex*, 204.

[253] Galileo Galilei, *[Discourses and Mathematical Demonstrations Concerning] Two New Sciences* (First published in 1638 in Leyden, tr. Stillman Drake (Madison, WI: University of Wisconsin Press, 1974) 49.

Influence of Archimedes

Marshall Clagett, late professor of the history of science at the Institute for Advanced Studies at Princeton, was convinced that the "most important achievement" of Archimedes was the "extraordinary influence" he had in the development of mathematics and physics in the sixteenth and seventeenth centuries.[254] And Giorgio de Santillana, a physicist who taught the history and philosophy of science at the Massachusetts Institute of Technology, said that both Galileo and Newton were brought up on Archimedes and Euclid.[255]

Santillana found "perfection" in all the works of Archimedes, his influence shaping modern science: "Archimedes is the founder of mathematical physics as we mean it today," he wrote. "Notwithstanding many efforts in the intervening centuries, mathematical physics did not start again until his books were translated and published in the sixteenth century. It is to him that Galileo and Stevinus [the self-made Dutch scientist Simon Stevin] owe their techniques and their frame of thought."[256]

In 1986, Wilbur Knorr, professor of the history of Greek mathematics at Stanford University, had this to say about Archimedes:

"Among the ancient scientists none has attained greater fame, either in his own time or in ours, than Archimedes of Syracuse. His immense achievements in pure geometry extended into the field of mechanics where his theoretical findings had consequences for the design of mechanisms. Siege engines constructed under his supervision held the Roman fleet at bay for three years during the ill-fated campaign which despite his efforts ended in the city's fall and his own murder in 212 B.C."[257]

Another American scientist, George Simmons, professor of mathematics at Colorado College, has nothing but the highest praise for Archimedes, raising him almost to the heavens. He said, "[I]f we consider what all other men accomplished in mathematics and physics, on every continent and in every civilization, from the beginning of time down to the seventeenth century in Western Europe, the achievement of Archimedes outweighs it all. He was a

[254] Clagett, *Greek Science in Antiquity*, 82.

[255] Giorgio de Santillana, *The Crime of Galileo* (New York: Time, 1962) 185.

[256] Giorgio de Santillana, *The Origins of Scientific Thought: From Anaximander to Proclus, 600 BC to AD 500* (New York: New American Library, 1961) 239–240.

[257] Wilbur Richard Knorr, *The Ancient Tradition of Geometric Problems* (First published in 1986, New York: Dover, 1993) 151.

great civilization all by himself." Simmons also described the discovery Archimedes made of the formula for the volume of a sphere as "one of the greatest mathematical achievements of all time." Archimedes did that in his work, "On the Sphere and Cylinder." Simmons explained that Archimedes made his discovery "by means of a long and rigorous argument of classic perfection."[258]

In 2010, Mario Geymonat, an Italian classical scholar, argued that Archimedes was essential in the making of the Renaissance, being "one of the main driving forces" in the development of science in the sixteenth century and after. Geymonat explained the influence of Archimedes as follows:

"During the Renaissance, the discovery, translation, and in-depth study of Archimedes' works gave an extraordinary impulse to the foundation and development of all aspects of modern science, both in methods of research and in geometric concepts and has constituted an important basis for future developments in mathematics and geometry."[259]

Archimedes influenced major scientists of the Renaissance like Leonardo da Vinci, 1452–1519; Niccolo Tartaglia, 1500–1557; Federico Commandino of Urbino, 1509–1575; and, of course, Galileo. Tartaglia was a mathematician and military engineer. Commandino translated several of the works of Archimedes into Latin. In fact, Archimedes was so important in the development of mechanics in the Renaissance that scholars speak of the Archimedean Renaissance in mechanics.[260]

Archimedes also influenced the Dutch mathematician and engineer Simon Stevin, 1546–1620. Like Archimedes, Stevin made original contributions to statics and hydrostatics. In mechanics, he was the "first true successor of Archimedes in the Renaissance."[261]

In his assessment of the Archimedes palimpsest, Stanford professor Netz wrote in 2011: "Had we known nothing else by Archimedes, this would have been enough to establish him as the most important scientist of antiquity and the one to come closest to modern science." The palimpsest included seven

---

[258] Simmons, *Calculus Gems*, 43, 242.

[259] Geymonat, "Archimedes in Ancient Roman World," in *The Genius of Archimedes*, 361–368.

[260] Teun Koetsier, "La theorie des machines au XVIe siecle: Niccolo Tertaglia, Guidobaldo del Monte, Galileo Galilei," Corpus, 39, 2001, 155–189; Paolo Palmieri, "Breaking the Circle: The Emergence of Archimedean Mechanics in the Late Renaissance," *Archive for History of Exact Sciences*, 62, 2008, 301–346; Maarten van Dyck, "Gravitating Towards Stability: Guidobaldo's Aristotelian-Archimedean Synthesis," *History of Science*, 2006, 373–407.

[261] Teun Koetsier, "Simon Stevin and the Rise of Archimedean Mechanics in the Renaissance," in *The Genius of Archimedes*, 85–111.

of the treatises of Archimedes: *On the Equilibrium of Planes*, two books; *On Floating Bodies*, two books; *Method*; *On Spiral Lines*; *On the Sphere and Cylinder*, two books; *Measurement of a Circle*; and *Stomachion*.

These works and, especially, the *Method*, document Archimedes's mathematical physics. This meant, Netz wrote, the "radical idealization of physical objects through pure mathematical terms." Netz sees this approach in Greek science as "distinctively Archimedean." The mathematical physics of Archimedes became the "dominant tradition of science." "Archimedes' two major achievements—mathematical physics and the calculus," wrote Netz, "proved to be immeasurably important for the scientific revolution."[262]

The Greeks almost specialized in geometry, the diagram being the main tool of their mathematics. They made advances in number theory—showing that there are infinite prime numbers. Related to their interest in numbers, Archimedes invented combinatorics, counting and figuring out the number of possible solutions to a problem. This study of probabilities rose to a high level under Hipparchos. I will say more about him in chapter 6.

Archimedes was the most important player in the probability theory. In fact, one of the three works overwritten in the Christian prayer book, *Stomachion*, is, according to Netz, the "earliest evidence, anywhere, of the science of combinatorics."[263] *Stomachion* was such a difficult game that it triggered a bellyache, hence the name "Stomach." The challenge was to calculate how many ways one could use fourteen geometrical diagrams to make a square. Archimedes solved this game but the condition of the surviving *Stomachion* was so bad that it proved impossible to read more than a few lines of Archimedes.

Netz asked some combinatorial mathematicians from Stanford and the University of California-San Diego as well as a computer scientist from Chicago to solve the Archimedes problem. They calculated that there are 17,152 ways fourteen geometrical figures can be arranged in making a square.

Archimedes calculated the circumference of the circle by using two 96-sided polygons, one inside and the other outside the circle. He then estimated the perimeters of these polygons and, therefore, the circumference of the circle squeezed between them.

[262] Reviel Netz, "The Place of Codex C in Archimedes Scholarship" in *The Archimedes Palimpsest*, ed. Reviel Netz et al., 2 vols (New York: Published for the Walters Art Museum by Cambridge University Press, 2011) I, 267–268.

[263] Netz and Noel, *The Archimedes Codex*, 260.

He calculated the "circumference: diameter" to be something like 3.140845 and 3.142857. In the early eighteenth century, this value took the name of Π (pi) and it now stands at 3.14159.

Archimedes also studied the lever and the center of gravity and, in this ingenious way, he founded mechanics. His investigation of the equilibrium of floating bodies established the science of hydrostatics and naval architecture.

In general, Archimedes measured curves, shaping calculus in the process. He invented combinatorics, which is at the core of our understanding of the theory of probability. These, among many other great mathematical and technological achievements, form the underbelly of computer and imaging science that is, according to Netz, fundamentally Archimedean. It was that science that helped in the decipherment of (1) the hidden and almost destroyed works of Archimedes; and (2) the inscriptions on the plates of the Antikythera Mechanism.

Chapter 5

An Awesome Laptop Computer from the Greeks

The twentieth-century world of science was shocked with the discovery, in 1900, of the Antikythera computer/Tablet in the waters of the Aegean island of Antikythera. Scientists spent nearly fifty years guessing its origins and science. This chapter discusses the many twists and turns that led to a full understanding of the significance of the world's first computer and the pivotal role played by Derek de Solla Price. His 1974 *Gears from the Greeks* proved this: a machine of mathematical, astronomical, and engineering genius was the earliest device to predict the eclipses of the Sun and the Moon—and much more.

The Shipwreck

On Tuesday, April 4, 1900, sponge divers from Syme, a small island near Rhodes, discovered a shipwreck full of precious ancient Greek treasures. The shipwreck was down deep in the waters of the tiny island of Antikythera.

Sponge diving had always been by far the most dangerous occupation in the Greek world. Yet sponges were so valuable and, often, irreplaceable. For this reason, there were always sponge hunters. They did their life-threatening work because they also earned a good living, which is why sponges were often described as "white gold."

In ancient Greece, sponges were used as protective material in the head and leg gear of warriors. Sponges remain to this day irreplaceable to the manufacturers of weapons and porcelain.

The sponge divers of Syme and the people of other Dodecanese islands were, in 1900, still living under Turkish rule. This gave the sponge divers fishing rights in North African waters. The Syme sponge divers sailed thirty to forty miles from the African coast, starting from Egypt to Morocco. There, in the deep waters of the Mediterranean and on the sides of underwater hills, sponge divers harvested precious sponges for sale to customers in London, Venice, and other international markets.

The usual route for the two ships of sponge divers from Syme for the North African coast was through the islands of Cythera and Antikythera. This was a convenient but unusually unsafe route. Countless ancient ships sank between these two islands. In 1802, Lord Elgin lost the marble treasure he had cut off from the Parthenon. His ship, *Mentor*, carrying the stolen Parthenon sculpture, sank in the harbor of Cythera. It took Elgin until 1804 and the employment of divers from different Greek islands to recover his loot.

Storms affecting the sea around Antikythera are unpredictable but always hazardous. Antikythera is a tiny one-mile-long island south of Cythera. Cythera, the southernmost of the Ionian Islands, was essential to the sponge divers for replenishing their supplies. They worked at sea for seven months. They usually departed from Syme in early April and returned in late October.

Two men, Photes Lindiakos and Demetres Kontos, made the 1900 sponge fishing possible. Lindiakos funded the journey for sponges and Kontos was the captain of the two ships. When the two ships found themselves at Antikythera waiting out the dying of a storm, one of the sponge divers, Elias Lykopantes (known also by the nickname Stadiates), discovered the remnants of an ancient ship some fifty-eight meters below the surface of the sea.

In great astonishment and fear, Lykopantes swam as quickly as he could to the waiting sponge divers. He was holding a bronze human hand. He said, "Captain, I found this hand down there." Captain Kontos, experienced man of the sea and familiar with ancient Greek antiquities, immediately dove to the location of the shipwreck and confirmed the discovery of the treasure. He saw what was left of the cargo hold of a wrecked sunken ship. He recognized statues of horses, statues of men and women, and lots of vases and other material remnants of Greek civilization.

Captain Kontos telegraphed the Greek government of the archaeological discovery in the waters of Antikythera. The Greek government did not allow its near bankruptcy to delay action. It sent a warship to Antikythera and hired the sponge divers to retrieve the ancient treasure from the sea. The urgency of the task was so high that divers worked long hours during the treacherous winter months—all the way from November 1900 to September 1901.

Both Photes Lindiakos and Demetres Kontos moved their families to Cythera to supervise the recovery of the ancient treasure. Yet despite their efforts to protect the archaeological treasure and the health of the sponge divers, one of the eight divers died and two were paralyzed. The conditions of diving repeatedly in the frozen and stormy waters of the island were harsh and dangerous. Sponge divers sacrificed everything in their work. They saved the Antikythera computer from oblivion.

In late 1900, divers brought up a beautiful bronze statue of a naked young man, which eventually received the name of the Antikythera Youth. The naked youth caused quite a stir in Athens. According to Helen Kladake-Vratsanou, historian of Syme and the sponge divers, Greeks flocked to one of the ministries where officials kept the bronze statue.

The visitors were full of emotions and wanted to see the statue at close range. She says the experience of multitudes of Greeks paying their respect to the Antikythera Youth was similar to the religious experience of crowds venerating icons.[264]

Experts surmised that the shipwreck probably happened during the chaotic period of the Roman civil wars. The large Greek ship sank in the first half of the first century BCE. We know the ship was Greek because the surviving drinking cups have the names of their Greek owners on them.

The ship was sailing from Rhodes to Rome. Probably hired by Romans, the forty-meter-long ship carried looted Greek treasure: about fifty bronze and marble statues, and lots of jewelry, amphorae, and coins.

Some of the amphorae came from Rhodes. The silver coins originated in Pergamon (Pergamum), a Greek kingdom in northwest Asia Minor, and the bronze coins were from Ephesos, a Greek polis about 100 miles south of Pergamon.

Some of the sculpture on board the doomed ship possibly came from Pergamum, Ephesos, and Delos. The bronze coins from 70 to 60 BCE also

[264] Helen Kladake-Vratsanou, "Syme, Sponge Divers and the Shipwreck," in Kostes Stekas, ed., *The Antikythera Mechanism: Researchers Unlocking the Secrets of the Most Ancient Mechanical Universe* (Athens: Kostes Stekas, 2016) 65–70 (in Greek).

dated the ship. The statue of the Antikythera Youth was possibly sculpted by Euphranoros around 340–330 BCE. It is a masterpiece. The handsome ephebe—young man—may well represent Perseus holding the head of the monster Medusa or Paris offering an apple to Aphrodite.[265]

Hidden among these works of art and culture was a small metal object that looked like a stone. After cleaning and lengthy examination, observers noticed Greek letters and what appeared to be the teeth of gears. Archaeologists of the National Archaeological Museum in Athens named the mysterious device Antikythera Mechanism.

From a Star Catcher to a Mechanical Universe

At first, the museum officials catalogued the small, corroded mechanism as item 15087. They put it away, dubbing it *astrolabe*, which in Greek means "star catcher." Astrolabes helped figure out the position of the Sun and the stars in the sky. They were used for the measurement of angular distances of the Sun and the stars from the Earth. This information was relevant for figuring out the length of the day, and the rise and setting of the Sun. They also gave clues for astronomical predictions but were not useful for navigation.

It wasn't until archaeologist Spyridon Stais, 1859–1932, saw an inscription in ancient Greek on one of the dials of the ancient device that museum archaeologists started a more careful examination of the corroded treasure. The Greek letters in the inscriptions dated from 150 to 100 BCE.[266] Other experts noticed the perfectly cut triangular gear teeth, unheard of before this discovery. They took the metal device back into the laboratory, named it the Antikythera Mechanism, and began to study it more intensely.

On May 23, 1902, the Athenian newspaper *Asty* (City) published the first report of the discovery of the geared device. That article reported its date to

---

[265] Nikolaos Kaltsas, "The Treasure of Antikythera," in Stekas, ed., *The Antikythera Mechanism* 79–84.

[266] There's also an earlier date for the origins of the Greek computer (late third century BCE). This estimate comes from the 2014 work of two scientists: Christian Carman, professor of the history of science, National University of Quilmes, Buenos Aires, Argentina, and James Evans, professor of physics, University of Puget Sound, Tacoma, Washington. They studied the eclipse predictor Saros dial of the Antikythera Mechanism. They assumed the Greek designers of this computer borrowed Babylonian arithmetic to make sense of the eclipse predictions of the Saros dial. They concluded that the "eclipse prediction scheme on the Antikythera Mechanism really does derive from the Babylonian example." The Saros dial, they said, started predicting solar and lunar eclipses in 204 BCE. This brings the "epoch year" for the manufacture of the Antikythera Mechanism closer to 204 BCE: Christian C. Carman and James Evans, "On the epoch of the Antikythera mechanism and its eclipse predictor," *Archive for History of Exact Sciences*, (2014) 68:693–774.

range from the second century BCE to the second century of our era. The article also described the device as an unusual example of an astrolabe. The studies continued. In 1905, Konstantinos Rados, a naval historian, said the Antikythera Mechanism was too complex to be an astrolabe.

With the possible exception of the German philologist Albert Rehm, 1871–1949, who in 1907, sided with Rados, most scientists who studied the mechanism repeated the largely misleading statements of most of their Greek colleagues that the ancient device was an astrolabe.

Rehm suggested the Antikythera gear assembly resembled the Sphere of Archimedes that Cicero saw and described sometime during the first half of the first century BCE. This sphere was one of two of Archimedes's mechanical devices the Roman general Marcellus brought with him to Rome after his troops plundered Syracuse and assassinated Archimedes in 212 BCE. Marcellus placed the first of these devices, a celestial sphere, in the temple of Vesta, the goddess of the hearth similar to the Greek goddess Hestia.

Cicero learned about the planetarium of Archimedes from a friend he called Gallus: The planetarium or sphere reproduced the eclipses of the Sun and the Moon and the movements of the heavens, the Sun, and the Moon, including those of the planets one could follow with the naked eye: Aphrodite (Venus), Hermes (Mercury), Ares (Mars), Kronos (Saturn), and Zeus (Jupiter). According to Cicero, the same eclipse of the Sun happened on the globe as would actually happen in the sky. Cicero was certain that "there was more genius in that Sicilian [Archimedes] than human nature seems able to encompass ... the invention of Archimedes deserved admiration on this account, that he had discovered how to reproduce the unequal velocities and different orbits [of the Sun, the Moon, and the planets] in a single turning mechanism. When Gallus set this globe in motion, it came about that the moon was as many revolutions behind the sun on that bronze instrument as in the heavens themselves, and therefore there was that same eclipse of the sun in that sphere, and the moon then passed into the region of the Earth's shadow."[267]

Cicero also repeated his astonishment of observing Archimedes's mechanical universe at work. He said: "When Archimedes exhibited the motions of the moon, the sun, and the five planets on a sphere, he did what the god of Plato, who is represented in the *Timaeus* as constructing the universe, did. He made it possible by a single rotation to control motions of most diverse velocities. Now if this cannot be done without a god in the case of the universe, Archimedes

[267] Cicero, *The Republic* I.14.21–22, tr. Clinton Walker Keyes (Loeb, 1928).

surely could not have been able to imitate those same motions on this sphere without divine genius."[268]

Cicero knew Poseidonios, a philosopher-engineer in Rhodes, who, like Archimedes, had built a mechanical universe. "Our friend Posidonius," says Cicero, "has ... made a globe which in its revolution shows the movements of the sun and stars and planets, by day and night, just as they appear in the sky."[269]

Cicero's reports are important testimony that the Greeks had models of the mechanical universe in use in the first century BCE. This cannot be underestimated in the curious absence of additional written evidence documenting the historical existence of the Antikythera machine or, more appropriately, Antikythera computer-like devices. I take this up in chapter 6.

Rehm worked on the device for several years. He published on the Antikythera Mechanism but left out his most interesting ideas, which remained in his research notes.[270] In the 1920s, another researcher, Ioannis Theophanides,1877–1939, a Greek rear admiral, also studied the device. He concluded that it was a planetarium.

For more than half a century, the interest in the Antikythera Mechanism waned and became part of the backlog of unexplained mysteries of the ancient world. Its fragments were cleaned and, occasionally, photographed.[271]

Derek de Solla Price and the Tablet

In 1958, Derek de Solla Price, a British historian of science and experimental physicist, studied the fragments of the Antikythera Mechanism. He then spent two years at the Institute of Advanced Studies at Princeton before becoming professor of the history of science at Yale University in 1960.

In 1959, he wrote an article about "An Ancient Greek Computer" for *Scientific American.* He spoke of the high scientific technology of the Tablet and assured his readers, "Nothing like this is preserved elsewhere." He likened the Antikythera Mechanism to the great astronomical cathedral clocks built in

[268] Cicero, *Tusculan Disputations* I.63.

[269] Cicero, *On the Nature of the Gods* 2.87–89, tr. Horace C. P. McGregor (New York, Penguin Books, 1972).

[270] I am grateful to Tony Freeth for this suggestion, July 10, 2018. Freeth, a British mathematician, organized the 2005 scientific team for the study of the Antikythera Mechanism. I discuss his contributions in the text.

[271] A. Jones, "Historical Background and General Observations," *Almagest*, 7–1, May 2016, pp. 37–66.

Europe during the Renaissance or to a "modern analogue computer." He saw the Tablet as "the venerable progenitor of all our present plethora of scientific hardware." "It is a bit frightening," Price admitted, "to know that just before the fall of their great civilization the ancient Greeks had come so close to our age, not only in their thought, but also in their scientific technology."[272]

Arthur C. Clarke, the world-famous science fiction author, read Price's article about the Antikythera Mechanism and went to Athens in 1965 to see the ancient device. He had a letter of introduction and support from an admiral. He visited the National Archaeological Museum in Athens three times before the museum staff located the fragments of the Antikythera Mechanism. They kept those fragments in a simple cigar box. Clarke was disappointed that the Greeks were so careless with such a great treasure of their ancient achievements.

Fragment C: showing an inscription and parts of the two celestial circles: the Zodiac depicting 12 constellations (inside) and 365-day year calendar (outside). Courtesy Tom Malzbender and Hewlett Packard.

Far more important than his annoyance with the Greek bureaucracy was Clarke's reaction to the fragments of the mechanism. Like Price, he fell into deep thought. "Looking at this extraordinary relic is a most disturbing experience," he wrote. "Few activities are more futile than 'what if ...' type of

[272] Derek de Solla Price, "An Ancient Greek Computer," *Scientific American*, 200 (6) (1959) 60–67.

speculation, yet the Antikythera mechanism positively compels such thinking. Though it is over two thousand years old, it represents a level [of achievement], which our technology did not reach until the eighteenth century... . If the insight of the Greeks had matched their ingenuity, the industrial revolution might have begun a thousand years before Columbus. By this time [1975] we would not merely be pottering around the moon; we would have reached the nearest stars."[273]

True, but not because the Greeks missed the breakthrough into experimental science. The Greeks employed experimental science in stitching the computer together. The reason students of the Antikythera computer refuse to contemplate why it nearly disappeared comes straight from metaphysics and history. **The Christians made the Greek industrial revolution impossible.** I explain that dark page of history in chapter 9.

In 1971, Price came back to the Antikythera fragments. He read a report from the Oak Ridge National Laboratory in Tennessee about gamma rays revealing the insides of metals. He wrote to the director of the Tennessee lab, who put him in touch with the Atomic Energy Commission in Athens. The AEC had a nuclear physicist, Charalambos Karakalos, who responded to Price's request to X-ray the Antikythera fragments. In the summer of 1972, Karakalos took hundreds of pictures of the fragments.

Using those images and dogged careful analysis of the device itself, Price took years to master the intricacies of the ancient machine. His research convinced him that the Antikythera Mechanism was "one of the most important pieces of evidence for the understanding of ancient Greek science and technology." He explained why: The complexity of its gear design demonstrates a more precise picture of the advanced level of Greco-Roman "mechanical proficiency" than that coming out of any of the surviving textual evidence. He found the Antikythera Mechanism was a "singular artifact... . The oldest existing relic of scientific technology, and the only complicated mechanical device we have from antiquity. [It] changes our ideas about the Greeks and makes visible a more continuous historical evolution of one of the most important main lines [of Greek science and technology] that lead to our civilization," he wrote.[274]

---

[273] Arthur C. Clarke, "Technology and the Limits of Knowledge," in *The Frontiers of Knowledge*, ed., Brooke Hindle (Garden City, New York: Doubleday, 1975) 120–121.

[274] Derek de Solla Price, "Gears from the Greeks: The Antikythera Mechanism—A Calendar Computer from ca. 80 B.C." in *Transactions of the American Philosophical Society*, 1974, 64 (7) 13.

Fragment A: the largest fragment of the Antikythera Mechanism. It shows the Cosmos of the Sun, Moon, and the planets. Twenty-seven of the total thirty gears of the computer are in this governing Cosmos fragment. Courtesy Tom Malzbender and Hewlett Packard.

Price's *Gears from the Greeks* is a masterful account of how he decoded the Antikythera Mechanism. This study was the first scientific examination of the approximately 2,060-year-old device.[275] After his groundbreaking study, it became known as the world's first computer.

[275] Other studies I found fundamental in my investigation of the Antikythera Mechanism include the two reports published in the British science journal *Nature* (November 30, 2006, and July 31, 2008) by a few international researchers and an article on "Fragmentary Knowledge" by John Seabrook in *The New Yorker* of May 14, 2007.

One of Price's principal discoveries was that in addition to its collection of precisely interlocking gears, it also had a differential gear, the first ever created, which enabled the Antikythera Mechanism to show the movements of the Sun and the Moon in "perfect consistency" with the phases of the Moon. "It must surely rank," Price said of the differential gear, "as one of the greatest basic mechanical inventions of all time."[276]

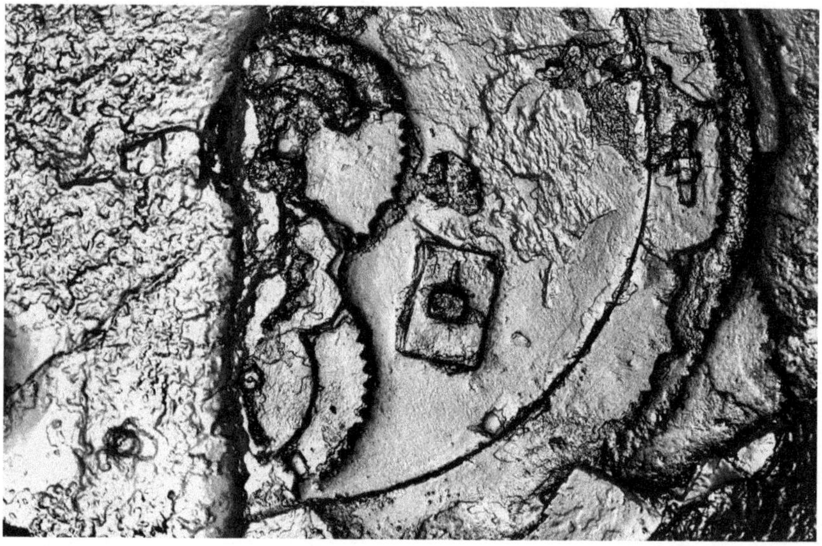

Back side of fragment A. See the interlocking gears. Courtesy Tom Malzbender and Hewlett Packard.

Price showed that the differential gear of the Antikythera computer preceded the geared inventions of Leonardo da Vinci and other Renaissance inventors. The differential gear reappeared in 1575 in a clock made by Eberhart Baldewin in Kassel, Germany.

Price noted that the experts judge the Greeks from scraps of building stones, buildings, statues, coins, pottery, and a few selected written sources. Yet,

[276] Price, "Gears from the Greeks," 60. Tony Freeth said to me (July 10, 2018) Price was wrong about the differential gear: "Price's so-called 'differential' was wrong—as I showed in [the article published in the science magazine] *Nature* [30 November] 2006. The gears that constitute the system that Price thought were a differential in fact serve another more subtle purpose." Freeth gives credit to Price for his discovery of the "Metonic Cycle, originally discovered by Rhem in the inscriptions embedded in the gearwheels. Price's Metonic gear train was of fundamental importance. Price got all the other gear trains wrong. It was Michael Wright and myself, who discovered the actual functions of the gear trains." In the text, I explain why this reasoning is unconvincing. The Antikythera computer has a differential gear. Price's genius is all over the decoding of the Antikythera computer.

when it comes to the heart of their lives and culture, how they did their work in agriculture, how they built the great temples like the Parthenon, what kind of mechanical devices they had for doing things in peace and war; in sum, what their applied science produced in the field of technology, we have practically nothing from the Greek past.

"Wheels from carriages and carts survive from deep antiquity," he said, "but there is absolutely nothing but the Antikythera fragments that looks anything like a fine gear wheel or small piece of mechanism. Indeed, the evidence for scientific instruments and fine mechanical objects is so scant that it is often thought that the Greeks had none."[277]

Price died in 1983 and did not have enough time to complete his studies of the geared astronomical mechanism or update his 1974 *Gears from the Greeks*. Nevertheless, his work began to bear fruit.

Next to Price, the other scientist who also spent his life studying the Antikythera Mechanism was Michael Wright, who worked at the Museum of Science in London. The spark for Wright's interest in Greek technology came from Price. But after engaging in his own research, Wright was slowly disillusioned with Price's research and questioned many of his findings and theories. He charged that the work of Price was "damned by his misinterpretation of the internal arrangement [of the mechanism] and his imposition of wishful-thinking in conflict with observable detail."[278]

Wright said that the Antikythera artifact reproduced the motions of the Sun and the Moon with epicyclic gearing based on additional gears. A Greek physicist and historian of Greek science at the University of Athens, Antonios Pinotsis, questioned that assumption, siding with Price. He explained that the epicyclical mechanism or epicyclic gearing is much more complicated and difficult to use in the small volume of the Antikythera Mechanism. In contrast, the differential gear is better suited to the Antikythera Mechanism as well as being "the easiest and shortest way to achieve the resonance of different rotations."[279]

Wright concluded there was very little that was unusual about the Antikythera device, which he never described as a computer. Instead, he called it a planetarium and the "world's oldest known geared mechanism." He also

[277] Price, "Gears from the Greeks," 51.

[278] M. T. Wright, "The Antikythera Mechanism reconsidered," *Interdisciplinary Science Reviews*, 2007, Vol. 32, No 1, 27–43.

[279] A. D. Pinotsis, "The Antikythera mechanism: who was its creator and what was its use and purpose?" *Astronomical and Astrophysical Transactions*, Vol. 26, Nos. 4–5, August–October 2007, 211–226.

avoided ascribing the invention of the first "geared mechanism" to the Greeks, though he credited Hipparchos whose astronomical model was at the core of the Antikythera Mechanism. He described the Antikythera Mechanism and its technology as "ancient," but not necessarily Greek in origin. Wright started his own research into the Antikythera Mechanism in 1990. He teamed up with a scientist from Australia, Alan Bromley, an expert in the history of computers. They went to Athens and started their intensive study of the Antikythera fragments. They took hundreds of their own X-ray pictures of the fragments during their three-year study. Wright had put most of the labor and skills into this project. However, in February 1994, they parted ways and Bromley took the pictures and left for Australia. He died several years later without publishing his research. Wright acquired many of the photos from Bromley's estate.

Finding fault with Price's model, Wright even decided to build his own model of the Antikythera Mechanism to disprove the predictive power of the device that Price postulated. In his estimation, it was a planetarium, not a computer. He published his research in 2007. Much of Wright's position—biased by his hatred for Price—has proved baseless. Price was an outstanding scientist and scholar with two doctorates, one in physics and another in the history of science with emphasis on the history of scientific instruments and astronomical instruments in particular. Price also worked with Joseph Needham, a top historian of Chinese science. He collaborated with Needham and a Chinese historian, Wang Ling, on a review of Chinese astronomical clocks. So, Price was well prepared for handling the intricacies and importance of the Antikythera Mechanism.

Seeing the Inside World of the Tablet

But even before the publication of Wright's work, the controversy had been raging for years over the origins and nature of the Antikythera Mechanism. In 2005, the British mathematician Tony Freeth put together an international group of scientists to get to the bottom of the Antikythera Mechanism by using the latest imaging technologies. Freeth collaborated with Mike Edmunds, an astronomer from Cardiff University.

Freeth convinced X-Tek Systems from England and Hewlett-Packard from the US to donate their high-tech imaging technologies. Armed with this international support, he applied for permission to the Greek museum to study the Antikythera Mechanism; the museum rejected his initial request. Undeterred, Freeth then brought to his project three Greek scientists: John Seiradakis, late professor of astronomy at the Aristotle University of Thessalonike; Xenophon Moussas, professor emeritus of astrophysics at the University of Athens; and Agamemnon Tselikas, director of the Center for History and Paleography in Athens. However, adding these Greeks to Freeth's team did not change the policy of the museum.

It wasn't until Moussas met with the deputy minister of culture, Petros Tatoulis, that permission was granted. Moussas had influence. He also had a grasp of the seminal importance of the Antikythera computer. He argued that conclusive research into the nature and origins of the Antikythera Mechanism would be of immense value. Greece would then reaffirm its culture as the foundation of modern science and technology. Tatoulis ordered the museum to cooperate with Freeth and his team.

In September 2005, Freeth and his technicians started their imaging work at the museum. They examined most of the eighty-two bronze fragments and thirty gearwheels of the Antikythera Mechanism. Twenty-seven gears exist in the largest fragment A, making up the face of the computer. One gear exists in each of the fragments B, C, and D. One of the technicians, Tom Malzbender, who was also leading the Hewlett-Packard team, came to Athens from California. He took pictures of the Antikythera fragments with a flashbulb dome, which emitted lighting from different angles. This technique, Polynomial Texture Mapping (PTM), enhanced surface detail and revealed previously buried surface inscriptions.

Then X-Tek Systems used its state-of-the-art technology, the Bladerunner, to probe the insides of the device by microfocus X-ray Computed Tomography (CT) like a scan of a human in a hospital. All in all, the fragments of the computer were exposed to X-ray bombardment unprecedented in intensity and sophistication. Out of thousands of images of the interior and exterior of the ancient machine, there emerged the first complete portrait of the architecture and engineering of this ancient device. They got more precise images of the inscriptions on the front and back bronze plates of the computer, which seemed to indicate a kind of guide for its operation.

The inscriptions on the fragments needed to be translated. Moussas hired Yanis Bitsakis, a physicist working at the Center for History and Paleography in Athens, to decipher them. The Greek letters of the inscriptions were very small; some were 1.2 millimeters or one-twentieth of an inch.

Inscription, Polynomial Texture Mapping photograph of the surface of the Greek computer, Courtesy Tom Malzbender and Hewlett Packard.

Bitsakis would go through the computer images of the fragments and pass along any image with lettering to Agamemnon Tselikas for translation. The first inscription Bitsakis and Tselikas translated was on the back of the Antikythera Mechanism. It said: ΕΛΙΚΙ ΤΜΗΜΑΤΑ ΣΑΕ, "the spiral [ΕΛΙΚΙ] is divided into 235 sections."

This meant that one of the back dials was a spiral representing the nineteen-year Metonic Moon and Sun calendar of 235 months. Another back calendar dial of eighteen years and 223 lunar months named Saros predicted the eclipses of the Sun and the Moon. The Metonic calendar enclosed the two- and four-year Olympics dial and seventy-six-year Kallippic dial and the Saros dial enclosed the Exeligmos (fifty-four-year) dial. More on Saros below.

I will examine the front bronze plate and its inscriptions in more detail at the end of chapter 7. Those inscriptions were about the days of the year, noting that the Zodiac ran clockwise around the dials. Furthermore, the inscriptions explained the Cosmos: the heliacal risings and settings of constellations: Capricorn, Aquarius, Pisces, Aries, Hyades, Taurus, Scorpio, Pleiades, Gemini, Eagle, Arcturus, Sagittarius, Cancer, Orion, Dog, Leo, and Virgo. They also described the movements of the five planets, including those of the Sun and the Moon.[280]

Inscriptions on the back plate describe the dials, pointers, and other external parts of the Antikythera computer. The "best preserved" passages of the inscription text dealt with features of the computer, now lost. These lost parts were pointers with small spheres showing the Sun and the planets on the front dial. There was also a dial on the upper back plate modeling the extremely accurate seventy-six-year Kallippic calendar of the nineteen-year Metonic cycle.[281] The seventy-six-year calendar is four times the nineteen-year cycle. More on the Kallippic calendar below.

The largest wheel—fragment A—became known as the Wheel of the Sun god Helios because it drives a pointer showing the movement and position of the Sun on two circular concentric dials, the Zodiac being the inside dial and the 365.25-day solar year being the outside dial. Moreover, other front dials showed the movement and position of the Moon and the planets in the Zodiac. They also revealed the phases of the Moon.

Freeth's team finished their work and published their findings in the July 2008 issue of *Nature*. They stated the Antikythera Mechanism was "an extremely rare original document that gives us critical information about the astronomy and technology of its era."[282]

The Fingerprints of Hipparchos

One of the scientists who contributed to the sophisticated technology of the Antikythera computer was Hipparchos. His fingerprints are all over the ancient machine. His astronomical ideas dominated the second century BCE

[280] M. Anastasiou et al., "The Front Cover Inscriptions," *Almagest* 7–1, May 2016, pp. 250–297.
[281] Y. Bitsakis and A. Jones, "The Back-Cover Inscriptions," *Almagest* 7–1, May 2016, pp. 216–249.
[282] "Supplementary Notes," 5.

as much as Archimedes's mathematics and engineering did the third century BCE. Hipparchos invented plane trigonometry and made astronomy the predictive mathematical science it is today. He computed eclipses. He calculated the distance between the Earth and the Moon to be sixty Earth radii. His measurement of the distance from the Earth to the Sun was well below the correct value. In addition, according to the astronomer Ptolemaios writing about three centuries later, Hipparchos discovered that the sphere of the fixed stars moves very slowly. This circular motion of the fixed stars from west to east takes place around the poles of the ecliptic, which is the annual circular path of the Sun. The discovery that the fixed stars move in a circular motion, paralleling the ecliptic, is an achievement of Hipparchos.[283]

We now describe the discovery of Hipparchos that the fixed stars move as the "precession [movement] of the equinoxes." Equinox is the point where the ecliptic intersects the celestial equator. The slow movement of the fixed stars, or the precession of the equinoxes in relation to the circular arc of the motion of the Sun, gives us two ways to define what is "year."

The first, tropical year, is the period the Sun takes to travel from the vernal or spring equinox to the same spring equinox. The second, sidereal year, covers the travel of the Sun from one fixed star back to the same star. This journey is slightly longer than going back and forth to one equinoctial point because the stars are not stationary. The time difference between these two "years" is minimal, however: one sidereal year equals 1.000039 tropical year.

Hipparchos left a list with all his astronomical observations, including the observations he borrowed from Babylonian and Greek astronomers. His only surviving book is a commentary on the works of Eudoxos and Aratos. Most of what we know of him comes from the *Almagest* of Ptolemaios.

Pliny the Elder said that Hipparchos "can never receive sufficient praise. No one has done more to prove that the stars are linked to man and our souls are a part of heaven." Pliny thought nearly divine the work of Hipparchos of listing the stars and constellations, giving their names, positions, and brightness in the sky. That way Hipparchos, said Pliny, gave us clues of whether stars are born and die; decrease or increase in size. In other words, Hipparchos "left the sky as a legacy to all men—if anyone should be found to claim this inheritance."[284]

---

283 Ptolemy, *Almagest* III.1, VII.1–3.

284 Pliny the Elder, *Natural History* II.95, tr. John F. Healy (London: Penguin Books, 1991).

Could it be that Hipparchos, whose astronomical model permeates the design of the Greek computer, had created a prototype, which then inspired the manufacturing of that mechanism found off the island of Antikythera? From about 140 to 120 BCE, he had his laboratory in Rhodes. More than other Greek astronomers, he made use of the data of Babylonian astronomers. But like the rest of the Greek astronomers, he employed geometry in the study and understanding of astronomical phenomena.

The connection of Hipparchos to the Antikythera Mechanism is in the front bronze plate of the device where pointers displayed the positions and speed of the Sun and the Moon in the Zodiac.[285] Hipparchos knew the Moon moved around the Earth at different speeds. When the Moon is close to the Earth, it moves faster than when it is farther from the Earth when it slows down. This is because the Moon's orbit is elliptical, not the perfect and eternal circular movement the Greeks associated with the stars. Perhaps Hipparchos had read Aristarchos and accepted his heliocentric theory. Hipparchos resolved the difficulty of lunar motion around the Earth with his epicyclic Moon theory, which superimposed one circular motion of the Moon onto another, the second movement having a different center.

Whoever made the Antikythera Mechanism modeled it on the ideas of Hipparchos by creating a series of gears that replicated the orbital cycles of the Moon in relation to the Earth and Sun. He did that with one gearwheel sitting on top of another but located on a different axis. A pin-and-slot mechanism, modeling the lunar theory of Hipparchos,[286] takes under consideration the non-circular or elliptical orbit of the Moon around the Earth. A pin originating from the bottom wheel enters the slot of the wheel above it. When the bottom wheel turns, it also drives around the top gearwheel. However, the wheels have different centers and, therefore, the pin slides back and forth in the slot, which enables the speed of the top wheel to vary while that of the bottom wheel remains constant.

Price proposed that the engineer who created the Antikythera Mechanism was either Andronikos Kyrrhestes, who built the Tower of the Winds in Athens in the second century BCE, or Poseidonios and Geminos, the astronomers of Rhodes who flourished from the late second and first centuries BCE.

---

[285] "Supplementary Information," 2–3 (*Nature*, July 31, 2008).

[286] "Supplementary Information," 23–26 (*Nature*, November 30, 2006).

Poseidonios and Stoicism

Poseidonios (c. 135–c. 51 BCE) was born in Apamea, Syria, by the River Orontes. He studied in Athens and toured much of the known world to study natural phenomena. Strabo admired his natural science and geographical knowledge.[287] Poseidonios was a great polymath. He moved to Rhodes because Rhodes had a reputation and tradition conducive to the sciences.

He became a citizen of Rhodes and served his adopted country in senior executive political positions. Powerful Romans visited him for his friendship and learning. General Pompey stopped by Rhodes in 66 and 62 BCE to see him. Cicero studied under him and reported that Poseidonios constructed a sphere that looked like the planetarium of Archimedes: being a mechanical universe where the Sun, the Moon, and the planets moved just like they moved in the sky.[288]

Poseidonios, painting by Evi Sarantea. Courtesy Sarantea.

[287] Strabo, *Geography* books 1–3.

[288] Cicero, *The Nature of the Gods* 2.87–89.

Poseidonios was more than an astronomer and an engineer. Only fragments of his writings survive, however. He was a natural philosopher in the tradition of Aristotle. He hoped his writings would educate people about the Cosmos and human behavior.[289] Seneca the Younger, c. 8 BCE–65 CE, Roman philosopher and teacher and advisor to Emperor Nero, thought Poseidonios left a useful legacy of knowledge for both secular and divine affairs; not merely theories but practical advice for action leading to a good and ethical life.[290]

Poseidonios was eclectic in his philosophical ideas. He probably admired Plato and emulated Aristotle. But his age, late second and early first centuries BCE, had more than Plato and Aristotle as models of philosophy and politics. Stoicism was the new paradigm. Stoicism captures an alternative political system to the polis, which means different politics than those of Plato and Aristotle. Yet the natural ideas of Stoicism[291] were vaguely similar to the Hellenic traditions of natural philosophy and science, including those of Plato and Aristotle.

Poseidonios probably embraced certain aspects of Stoicism and rejected others. The founder of Stoicism was Zenon of Kition, Cyprus. Zenon started preaching in the Athenian agora in the early third century BCE.

Stoicism borrowed logos (reason) from Herakleitos, making it the universal force among people and the Cosmos. Fire was especially close to divine reason, a tradition the Stoics also adopted from Herakleitos.[292] Herakleitos, "a god-like man, the inspired glory of ancient Ephesos,"[293] saw the stars as bowls of fire and fire as logos, the wisdom one gains in understanding the Cosmos. Man was inextricably tied to nature.[294]

However, Stoics like Zenon broke with the Greek tradition of the polis and the philosophical ideas of Plato and Aristotle.[295] Stoicism degenerated to the popular talk of the shopping mall: an incoherent, flexible formula of fears, ideas, superstitions, abstractions, theologies, and wishes satisfying almost everybody, especially those in power. It mixed up religious practices from outside the

---

[289] I. G. Kidd, ed., *Posidonius: The Translation of the Fragments*, Vol. III (Cambridge: Cambridge University Press, 1999) 3–28.

[290] Seneca, *Epistles* 104.21–24., ed. and tr. R. M. Gummere (Loeb, 1917–25).

[291] Laertios, *Lives of Eminent Philosophers* 7.132–160; A. A. Long and D. N. Sedley, eds. and trs., *The Hellenistic Philosophers* (Cambridge: Cambridge University Press, 1987), I, 158–437.

[292] A. A. Long, *Hellenistic Philosophy: Stoics, Epicureans, Skeptics*, second edition (Berkeley: University of California Press, 1986), 145–147.

[293] *The Greek Anthology* 2.355.

[294] Herakleitos, *Fragments* 194–227, in *The Presocratic Philosophers*, ed. and tr. Kirk et al.

[295] I am grateful to my Greek philosopher friend Kostas Kalimtzes for this insight on Stoicism.

Greek world with the demands and superstitions of tyrants. Overall, Stoicism was a religion for the masses.

Seneca hinted that Stoicism made people like dogs tied to a moving cart. The dogs had the option of walking behind the cart or being dragged by the cart.[296] Another Stoic philosopher, Epictetus, a Greek from Asia who was born into slavery, c. 55–c. 135, was equally harsh. He said life was tough: some people have had to hold the chamber pot for others.[297] Some critics reproached Stoicism for "painting resignation in the colors of fervor; for sublimating human powerlessness by transforming it magically into purely inner liberty." No doubt, Stoicism was an ideology for hard times.[298]

Stoic philosophers plundered the ancient Greek cosmological tradition. They peppered their speeches and writings with Greek powerful concepts like logos or *pneuma* (breath, mixture of cold and hot, air and fire). But these ideas became obscure, though pretty and meaningless Stoic decoration for the few. In contrast to the eternity of the Platonic and Aristotelian Cosmos, the Stoic Cosmos was going out periodically in a big fiery bang of flames, only to reconstitute itself and start all over again.

The real purpose of Stoicism, however, was political. It preached the accommodation of the people to the emerging ruling class of undemocratic states. Stoicism, says Tim Whitmarsh, professor of Greek culture at Cambridge University, "imagined a universe governed for the good by a single divine force." This proved "congenial to a Roman imperial mentality from ancient times."[299] Stoicism benefited the Roman Empire in the Mediterranean.[300]

Zenon denounced Greek traditional political values. He had no use for Plato's *Republic*. He wrote his own *Republic*, in which he envisioned a global empire running the world. Ploutarchos put it diplomatically this way: Zenon's *Republic*, he said, would favor the undoing of Greek institutions. Its main point is that Greek household arrangements should not be based on poleis or smaller communities, each one living by its own laws. Zenon wants us to consider all human beings from all over the world as our fellow citizens and

[296] Seneca, *Epistles* 107.11.

[297] Epictetus, *The Discourses* 1.2.8–10, ed. Christopher Gill and tr. Robin Hard (London: Everyman, 1995).

[298] Jacques Brunschwig, "Stoicism," in *Greek Thought*, 977–996.

[299] Whitmarsh, *Battling the Gods*, 235.

[300] Brent D. Shaw, "The Divine Economy: Stoicism as Ideology," *Latomus*, Vol. 44 (No 1, January–March 1985), 16–54.

residents following an identical way of life. In other words, Zenon wants the Greeks to become a grazing herd of cattle.[301]

Zenon also attacked the Greek family and role of women in Greek society, making all women prostitutes—property shared among men. In addition, he did not want to see the construction of any more temples, gymnasia, and law courts in the Greek poleis.[302]

Poseidonios rejected Zenon's extreme anti-Greek ideology. Galen says the Stoics attacked Poseidonios because their allegiance was to their dogmas rather than to their country. Poseidonios, however, betrayed the Stoics rather than the truth and Greece.[303] He remained Greek in his philosophy and science. Galen also says Poseidonios made novel and important contributions to history, geography, meteorology, cosmology, psychology, and mathematics.[304]

Geminos and the Phenomena

Poseidonios passed on his knowledge to his students, especially Geminos, who was also an astronomer. Geminos flourished in Rhodes in the first century BCE. His book, *Introduction to the Phenomena*, is a general overview of Greek astronomy, covering four hundred years of Greek astronomical thought. It connects Hipparchos (second century BCE) and Ptolemy (second century CE).

The book of Geminos summarized the facts and theory of the heavens of his time. Like Aratos, Geminos thinks the geocentric Cosmos is spherical and divine, hardly amenable to human intervention, but perfectly amenable to reason. Geminos describes the Zodiac, the constellations, the celestial sphere, the phases of the Moon, the risings and settings of the fixed stars, the eclipses of the Sun and the Moon, and the Greek practice of reckoning time.

Introduction to the Phenomena includes ideas that resemble the inscriptions in the Antikythera Mechanism on the names of the months, which years had thirteen months, which month would be repeated in those years, and which months had thirty and which had twenty-nine days.

---

[301] Plutarch's, *On the Fortune of Alexander* 329a–b.

[302] Laertios, *Lives of Eminent Philosophers* 7.32–33.

[303] Galen, "The Soul's Dependence on the Body" 819–820 in *Selected Works*, tr. P. N. Singer (New York: Oxford University Press, 1997) 174–175.

[304] *The Encyclopedia of Ancient Natural Scientists* 691–692.

The scientists who studied the Antikythera Mechanism, reading its inscriptions, saw the hand of Geminos in the Antikythera Mechanism.[305] The inscriptions of the Metonic dial in the back plate of the Antikythera device imply a calendar similar to the calendar of Geminos, which originated in Corinth and Epiros.[306]

Archimedes is another plausible creator of the world's oldest computer. While his death in 212 BCE makes his candidacy impossible as the device has been reliably dated from the second century BCE, there's little doubt that his work was the primary inspiration for the actual manufacturer of the Antikythera Mechanism.

According to Pappos of Alexandria, Archimedes wrote a book, now lost, on building astronomical devices in the shape of spheres. Pappos said that sphere makers "construct a model of the heavens [and operate it] with the help of the uniform circular motion of water."[307] In such cases, then, those planetariums worked with waterpower. The Antikythera Mechanism was not powered by water but by a hand crank. It seems clear, however, that the work of Archimedes served as one of the primary influences in its creation.

Was the Oldest Computer Manufactured in Rhodes?

Pinotsis says the mechanism was most likely manufactured on the island of Rhodes. From his study of the coins of Rhodes, he noticed an interesting correlation between the ray-crowned head of Helios and the advances in astronomical knowledge on the island. In the Alexandrian Era, Rhodes was a center for science and, especially, astronomical studies, including "naval and military technology."[308] Famous scientists, philosophers, and strategists lived and flourished on the island of Helios. They were one of the reasons why Rhodes became one of the earliest and most important cultural centers of ancient Greece, including a science laboratory that possibly gave rise to the technology of the Antikythera computer. Its Olympics dial records the name of the games in Rhodes: *Halieia* (in honor of the Sun god, Helios). This strengthens the

[305] Freeth et al., "Calendars with Olympiad display and eclipse prediction on the Antikythera Mechanism," *Nature*, July 31, 2008, 454: 614.

[306] M. Anastasiou et al., "The Back Dial and Back Plate Inscriptions," *Almagest* 7–1, May 2016.

[307] Pappos, *Mathematical Collection* 8.1–5.

[308] Antonios D. Pinotsis, "Astronomy in Ancient Rhodes" (unpublished paper, 2008; Section of Astrophysics, Astronomy and Mechanics, Department of Physics, University of Athens); Pinotsis, "Kleovoulos of Lindos: The Precursor of Science in Ancient Rhodes," *Dodecanesian Chronicles*, Vol. 18, 2005, 332–351 (in Greek). For other works of Pinotsis, see the bibliography.

prospect that Rhodes was the home for the "prototype" engineering mechanism that became the Antikythera computer.[309]

The first meteorological observations necessary for Greek star calendars took place in Rhodes. The Greeks called star calendars *Parapegmata* (singular: *Parapegma*). These star calendars included information of astronomical and meteorological phenomena useful to farmers and possibly sailors. Such phenomena included equinoxes, solstices, and the risings and settings of stars and constellations just before the rise and a little after the setting of the Sun. These astral risings and settings took place once a year.

The surviving Antikythera Mechanism is only a fraction of the original. Only 20 percent of the front plate survives, including 90 percent of the Zodiac and part of the *Parapegma*—showing thirty-six risings and settings of the stars. Twenty-five percent of the back place survives, showing 33 percent of the Metonic and Saros cycles.[310] The surviving fragments of the Greek computer give us less than 50 percent of the *Parapegma* inscriptions.[311] Nevertheless, the *Parapegma* of the front plate of the Antikythera computer gives the names of the following seven constellations: Sirius, Arcturus, Pleiades, Hyades, Lyra, Aquila, and Orion. In the back plate inscription we read Aphrodite, Helios, golden sphere of Helios, sphere of the Cosmos, spiral of 235 [months of the Metonic calendar of 19 years], 76 years, and 19 years [of the Metonic calendar] and 223 [months of the Saros eclipse predicting dial] and elliptical [trajectory of the Moon around the Earth].

In other words, the *Parapegma* did what the Greeks expected it would do: list in alphabetical order astronomical phenomena of constellations and stars rising and setting in relation to the rise and setting of the Sun.[312]

Stunning, Enigmatic, and Complicated

The Antikythera Mechanism was pregnant with the original version of advanced technology that made our culture. It combined the precision and accuracy of Swiss watches 1,500 years before those watches came into being.

[309] Paul A. Iversen, "The Calendar on the Antikythera Mechanism and the Corinthian Family of Calendars," *Hesperia* 86 (2017) 129–203. Tony Freeth argues Epiros, rather than Rhodes, was the most likely place for the construction of the Antikythera computer (personal communication, July 10, 2018).

[310] Xenophon Moussas, "Antikythera Mechanism as evidence for Hellenistic technology excellence," in *Hellenistic Alexandria: Celebrating 24 Centuries*, ed. Christos S. Zerefos and Marianna V. Vardinoyannis (Oxford: Archaeopress Publishing, 2018) 209–225.

[311] Anastasiou, *The Antikythera Mechanism*, 24–25, 86–91, 103–118.

[312] Y. Bitsakis and A. Jones, "The Front Dial and Parapegma Inscriptions," *Almagest* 7–1, May 2016, pp. 68–137.

The scientists and engineers who decoded the Antikythera computer were delighted with the hidden technology embryo of the ancient device. They concluded that the Antikythera computer was the most sophisticated technology in the Mediterranean for more than a millennium. Their praises buttressed the findings and praises of Price for the Greek computer. But they refrained from elaborating on his moral and scientific stance that the West needs to rethink its treatment of the Greeks, especially its obsolete view that the Greeks had no advanced technology.

If the students of the Antikythera Mechanism after Price were reluctant to take up his cause, they spoke loudly about the technological genius embedded in the Antikythera device. One started with the inscriptions wrapping the device. The inscriptions mirrored the Greek in use in the second century. And the Greek symbols announcing eclipses do so with economy and forcefulness.

Detail from slice of the back side of fragment A. Notice the letters and the signs forming the signature for a Moon and Sun eclipse prediction. Sigma (Σ) stands for (ΣΕΛΗΝΗ, Selene) Moon and H (eta) is for (ΗΛΙΟΣ, Helios) Sun. Eclipse prediction symbols are looking like anchors. These anchors abbreviate two letters of the word hour: (hora) ωρα (ω/ρ). Next to Σ and H we see the symbols for the hour the eclipses are expected to take place.
Courtesy Tom Malzbender and Hewlett Packard.

The deciphering of the inscriptions added light to the scientific consensus that this Antikythera artifact was unusual; in fact, it was spectacular. Jo Marchant, a journalist and former editorial staff member of the science journal *Nature*, captured the widespread sentiment of the scientific community by repeating Price: that the Antikythera Mechanism is "by far the most advanced scientific instrument to survive from antiquity."[313] The BBC called the Antikythera Mechanism a "priceless artifact."[314] John Noble Wilford, a reporter for *The New York Times*, described it as "a surviving marvel of ancient Greek technology."[315]

Tony Freeth was extremely pleased with the results of the decoding of the Greek computer. He praised it this way: "It's an absolutely unbelievably stunning and sophisticated idea. I don't know how they thought of it. We're just following in the tracks of the ancient Greeks."[316]

According to Xenophon Moussas, the Greek computer is "the epitome of Greek philosophy, *techne*, and science." It is also "a miracle of science, technology and philosophy." Indeed, it was a complicated astronomical device that worked on the basis of theories of physics and astronomy. These theories, Moussas says, mirrored exact mathematics, reproducing natural phenomena like the movements of the planets in relation to the Earth. In addition, Moussas argues that such theories followed the "theory and practice of Pythagoras." He is right. Pythagoras said mathematics was the proper language for the study and understanding of the natural world and the Cosmos. In other words, geometry (triangles, circles, and epicycles and other mathematical constructions) describes the heavens. With the understanding geometry explains how the heavens work, it's easy to translate our geometrical constructions, including circles, "into gears moving dials on panels."[317]

In 2014, Magdalini Anastasiou, a physicist at the Aristotle University of Thessalonike, wrote her doctoral dissertation on the Antikythera computer. After studying the science, technical complexity, and reconstruction of the device for several years, she found something magical and perfect about its conception and design. More than that, like Moussas, she admitted the computer

---

313 Jo Marchant, "Was ancient Greek 'computer' an astronomical tool?" *New Scientist*, July 30, 2008.

314 BBC News, "Olympic link to early 'computer,'" July 30, 2008.

315 John Noble Wilford, "Discovering How Greeks Computed in 100 B.C.," *The New York Times*, July 31, 2008.

316 Quoted in Marchant, *Decoding the Heavens*, 240.

317 Xenophon Moussas, *The Antikythera Mechanism*, 15–43, 103. Also, see Xenophon Moussas's article in Greek: "The Mechanism of Antikythera and the Pythagoreans," *Socrates*, Vol 1 (2015) 77–86.

was a perfect mechanical construction.[318] In a similar spirit, Alexander Jones, historian of science at New York University who investigated the inscriptions of the Antikythera computer, reported, in 2016, that, "In this very small volume of messed-up corroded metal [of the Antikythera computer] you have packed in there enough knowledge to fill several books telling us about ancient [Greek] technology, ancient science and the way these interacted with the broader culture of the time. It would be hard to dispute that this is the single most information-rich object that has been uncovered by archaeologists from ancient times."[319]

Bruce Dorminey, a contributor to *Forbes* magazine, interviewed Jones. He was so impressed by what he learned about the Antikythera device he concluded its "ultimate purpose" was helping in the teaching of science, technology, engineering, and mathematics.[320] Jones may be right that such a sophisticated device was also used to educate the Greek elite about the heavens.[321] In other words, the computer, designed like a working mechanical universe, was possibly a teaching tool illustrating the astronomical and mathematical knowledge of how the world works. This is all well and good and the computer might have had didactic purposes. Yet Jones was not completely satisfied with it. He found technical problems with its design. Its "triangular gear teeth" were not efficient; its astronomy was problematic. But more important than that was Jones's rejection of the practical uses of the computer. "I do not believe that its purpose was to compute data or make predictions to be applied in some practical context," he wrote.[322] In addition, Jones indulged in putting down Derek de Solla Price,[323] the scientist whose pioneering investigations set the foundations for our understanding of the fantastic importance of the Greek computer. Such sophisticated yet misleading assessments are spreading the metaphysics of raising doubt about or covering up Greek achievements. The game most modern scientists play

[318] Magdalini Anastasiou, *The Antikythera Mechanism: Astronomy and Technology in Ancient Greece* (in Greek, doctoral dissertation, University of Thessalonike, June 2014), 10, 16, 39–40, 137.

[319] Quoted by Sarah Kaplan, "The world's oldest computer is still revealing its secrets," *The Washington Post*, June 14, 2016.

[320] Bruce Dorminey, "Antikythera Mechanism May Have Been World's First 'STEM' [Science, Technology, Engineering, and Mathematics] Project," *Forbes*, June 30, 2016.

[321] Alexander Jones, *A Portable Cosmos* (New York: Oxford University Press, 2017) 236.

[322] Jones, *A Portable Cosmos*, 234–235.

[323] Jones, "Like Opening a Pyramid and Finding an Atomic Bomb: Derek de Solla Price and the Antikythera Mechanism," *Proceedings of the American Philosophical Society*, Vol. 162, No. 3, September 2018, 259–294.

is they want everything to themselves: science and technology, they say, are for the most part exclusively products of modernity. Price demolished that delusion. He demonstrated that the Antikythera Mechanism was a computer of scientific technology, which the Greeks created in order to serve astronomers, farmers, priests, and politicians and, above all, for bringing the heavens down to Earth.

Of course, the scientists who worked on the fragments of the Antikythera device saw the world differently. In early June 2016, they began seeing it like a "kind of philosopher's guide to the galaxy."[324] Elizabeth Rayne, writer for a popular astronomy magazine, liked the metaphor of the galaxy so much she described the Antikythera computer as a "miniature galaxy." She praised Greece for giving so much to civilization. She credited the little galaxy for giving us "insight into the advanced intelligence of a civilization that can now add computers to its already impressive resume as the birthplace of philosophy, democracy and, of course, astronomy."[325]

Another scientist who praised the technological genius of the ancient computer was Mike Edmunds, professor of astronomy at Cardiff University, UK. Edmunds is also a member of the Antikythera Mechanism Research Project, hence intimate with all research evaluating the device's history and significance. On July 20, 2016, he said the Antikythera computer has rightly "captivated the world." It charted and tracked the planets (Mercury, Mars, Venus, Jupiter, and Saturn). But its tracking displayed the planets in such a fashion that included their "rather irregular 'wanderings' about the sky." The computer's inscriptions explained the sky to the users. "The [Antikythera] mechanism is basically an astronomical device, which bears witness both to the Greeks' astronomical knowledge and their extraordinary, and rather unrecognized, mechanical design skills," wrote Edmunds.[326]

Yet, it has taken more than a century for scientists to take the Antikythera computer seriously. The sluggish Greek museum bureaucrats in charge of the remnants of the device had little interest in Hellenic treasures that do not catch the eye of the tourists. They simply stored the twisted and corroded lump of

[324] Nicholas Paphitis, "Decade of labor reveals philosopher's guide to the galaxy," Associated Press, June 9, 2016.

[325] Elizabeth Rayne, "The Antikythera mechanism is officially the oldest computer ever," *Blastr*, June 23, 2016.

[326] Mike Edmunds, "Does the Antikythera Mechanism have a hidden purpose?" *Daily Mail*, July 20, 2016.

bronze in the museum for decades. All this changed with Price, who described the Antikythera Mechanism as a "singular artifact" and an "astronomical or calendrical calculator ... the most enigmatic, most complicated piece of scientific machinery known from antiquity."[327]

Decoding the Tablet

While the scientific community was able to decode the inner workings of the Antikythera computer, the archaeological community has not yet shed much light on its origins. This may be changing because, in 1976, but especially since 2014, underwater archaeological explorations have been targeting the sea floor[328] where, in 1900, sponge divers found the original astronomical device. These underwater investigations have unearthed human remains, coins, amphorae, small statues, gold rings, gold earrings, and glassware. In one of the underwater investigations in the fall of 2016, scientists discovered under sand and broken jars fragments from a human skeleton: part of a skull, including three teeth, and bones from arms, ribs, and femurs.

Could it be that these 2,000-year-old bones belonged to the Greek owner or a sailor of the large ship doomed by a storm in the first century BCE? Hannes Schroeder, a biologist from the Natural History Museum of Denmark, is hopeful he will be able to sequence DNA from the bones to answer these questions.[329] Marine archaeologists revisited the shipwreck site in the fall of 2017. This time they discovered a bronze hand, a small bronze disc decorated with a bull, and wooden planks. The disc could be a missing part of the Antikythera computer.[330]

We may never know who actually invented the Antikythera computer. We may also never know where it originated. What matters is that the Antikythera Mechanism represents the scientific technology of the ancient Greeks. From this realization we understand their role in the creation of modern science. The next chapter highlights how and why the Greeks built a modern-like world.

---

[327] Price, "Gears from the Greeks," 5.

[328] Christopher Joyce, "Ancient Shipwreck Off Greek Island Yields a Different Sort of Treasure," *National Public Radio*, June 28, 2016.

[329] Jo Marchant, "Human skeleton found on famed Antikythera shipwreck," *Nature*, September 19, 2016.

[330] Nicholas St. Fleur, "Bronze Arm Found in Famous Shipwreck Points to More Treasure Below," *The New York Times*, October 5, 2017.

Chapter 6

The Golden Age of Greek Science and Technology in Alexandria, Egypt

The Antikythera computer was the culmination of advanced mathematics, astronomy, metallurgy, and engineering. It incorporated the philosophy and science of Aristotle, the gears of Ktesibios, the mathematics and mechanics of Archimedes, and the astronomical ideas of Hipparchos. The Antikythera computer and the infrastructure of technology that made it possible were the products of the golden age of science in the Alexandrian Era, from the late fourth century BCE to the second century of our times. This chapter explains the origins and significance of the Alexandrian Era: how the Greeks developed the first model for the emergence and cultivation of advanced science and technology. The greatest achievement of that model was the Antikythera computer.

The Island of the Sun God Helios and the Tablet

In chapter 5, I reported that the Antikythera geared device was probably made in Rhodes or, quite possibly, Corinth or, more likely, in one of the daughter-poleis of Corinth in northwestern Greece, Kerkyra, Epiros, and Syracuse, Sicily—in southern Magna Graecia, Italy.

Rhodes and Syracuse are the most attractive possibilities for giving birth to the Antikythera Mechanism–like devices. In fact, the Greek computer came from both. The advantage of Rhodes is that, in all its life, it was "a home of Greek civilization, Greek learning, and Greek art."[331] And, second, as I documented in chapter 5, the fingerprints of two Rhodian astronomers, Hipparchos and Geminos, are all over the Antikythera computer, which may be the original model or copy of another standard astronomical model dating from the third century. Were that to be the case, Hipparchos and Geminos also mirrored earlier advancements in science.

The history of Rhodes goes back to the very beginnings of Greek culture. Strabo says that before Rhodes had the name Rhodes, it was known as *Ophioussa* (Serpent-island) and *Stadia* (Strong). Then it became *Telchinis*, the country of Telchines, who colonized it. These Telchines, Strabo recounts, were controversial people. Some described them as "maligners" and "sorcerers" who mixed sulphur with water from the sacred River Styx to damage and kill animals and plants. However, there was an alternative view of Telchines as superb craftsmen maligned by competing workers. Telchines came to Rhodes from Crete. They were the first craftsmen who worked iron and brass. In fact, the Telchines were so ancient that, according to tradition, they manufactured the scythe for Titan Kronos, father of Zeus.[332]

In time, the Telchines withered away or intermarried with a new group of Greek invaders of Rhodes known as the *Heliadai*, children of the Sun god Helios. It was also during the time of the *Heliadai* that Athena was born in Rhodes from the head of her father Zeus.

According to Pindar, Helios urged his children to build for Athena: "a shining altar and burn a sacred offering to gladden the heart of Zeus and Athena holding in her hand the spear-bolt." Zeus gathered a blond cloud and rained lots of gold upon them. Athena, the bright-eyed goddess, taught them craftsmanship, which increased the skill of their hands. The people of Rhodes sculpted works of superior cunning. They looked like they were breathing and moving.[333]

Eventually, the son of Herakles, Tlepolemos, brought his followers to Rhodes, the island of Helios. Homer tells us that Zeus blessed them with

---

[331] M. Rostovtzeff, "Rhodes, Delos and Hellenistic commerce" in *The Cambridge Ancient History*, Vol. VIII, *Rome and the Mediterranean 218–133 B.C.*, ed. S.A. Cook et al. (Cambridge: Cambridge University Press, 1965) 641.

[332] Strabo, *Geography* 14.2.7.

[333] Pindar, *Victory Songs* Olympian 7.39–57.

gold and wealth.[334] Thus, Rhodes had the mythic and historical tradition of technological achievements. In the Alexandrian Era, the island of Helios was a center for science and, especially, astronomical studies. Famous scientists and philosophers lived and flourished in Rhodes. They were one of the reasons why Rhodes became one of the earliest and most important cultural centers of ancient Greece. In the early third century BCE, Rhodes showed off its technological achievement and power with a colossal bronze statue of the Sun god Helios, patron god of Rhodes. It was sculpted in the years 294 to 282 BCE by Chares of Lindos, a great polis of Rhodes. Chares was a student of Lysippos, who did portraits of Alexander the Great. Writing about three centuries after the construction of the Colossus of Rhodes, Pliny the Elder thought it was a great achievement.[335] In addition, the first meteorological observations necessary for Greek calendars or *Parapegmata* took place in Rhodes.[336] This tradition of technology was not limited to Rhodes.

Marriage of Craftsmanship and Theory

As explained in chapter 3, Plato loved more than theory. He admired the craftsmanship of Hippias; his technical skills enabled him to do just about everything he needed. He engraved his own ring, made his own shoes, wove his own cloak and tunic, and plaited the belt he wore around his tunic. This was also a man of words and knowledge.[337] Plato also respected men who worked with their hands because they are careful in what they do: using their skills and organization to give their products the best possible shape, appearance, and quality.[338]

Moreover, Plato admired the mathematical nature of craftsmanship. Without counting, measuring, and weighing, Plato said, arts and crafts would be pretty much worthless. People would have to resort to conjecture and guesses in dealing with each other and in doing things. Plato had a special liking for

334 Homer, *Iliad* 2.653–670.

335 Pliny the Elder, *Natural History* 34.41.

336 Antonios D. Pinotsis, "Astronomy in Ancient Rhodes" (unpublished paper, 2008; Section of Astrophysics, Astronomy and Mechanics, Department of Physics, University of Athens); Pinotsis, "Kleovoulos of Lindos: The Precursor of Science in Ancient Rhodes," *Dodecanesian Chronicles*, Vol. 18, 2005, 332–351 (in Greek). For other works of Pinotsis, see the bibliography.

337 Plato, *Hippias Minor* 368b–e.

338 Plato, *Gorgias* 503d–504a, ed. E. R. Dodds (Oxford, 1990).

building. He explained the "superior craftsmanship" of construction to the frequent use of instruments and measures. The woodworking craft in shipbuilding and house building, for example, was accurate thanks to the tools of the craftsmen: straightedge, compass, a mason's rule, a line, and the carpenter's square.[339]

Aristotle also admired craftsmen and inventors for their useful devices and wisdom.[340] In fact, of all the social classes in a polis, he considered the class of mechanics the most essential. No polis could exist without the mechanics practicing their arts and crafts. Of those arts and crafts, Aristotle said, some are "absolutely necessary" while others contribute to luxury or enrich life.[341]

One student or follower of Aristotle wrote a work on *Mechanics*, and that book is within the works of Aristotle. This technician focused on the lever, pulley, wedge, and windlass, not complicated machines but practical aids to doing necessary work like lifting water from wells and the sailing of ships. Yet, the author of this technical manual also tried to show the mathematical principles behind his machines.

Explosion of Objective Knowledge about the World

Greek thinkers since Thales sowed the ground for the final flowering and globalization of science and technology in the era of Alexander the Great. This lasted for about 300 years: from the death of Alexander the Great in 323 BCE until the death of queen Kleopatra of Egypt in 30 BCE. However, the Greek scientific revolution that produced the Antikythera Mechanism went beyond the borders of the Alexandrian Age for another 200 years, covering the second century of our era. The third and second centuries BCE marked the climax of the golden age of Greek science. The Greeks institutionalized their new way of constructing and construing the world.

According to Ploutarchos, Alexander put into practice the best ideas of Greek philosophers. He also abolished many abominable traditions among some of the non-Greeks in his empire. Alexander convinced the Hyrcanians (Jews)

---

[339] Plato, *Philebos* 55d–56c.

[340] Aristotle, *Metaphysics* 981b13–17.

[341] Aristotle, *Politics* 1291a1–4.

to adopt the custom of marriage; the Arachosians (living in Old Kandahar, Afghanistan) to farm; the Sogdians (people of present-day Tajikistan and Uzbekistan) to respect their parents, not kill them; the Persians to respect their mothers, not marry them; the Indians to worship Greek gods; and convinced the Skythians (present-day Ukrainians and Russians) to bury their dead, not eat them. In addition, Alexander united the world for the first time. But he rejected Aristotle's advice to treat the Greeks as the masters of barbarians.

Ploutarchos reports that Alexander implemented the following strategy for creating one world: He was certain he was a god-sent ruler and mediator for the whole world. He waged war against those he could not bring over by persuasion. He integrated people from all over the world. He instructed them to consider the inhabited world to be their native land, and his camp to be their acropolis for their defense. Alexander also expected his subjects to ally with good men and treat the wicked as strangers. The difference between Greeks and barbarians, he told them, was not a matter of arms and clothing but excellence. Wickedness marked the barbarian.[342]

Alexander's successors also spread Hellenic civilization throughout Asia and the Middle East while uniting Greece for the first time. The "extremely sudden extension" of the Greek world gave an opportunity to Greeks to earn a good living almost everywhere. This encouraged the use of reason "without any traditional constraints." And advanced educational institutions funded by Greek rulers gave a great boost to scientific research. Philosophy was so widespread that, when scientific disciplines started on their road to specialization, they carried with them a "valuable dowry of a well-developed consciousness of methods and problems, an awareness formed while scientific efforts had taken place within the context of philosophy."[343] In other words, Alexander's vision of Greek ideas and culture spreading East and West became a "marvelous reality" that triumphed for several centuries after his death, building the civilization of the world.[344]

---

[342] Ploutarchos, *De Alexandri Magni Fortuna aut Virtute* I 326c–329d in *The Hellenistic World from Alexander to the Roman Conquest*, Michael Austin, ed. and tr., second ed. (Cambridge: Cambridge University Press, 2006) 57–58 (I modified the translation).

[343] Albrecht Dihle, *Greek and Latin Literature of the Roman Empire*, tr. Manfred Malzahn (New York: Routledge, 1994) 284–285.

[344] Peter Bamm, *Alexander the Great: Power as Destiny*, tr. J. Maxwell Brownjohn (New York: McGraw-Hill, 1968) 22.

Alexander the Great. Silver tetradrachm, 315–294 BCE. Left side, silver coin is showing Alexander wearing ram's horns. The right side is depicting Zeus holding his sacred bird, the eagle. NM 1462. Courtesy Numismatic Museum, Athens.

The Greeks did spectacularly well in Egypt especially because of one of Alexander's top generals. This was Ptolemy, son of Lagos, 367–282 BCE. Alexander appointed him governor of Egypt. When Alexander died in 323 BCE, Ptolemy consolidated his power and, in 305 BCE, made himself the king of Egypt, taking the name Ptolemy I Soter (Savior), 305–283 BCE. He was fortunate to have the assistance of Demetrios of Phaleron, a student of Aristotle who was also author of philosophical works. Demetrios convinced him to replicate Aristotle's school in Alexandria, first of all, by building a library and a Mouseion, Shrine of the Muses, university and institute for advanced studies. Ptolemy was also a student of Aristotle. He encouraged Demetrios to go ahead with his Aristotelian proposal. At about 295 BCE, King Ptolemy founded the Mouseion for the study and cultivation of Greek culture, the sciences, and literature. The library also thrived very quickly. That way, the methods and science of Aristotle took deep roots in Alexandria, becoming the intellectual infrastructure of the golden age of Greek science.

Ptolemy I died in 283 BCE but his successor, Ptolemy II Philadelphos (of Brotherly Love), who ruled from 285–246, continued his father's tradition by lavishing money on the Mouseion and giving political support to its staff, famous scientists, poets, and scholars recruited from all over the Greek world. He also built the Pharos or Lighthouse of Alexandria. The Pharos of Alexandria was one of the Seven Wonders of the Ancient World. Built by Sostratos for King Ptolemy II Philadelphos, the Pharos was an example of great architectural and technological achievement. It lit up the harbor of Alexandria. Pharos was symbolic of the enlightenment of the Greek kings of Egypt. But there was more enlightenment emanating from the scholars advising the Ptolemies. They did independent research and writing. They advanced science and technology. They received handsome salaries and paid no taxes; they ate and lived in the Broucheion, which was part of the palace. The Ptolemies also established a library in the Mouseion: a main library of about 500,000 volumes in the palace and a sister library of probably 42,000 volumes in the temple of Zeus Serapis or *Serapeion*. One of the librarians, Kallimachos, compiled the *Pinakes*, Πίνακες, a 120-volume catalogue of the collections of the library. Staff of the library combed Greece for manuscripts. Books found in ships coming to the harbor of Alexandria were copied for the library.

The Greek kings of Alexander's empire, especially the Ptolemies of Egypt, created the foundations for a rational commonwealth characterized by scientific exploration, state-funded research, the scholarly study of earlier Greek culture, and the editing of the Greek classics. The scholars of Alexandria pioneered the "techniques of painstaking study and exegesis," which spread all over the civilized world. They continue to be the model for classical and scientific studies.[345]

Vitruvius praises the Library of Alexandria but he claims the Ptolemies stole the idea of founding a library from the Attalid kings of Pergamum. The Ptolemies established a library, according to Vitruvius, as a result of their "unbounded jealousy and avaricious desire" to be the best. The Attalids created a library because they were motivated by a "delight in literature." We don't know why Vitruvius was so biased against the Ptolemies, save for the fact that the Ptolemies resisted Roman imperialism until the death of Kleopatra, the last Greek Macedonian queen of Egypt in 30 BCE. As for the Attalids, they reigned over a small kingdom, which they willingly donated to the Romans in the late second century BCE.

Vitruvius also reports that the Ptolemies initiated literary competition among Greek scholars. In other words, the Ptolemies set up real cultural Olympiads. They consecrated "games in honor of the Muses and Apollo, and established prizes and honors for the successful writers of the day, in the same way as for successful athletes." Sometimes the victor of those games became the head of the Alexandrian Library. This was the case for Aristophanes who succeeded Eratosthenes in the top position at the Alexandrian Library around 195 BCE. Aristophanes was so proficient in Greek literature that in one cultural Olympiad he caught some of the competing poets plagiarizing other poets' work.[346]

The Mouseion and the Alexandrian Library were at the center of Greek society in Alexandria and the Greek world. Indeed, Alexandria surpassed Athens

345 Neil Hopkinson, ed., *A Hellenistic Anthology* (Cambridge: Cambridge University Press, 1996) 4–6.
346 Vitruvius, *On Architecture* 7. Preface 4–7.

in scientific and technological achievements. It became the center of knowledge and civilization for Greece and the world. Strabo visited Alexandria and was impressed by its wide streets crossing each other at right angles and suitable for horses and carriages. Alexandria had "magnificent" public buildings and palaces that covered a fourth to a third of the city. Alexandria was also "full of dedications and sanctuaries." The Gymnasium was the most beautiful building in Alexandria. The length of its porticoes was around 175 meters.[347]

The Greek writer Theokritos of Syracuse knew Alexandria well. He praised Ptolemy II Philadelphos for his wealth, military might, and wisdom. He reported Ptolemy II reined over Egypt, rich in soil and towns, and over regions of Syria, Asia, Phoenicia, Arabia, Libya, and Ethiopia. Ptolemy II was also the wealthiest king in the world.[348]

Another Greek writer, Athenaios, who lived in the second century, quotes a book on Alexandria written by Kallixeimos of Rhodes. Kallixeimos, 210–150 BCE, describes the great procession of 279 BCE. This was a procession of wealth and power the likes of which were rare at any time in the ancient world. The display of unfathomable riches was the work of Ptolemy II Philadelphos. Alexandrians must have been astonished by the exotic animals, carriages full of representatives of the gods, 57,600 soldiers and 23,200 cavalry men, and huge amounts of gold in statues, jewelry, and decorations. Kallixeimos explained this unrivaled wealth in gold as a gift of the Nile, which "streamed" with gold and unlimited amounts of food.[349]

Kallixeimos also reports that in the procession he saw a mechanized statue standing up and sitting down on its own. It held a gold libation bowl from which it poured libations of milk. The automated statue held a garlanded staff like the god Dionysos. The statue operated on a cart decorated with a canopy and four gilded torches.[350] We don't know who mechanized the statue, but in all likelihood it was Ktesibios, who lived in Alexandria during the reign of Ptolemy II. Another report on the splendor of Alexandria under the Ptolemies comes from Herodas, a writer of mimes who flourished during the reign of Ptolemy II. Herodas said all that the world produced existed in Egypt, and, most probably, Alexandria: wealth, athletic grounds, schools, the Mouseion,

[347] Strabo, *Geography* 17.1.8–10, tr. Horace Leonard Jones (Loeb, 1949).
[348] Theokritos, *In Praise of Ptolemy: Idyll 17* 71–110.
[349] Athenaios, *The Learned Banqueters* 5.201b–203e, tr. S. Douglas Olson (Loeb, 2006).
[350] Athenaios, *The Learned Banqueters* 5.198f3–199a1.

spectacles, power, peace, philosophers, money, young men, and countless beautiful women.[351]

The successor king, Ptolemy III Euergetes (the Benefactor), 246–221 BCE, continued the work of his father and grandfather. He persisted in the claims of his family that he and his wife Berenice II were descendants of Herakles and Zeus. He was a warrior king who won territory from Syria. His armies reached Ethiopia, where they met the *Troglodytai*, cave dwellers. They also went to Persia where they repatriated ancient Egyptian statues Persians had stolen from Egypt.

Gold octadrachm coin depicting King Ptolemy III Euergetes, 246-221 BCE. Courtesy NM Collection Demetriou, Numismatic Museum, Athens.

Ptolemy III was a ruler of all his subjects: Egyptians, Greeks, and Jews. Egyptian priests in particular appreciated the support of Ptolemy III for Egyptian institutions like organized priesthood and its temples with their gods. They were especially proud he brought back to Egypt the sacred Egyptian treasures Persia had looted from Egypt. They were grateful for his protection and care for the "famous sacred animals" of Egypt. They praised him, King Ptolemy III

[351] Herodas, *The Mimes and Fragments: Mimiamboi* 1.26–35, ed. and tr, Walter Headlam and A. D. Knox (Cambridge: Cambridge University Press, 1922).

Euergetes and his wife Berenice II, Benefactor Gods, for their generosity and respect. They treasured the policies of the Benefactor Gods. Those policies saved Egypt from draught and restored prosperity and peace.[352] Ptolemy III had a special attachment to two mighty gods: Poseidon and the Sun-Helios. In the golden coins showing Ptolemy III, we see the trident of Poseidon behind his head crowned by the rays of Helios. Ptolemy III was also lavish in his support of the Mouseion and the library.

Alexandria was not alone in scientific and cultural achievements. Pergamum also had a famous library dedicated to research, science, inventions, and learning. King Attalus I, 241–197 BCE, lavished the library with wealth and prestige. His dream was to make Pergamum a second Athens. He built the Pergamum Library next to the temple of Athena. He funded a replica of the Pheidias statue of Athena in the Parthenon for his library. During the reign of Eumenes II, 197–159 BCE, Egypt stopped exporting papyrus to Pergamum, no doubt the result of competition between Alexandria and Pergamum. Papyrus was essential for book production. Pergamum used this unexpected crisis to invent a better alternative to papyrus. This was a product from sheepskin known as *pergamene* from the name of Pergamum. *Pergamene* eventually came to be known as parchment, a technology that guaranteed hundreds of years of survival for books.[353]

The Greeks in Alexandria, Pergamum, and, possibly, other Alexandrian kingdoms and Greek poleis produced modern-like science and institutions. Miroslav Rostovtzeff, a great European historian of Greece and Rome in the twentieth century, summarized the achievements of the Greeks in the third and second centuries BCE.

Greek literature, art, and science, he said, remained Greek even after the death of Alexander the Great. It's wrong to call this a decadent or Hellenistic age. On the contrary, the Greek genius in the centuries after Alexander was just as creative as it had been in the centuries before Alexander. Greek civilization, in fact, spread over the world. City people spoke Greek. Cities in the post-Alexander Greek world, says Rostovtzeff, had a modern-like infrastructure of water supply, paved streets, healthy markers, schools, athletic stadia,

---

[352] An epigraphic account of Ptolemy III's Third Syrian War, document 268 and The Canopus Decree, document 271 in *The Hellenistic World from Alexander to the Roman Conquest*, ed. Michael Austin, second edition (New York: Cambridge University Press, 2007) 466–467, 470–475.

[353] John Ferguson, *The Heritage of Hellenism* (New York: Science History Publications, 1973) 99–100.

libraries, outdoor stone theaters, racecourses, public buildings for local assemblies, and beautiful temples and altars for the worship of several gods.[354]

Egypt under the Ptolemies, for example, had banks in administrative districts and villages. These royal banks lent money and regulated the currency and the economy. They invested funds and paid interest to depositors.[355] Ptolemy II was probably the richest king of late antiquity. His "immaculately harnessed economic machine"[356] enabled him to put up unparalleled displays of wealth and to fund unparalleled advancements in science and technology.

For several centuries, including those of the Alexandrian Era, Greek children went to school or received their education from a teacher. Girls also went to school, though perhaps in smaller numbers than boys. The educated classes read Homer and the same Greek books, went to the theater, and sent their children to the same wrestling schools and gymnasia. Children studied music, literature, and science—"a combination characteristic of Greece."

Finally, says Rostovtzeff, Greek education was the badge of civilization. Reading Homer, Plato, and Sophocles and enjoying the comedies of Menander was essential to being a citizen of the Alexandrian Age. Failure in this Greek education was the equivalent to being a barbarian.

The Greeks did not force non-Greeks to become Greek, much less take up their civilization. The culture of the Greeks, says Rostovtzeff, "owed its worldwide recognition mainly to its perfection." It helped, of course, that the Greeks had no problem non-Greeks participated and excelled in Hellenic civilization. "The conception of a certain quality peculiar to a civilized being—the conception embodied in the Latin word *humanitas*—dates from this age, and so does the idea of a civilized world."[357]

The nature of this civilization became stronger from the influence of Aristotle and Alexander the Great. Indeed, Aristotle and Alexander became its strongest pillars. Aristotle inspired Alexander with Greek national consciousness. He convinced Alexander he could unite and lead Greece and conquer Persia and the world. He taught him how to govern. He inspired him to make Greek culture global. In other words, Aristotle infused Alexander with the

---

[354] M. I. Rostovtzeff, *A History of the Ancient World*, tr. from the Russian by J. D. Duff, 2 vols. (Oxford: Clarendon Press, 1926) I, 379–381.

[355] Austin, ed., *The Hellenistic World*, documents: 296, 297, 299, 309, 316, and 319.

[356] Justin Pollard and Howard Reid, *The Rise and Fall of Alexandria* (New York: Viking, 2006) 82.

[357] Rostovtzeff, *A History of the Ancient World*, I, 379–381.

virtues of learning, thinking, and understanding Greek poetry, politics, history, science, philosophy, and ethics. Aristotle modeled Alexander after the heroes of Homer to be the best of men. As I have argued earlier in chapter 3, Aristotle gave Alexander the philosophical foundations of Greek culture. He inspired him with courage, justice, and the employment of reason.

Nevertheless, not everything was rosy between Aristotle and his pupil Alexander. Aristotle was probably troubled by Alexander's Oriental pretentions in forging one society out of barbarians and Greeks. To add insult to injury, in 327 BCE, Alexander put to death Kallisthenes, a relative of Aristotle serving on his staff as a historian. Kallisthenes triggered the wrath of Alexander by shaming him for his demands Greeks should behave like Persians in prostrating in front of him.[358]

Despite this potential conflict between Aristotle and Alexander, the influence of Aristotle on Alexander was lasting. Alexander accepted Aristotle's views that, with the Greeks with him, there was nothing he could not do. Alexander believed Aristotle when he said a united Greece could rule the world. Alexander also learned cosmology, geography, medicine, botany, and zoology from Aristotle. When he led his army into Persia, he surrounded himself with a variety of educated men and scientists.[359] Alexander built Alexandria in Egypt. His vision, fulfilled by his general and friend, Ptolemy, was Aristotelian: make Alexandria the political and intellectual capital of civilized world. But Alexander also influenced the world to hear Aristotle. Alexander and Aristotle were made for each other.

In 2004, Lucio Russo, Italian mathematician and historian of science, concluded that from the late fourth to the late second century BCE, the Greek-speaking countries brought into being "an explosion of objective knowledge about the external world."[360] In 2008, the Dutch classical scholar Orjan Wikander reached the same conclusion. In the period from 283 to 221 BCE, the Greek scientists of Alexandria brought about "an explosion of technological accomplishments." Ptolemy II Philadelphos and Ptolemy III Euergetes (Benefactor) supervised and funded the scientists who generated knowledge

[358] Quintus Curtius Rufus, *The History of Alexander* 8.5.5–21, tr. John Yardley (New York: Penguin Books, 1984); Arrian, *The Campaigns of Alexander* 4.10–14, tr. Aubrey de Selincourt (New York: Penguin Books, 1971); Ploutarchos, *Alexander* 55.9.

[359] N. G. L. Hammond, *The Genius of Alexander the Great* (Chapel Hill: The University of North Carolina Press, 1997) 5.

[360] Lucio Russo, *The Forgotten Revolution: How Science Was Born in 300 BC and Why It Had to Be Reborn*, tr. Silvio Levy (Berlin: Springer, 2004) 1.

and technologies. The Greek kings and scientists made the Mouseion "a hive of inventive activity."[361] The library was the heart of the entire enterprise.

In 2017, Dionyses Simopoulos, director emeritus of the Eugenideion Planetarium of Athens, said the Mouseion and Library of Alexandria were "lighthouses of knowledge." He is right. They were lighthouses of civilization. More directly, he meant the Mouseion and library expanded and advanced the knowledge of mathematics, astronomy, and philosophy. These were the landmarks of the scholarship of the Alexandrian Age, including the entire intellectual tradition and legacy of the classical era.[362]

The classical scholar Edward Alexander Parsons who studied the history of the Alexandrian Library called it the "glory of the Hellenic world." He is right. The Alexandrian Library was the glory of the Greeks. He said: "The Alexandrian Library was not only the largest and most important collection of Greek books that ever existed, but, in the days of its greatest glory, it was the greatest library in the entire world before the invention of printing."[363] According to the German scholar Rudolf Blum, the result of this ideal environment for the advancement of knowledge boosted civilization. Greek scientists made fundamental contributions to chronology, philology, history, literature, geography, mathematics, astronomy, and medicine.[364]

Experiments and Scientific Research

Philon of Byzantium was a product of the flowering of knowledge in the Greek world. Writing in the late third century BCE about mechanics, including the construction of weapons, he is emphatic that advancements in technology rely on theory and trial and error. He cited and probably followed the example of the sculptor Polykleitos that perfection comes "gradually in the course of many calculations." His experience came from working in Alexandria and being in contact with the master craftsmen of Rhodes. For example, catapults become

---

[361] Orjan Wikander, "Gadgets and Scientific Instruments," in *The Oxford Handbook of Engineering and Technology in the Classical World*, 785–799.

[362] Dionyses Simopoulos, "The Antikythera Mechanism," *Kathemerine*, June 11, 2017 (in Greek).

[363] Edward Alexander Parsons, *The Alexandrian Library: Glory of the Hellenic World* (Amsterdam: The Elsevier Press, 1952) 203.

[364] Rudolf Blum, *Kallimachos: The Alexandrian Library and the Origins of Bibliography*, translated from the German by Hans H. Wellisch (Madison, WI: The University of Wisconsin Press, 1991) 97.

more effective by changing the size of the hole through which the springs of the weapon, the twisted skeins, move. Doubling the size of a catapult required a larger diameter for the hole, which meant solving the interesting and difficult geometrical problem of doubling the cube.

Philon and other Greek engineers improved the catapults by geometrical and mechanical means. He describes how some of this happened. He also praises the Ptolemies, who funded science and technology experiments. Philon says the Ptolemies were ambitious men. They loved craftsmanship and power. Their metalworkers and engineers weaved craftsmanship into weapons. Philon explained the purpose of artillery was "the long-range shooting of missiles which will land with vigorous impact." But to get to that point he knew he had to have a definite method for figuring out the diameter of the hole for the twisted skeins. He was certain it was necessary to determine this diameter not accidentally or haphazardly but by testing. He complained the ancients failed to construct accurate catapults because their trials were limited. But the Alexandrian engineers, he said, succeeded in improving the catapults. The subsidies they received from the kings of Egypt made that possible. The kings were eager for fame and power. They were interested in the strategic importance of technology, including military weapons.

"For it is not possible to arrive at a complete solution of the problems involved merely by reason and by the methods of mechanics; many discoveries can, in fact, be made only as a result of trial," Philon said.[365]

In chapter 1, I discussed the role of Metis, Athena, and Hephaistos in Greek society. These were gods of science and technology. The stories Greeks knew about them inspired them to seek knowledge and invent and build tools for improving their lives. I quoted Homer to the effect that Hephaistos was a master of high-tech robots. Homer also says that Hephaistos fashioned weapons for the Trojan War hero Achilles. However, Hephaistos and Athena did more than splashy technology projects.

The wisdom associated with Metis and Athena links the mind to the hands, inspiring and guiding craftsmen in their useful work. The talent of the craftsmen is applied metis, what we would call applied science. The result of

[365] Philon of Byzantium, *Belopoeica* (Mechanics IV) 3. 5–7 in *A Source Book in Greek Science*, ed. Morris R. Cohen and I. E. Drabkin (New York: McGraw-Hill, 1948) 318–319. For the Greek text of *Belopoiika*, see: *Greek and Roman Artillery: Technical Treatises*, ed. and tr. E. W. Marsden (Oxford: Clarendon Press, 1971) 106–154.

that Metis tradition was philosophers like Aristotle; scientists and engineers like Euclid, Ktesibios, Archimedes, and Philon; and astronomers like Eudoxos, Aratos, Eratosthenes, Hipparchos, and Ptolemaios. Intelligence was behind the creation of a tremendous variety of weapons, plows, art, pots, clothes, shoes, houses, temples, sundials, speeches, books, and libraries.

In fact, the Greeks published their laws, offerings to the gods, and decisions on stone. They documented their culture with "acres of writing."[366] The result was the Antikythera computer. All this was also hard work.

For example, Erasistratos of the third century BCE said scientific research was a lifelong occupation. He explained: "Those who are completely unused to inquiry are, in the first exercise of their mind, blinded and dazed and straightway leave off the inquiry from mental fatigue and an incapacity that is no less than that of those who enter races without being used to them. But the man who is used to inquiry tries every possible loophole as he conducts his search and turns in every direction and so far from giving up the inquiry in the space of a day, does not cease his search throughout his life. Directing his attention to one idea after another that is germane to what is being investigated, he presses on until he arrives at his goal."[367]

Erasistratos did detailed anatomical studies. His model of the heart was like a water double pump: a twin-stroke pump with a double action.[368] Ktesibios probably had Erasistratos's mechanical heart model in mind in designing his double-piston water force pump.

Vitruvius knew this machine as *Machina Ctesibica*.[369] This was a great honor for Ktesibios. Vitruvius had no doubt he invented "the nature of wind-pressure and the principles of pneumatics." Pliny the Elder, a Roman natural philosopher and contemporary to Vitruvius, agrees. He says Ktesibios "discovered the theory of the pneumatic pump and invented hydraulic engines."[370]

By pneumatics, the Greeks meant wind or air power. Ktesibios used his insight of pneumatics to construct hydraulic machines, geared water clocks,

---

[366] Whitmarsh, *Battling the Gods,* 17.

[367] Erasistratos, *On Paralysis* 2 in Galen, *On Habits* ch 1, tr. G. E. R. Lloyd, *Greek Science After Aristotle* (New York: W. W. Norton, 1973) 86.

[368] Galen, *On the doctrines of Hippokrates and Plato* 6.6 in James Longrigg, ed., *Greek Medicine from the Heroic to the Hellenistic Age: A Source Book* (New York: Routledge, 1998) 94.

[369] Vitruvius, *On Architecture* 10.6–7.

[370] Pliny the Elder, *Natural History* 7.125, tr. John F. Healy (Penguin Books, 1991).

automata, and other machines powered by gears for measuring distances on land and sea. Vitruvius says the water machine of Ktesibios employed pistons and levers and moved water from a reservoir to water fountains. The water clock of Ktesibios had a rack and pinion, perhaps even gearwheels.[371]

Ktesibios developed his water clock in the early third century BCE, about the time that metal gearing, the cam, and the screw came into being in Alexandria.[372] Ktesibios also invented a water organ (*Hydraulis*). Athenaios says that in one of the lavish diners in Rome he and his friends (philosophers dining or *deipnosophistai*) heard "an extremely pleasant and enjoyable sound produced by a hydraulic organ coming from one of the neighboring houses." This was the organ crafted by Ktesibios of Alexandria, a barber.[373] A modern researcher calls this organ "the world's first keyboard instrument."[374] Thus, Ktesibios, a self-taught engineer, made fundamental advances to the gear technologies of the Greeks.

Erasistratos and Herophilos used surgical instruments. For example, they did abortions with the aid of an *embryotome* that cut the fetus to pieces. Erasistratos also invented and introduced the catheter for draining the bladder.[375]

Erasistratos, Herophilos, and Ktesibios, as well as other Greek scientists and engineers, especially Archimedes, used a great variety of measuring devices and instruments. These tools made a difference in advancing astronomy, mechanics, geography, medicine, chemistry, and craftsmanship.

Pliny the Elder credits Daidalos, a master craftsman and artist of the second millennium BCE, with inventing "woodworking, together with the saw, axe, plumb-line, gimlet, glue and isinglass."[376] Of course, Daidalos invented much more than the tools Pliny credited him with (chapter 1).

Leonidas of Tarentum, a writer of epigrams who flourished in the first half of the third century BCE, described the tools of a carpenter. Remember, a carpenter (*tekton*, τέκτων) in ancient Greece did much more than the carpenter does today. He was a craftsman but he could also be a physician, a sculptor, and

[371] Vitruvius, *On Architecture* 9.8.2–5; 10.7.1–3; 10.8.1–6.

[372] Andrew I. Wilson, "Machines in Greek and Roman Technology," in *The Oxford Handbook of Engineering and Technology in the Classical World*, 340; Orjan Wikander, "Gadgets and Scientific Instruments," in *Ibid.*, 787.

[373] Athenaios, *The Learned Banqueters* 4.174a–c, 5.174c1–2.

[374] James and Thorpe, *Ancient Inventions*, 116.

[375] Tertullian, "On the soul" 25.5 in *Greek Medicine*, 187, 189.

[376] Pliny the Elder, *Natural History* 7.198–199.

a master in any craft or art. This is because *tekton* encompassed the man working in wood as well as those working in technology. The carpenter of Leonidas of Tarentum was at the heart of craftsmanship and, therefore, engineering and technology. This carpenter dedicated his tools to Athena because Athena "gives grace to work." These tools included a grooved file; the plane, "the rapid devourer of wood"; the hammer that "strikes with both heads"; the rule stained with ochre; the drill-bow and rasp; the heavy axe with its handle, "the chief instrument of the crafts"; revolving augers and quick gimlets; screwdrivers; and the double-edged adze.[377]

A Greek farmer gifted his tools to Demeter: an ox-turning iron-tipped goad, a bag for measuring wheat and barley, a curved sickle, a winnowing fork, and his laced boots.[378] Greek farmers also used plows for the cultivation of the land. Farmers in Thessaly had a special plow that enabled them to dig deeper. They called that tool *mischos*. The other fundamental agricultural tool was the hoe or mattock. The hoe was usually two-pronged.[379] There was also a wide-bladed hoe for digging.[380]

The Greeks, according to Tassios, manufactured alloys that were solid and malleable. This meant they made better tools: astrolabes, dioptras, planetaria, more precise balances, and other instruments, including astronomical computers. Water clocks helped in measuring long time intervals more precisely than sundials. But in their simpler version of clepsydras, water clocks timed speeches made in the law courts.

Gearwheels made possible *hodometers* for measuring long distances on land and *dromometers* in measuring long distances on sea.[381]

Already in the fourth century BCE, the Greeks had high and exacting standards of quality control for metalworking and engineering. For example, a fourth-century inscription from Eleusis directs that the mixture of copper-tin for short metal rods (*poloi*) and their metal enclosures (*empolia*) connecting column drums ought to be made up of eleven parts copper and one part tin. The metal rods should be designed to rotate in their enclosures shaped like

---

[377] *The Greek Anthology* 6.205.

[378] *The Greek Anthology* 6.95.

[379] Theophrastos, *Causes of Plants* 3.20.8, ed. Kaktos Philological Team (Athens: Kaktos, 1998).

[380] Xenophon, *Oeconomicus* 17.12.

[381] Tassios, "Cross-Fertilization of Science and Technology in the Time of Archimedes," in *The Genius of Archimedes*, 347–359.

symmetrical cubes dug and fit into the central core of the connecting marble drums of the pillars. Furthermore, the inscription gives the exact specification for the shape, length, thickness, movement, and function of the parts of the metal mechanism. Copper had to have been imported from Marios, Cyprus, a reputable center for pure copper. Finally, the inscription prices the technical work and names the persons responsible for the delivery and installation of the metal parts in the twelve-column Stoa being erected in front of the Telesterion of Eleusis.[382] The Telesterion was a gigantic building hosting the ceremonies of the Eleusinian mysteries.

This technological progress and, especially, the availability of gearwheels, explains how the Greeks built the Antikythera Mechanism. By the early second century BCE, they had all that was necessary to construct the celestial machine. They had excellent astronomers, mathematicians, craftsmen, metallurgists, and engineers. Their metallurgical know-how dated centuries before the manufacture of the Antikythera computer. Workable gearwheels came into being during the third and second century BCE. Archimedes used gearwheels in his mechanical models of the universe. Tassios is right in saying, "Technically speaking, gearwheels are the main characteristic of the Antikythera Mechanism."[383]

Tassios also says the Greeks were "maniacs" for technology, by which he means they were passionate about things technical, such as craftsmanship. Their mythology, as we saw in the beginnings of this chapter and other parts of this book, was full of gods and heroes embodying extraordinary knowledge and powers. That mythology, which the Greeks considered early history, provided the metaphysics of zeal and success in technical work. Tassios reminds us that the Greeks had an insider's advantage to craftsmanship. One of their gods, Hephaistos, was a supreme metallurgist and engineer who designed and built Achilles's Shield and the robot Talos guarding Crete in the age of the Argonauts in the twelfth century BCE.

In addition, the Cyclopes, probably the earliest engineering geniuses in Greek culture, manufactured an almighty weapon, the thunderbolt, for Zeus to bring the war among the gods to an end.

---

[382] Society for the Study of Ancient Greek Technology and the Epigraphic Museum, eds., *Catalogue of an Exhibit of Ancient Greek Inscriptions of Technological Content* (Athens April 4– June 30, 2002, pp. 62–69 (in Greek); George Varoufakis, "The inscribed stele of Eleusis and quality control in antiquity," *Archaeology and Arts*, June 23, 2014.

[383] Tassios, "Prerequisites for the Antikythera Mechanism," in *The Antikythera Shipwreck*, 255.

Stephanos Paipetis, professor emeritus of mechanical engineering at the University of Patras, examined the technology in Homer and concluded, for instance, that Hephaistos built Achilles's Shield with advanced knowledge and technology of metals and materials. In other words, Hephaistos "possessed deep knowledge of the dynamic mechanical properties of laminated composite structures." The ships of Phaiakians that carried Odysseus from Phaiakia (Kerkyra) to Ithaca, Paipetis says, were governed by artificial intelligence.[384]

Husband-and-wife archaeologists from the University of Cincinnati, Sharon Stocker and Jack Davis, unearthed in May 2015 a tiny agate seal stone with great potential of throwing more light on the civilization of Bronze Age Greece. This thumb-size gem is engraved with a marvelous picture of a warrior who seems to come out of the *Iliad* of Homer, perhaps depicting Achilles killing Hector, with another dead soldier sprawled at his feet.[385] Of course, this cannot be, since this warrior died centuries before the Trojan War. So, the stunning reality about this gem is the detail it offers of the human body that, according to Davis,[386] does not appear until the classical age, about a thousand years later. This gem was part of the treasure-laden tomb of a warrior who died sometime between 1500 and 1450 BCE in Pylos, southwestern Peloponnesos, Greece. The treasure included four gold rings, precious stone beads, silver cups, ivory combs, a bronze sword, and several hundred other objects. One of them was an ivory plaque with a griffin, a mythical animal possessing a lion's body and an eagle's head and wings. This is why the Pylos warrior was dubbed "Griffin Warrior".

Like the fantastic image of the Pylos warrior, the gold rings mirror superior craftsmanship. The goldsmith[387] used "multiple sheets of gold." We find similar advanced knowledge and technology in the craftsmanship of the Shield of Achilles. The Pylos warrior died 243 years before Odysseus returned home to Ithaca in 1207 BCE. Nevertheless, the culture of the Griffin Warrior, captured in the tiny gem and the technology required for engraving it, informed the Greeks of the Homeric epics.

---

[384] Evaggelos Vallianatos, "The Shield of Achilles," *Huffington Post*, September 15, 2017.

[385] Sarah Gibbens, "Ancient Greek 'Masterpiece' Revealed on Thumb-Size Gem," *National Geographic*, November 7, 2017.

[386] Rachel Richardson, "Unearthing a Masterpiece," *UC Magazine*, November 6, 2017.

[387] Rachel Richardson, "Lord of the rings," *UC Magazine*, October 3, 2016.

Aside from the masterpiece qualities of the art and design of the combat scene on the gem, how did the artist work on such a tiny stone? Did he have a magnifying glass? You ask the same question when you see the gear teeth and inscriptions on the Antikythera Mechanism, the computer Greeks built about twelve and a half centuries after the death of the Pylos warrior.

The exquisite art of the Griffin Warrior fits nicely into the lavish and unforgettable images painted in the Bronze Age walls of homes at Akrotiri in the Aegean island of Thera. The island had more than beautiful art. Akrotiri had two-story houses, paved streets, running water, and toilets.

Bronze Age Greece then was a carrier of advanced civilization. The remarkable examples from surviving high quality technical works of the Mycenaean age, give one the impression that technology had found a friendly and fertile home in Greece. Naturally, and like the Parthenon, the Antikythera computer was the culmination of that legacy.[388]

The astronomer Ptolemaios probably owned an Antikythera Mechanism–like computer. He lived almost 300 years after the Antikythera machine came into being. Ptolemaios also flourished some 400 years after Erasistratos gave a classic definition of a scientist. Ptolemaios was so involved with his astronomical research that doing science was nearly divine to him. Investigating and decoding the heavens almost removed him from the affairs of the Earth, knowing not whether he was of the Earth or heavens. He said: "I know that I am mortal, a creature of a day; but when I search into the multitudinous revolving spirals of stars my feet no longer rest on Earth, but, standing by Zeus himself, I take my fill of ambrosia, the food of the gods."[389]

That is one way in which Greeks invented and developed science: being close to their gods in nature and the universe. For precious moments Ptolemaios shared in the immortality of the gods. Searching for answers in the heavens was getting the truth about the workings of the divine forces of nature. Thales was right that the Cosmos is beautiful[390] and full of gods.[391] But the truth about nature and the heavens was also essential for the working of society that relied

[388] Tassios, "The Greeks Were Maniacs for Technology," in Stekas, ed., *The Antikythera Mechanism* 151–158.

[389] Epigram attributed to Ptolemaios, *The Greek Anthology* 9.577.

[390] Diogenes Laertios, *Lives of Eminent Philosophers* 1.35 (κάλλιστον κόσμος).

[391] Aristotle, *On the Soul* 1.411a7.

on the seasons, the day, the month, and the year. Reckoning time was getting closer to understanding how nature works.

Machines from Nature for Useful Work

Greek mechanics and engineering continued for a long time. As late as the fourth century of our era, the Greek mathematician Pappos of Alexandria praised mechanics as "a science and an art," useful "for many important practical undertakings" as much as being prized by philosophers and mathematicians. Mechanics also is intimately connected to nature, especially the "material composition of the elements in the Cosmos." Mechanics, says Pappos, is theory and praxis. Theory includes "geometry, arithmetic, astronomy, and physics." Practical mechanics relies on "metal-working, architecture, carpentry," and "painting."[392]

In praising mechanics, Pappos mirrored a long tradition of science and engineering. Vitruvius, Heron of Alexandria, and Proklos had similar views. Vitruvius was Roman and Heron and Proklos were Greek. Vitruvius and Heron were two of the most distinguished observers of the Greeks during the time of the Antikythera Mechanism. Proklos was a later observer who flourished in the fifth century. But as a historian of science, his comments are very important.

Vitruvius was an eminently practical man with a comprehensive knowledge of Greek science and philosophy. He was an engineer, architect, and historian of science who lived through the transition of Rome from a republic to an empire. In fact, the first Roman emperor, Augustus, and his sister, Octavia, sponsored some of his work. Thus, Vitruvius lived during the first century BCE, when the Greeks were using the Antikythera Mechanism and other geared machines. He wrote extensively about architecture, machines, and inventions. He devoted the tenth book of his *De Architectura* to machines. He was also a careful observer of nature and society. He reported, for example, on water poisoned by lead, that drinking water was contaminated by white lead because of lead pipes. So, he advised Romans not to drink water coming out of lead pipes. Water, he said, "is much more wholesome from Earthware pipes than lead pipes." He warned: "In digging wells [for drinking water] we must not make light of science. The methods of nature must be considered

[392] Pappos, *Mathematical Collection* 8.1–5.

closely in the light of intelligence and experience."[393] Thus, Vitruvius is knowledgeable and trustworthy: he was certain that ideas of men like Ktesibios and Archimedes would benefit humankind. Vitruvius was also certain nature was the mother of invention and machines. He explained how the natural world becomes the model for the engineer. All machinery, he said, had something to do with nature and the movements of the stars. The "device" of nature is also responsible for the movements of the heavenly bodies. This explains the seasons and the ripening of the harvest. He was certain nature taught people to invent machines and handy implements in order to make their lives comfortable and lessen the burden of manual labor. This was especially true in food production and transportation of goods by land and sea. For example: "The equilibrium of balances and scales had been applied to free human life from fraud by the provision of just measures." In addition, Vitruvius explored military technologies that "serve both to protect against danger and to satisfy the needs of safety."[394]

The other trustworthy witness of the use of technology in Greek society at the time Greeks continued to build Antikythera Mechanism–like machines was Heron of Alexandria, who lived about a century after Vitruvius, that is, in the first century of our era. Heron pretty much agreed with Vitruvius that inventions of machines had a social function and usefulness among Greeks.

Heron was an engineering genius. He saved for posterity the technological legacy of the Greeks. He also made his own contributions to science and technology. He invented baroulkos, an engine with gears for lifting heavy things. He wrote extensively on geometry, physics, and mechanics. Reading his works,[395] I had the impression I was reading Euclid, Archimedes, and Philon. Indeed, Heron sprinkled his writings with the names of Greek thinkers from Homer to Hipparchos and Geminos.

His discussion of automation and how one builds automata (machines or mechanized operations and robots) is singularly attractive for his grasp of mathematics and mechanics. Engineering was at the heart of all the thought and practical wisdom and skills of Heron. He also advanced military technologies, optics, measurements, and mechanics, especially how to construct machines working with the power of water, steam, and compressed air. He was

---

393 Vitruvius, *On Architecture* 8.6.10, 12.

394 Vitruvius, *On Architecture* 9. Preface.15; 10.1.4–6; 10.10.1–6.

395 Wilhelm Schmidt, L. Nix, Hermann Schone and J. L. Heiberg, eds., *Heronis Alexandrini Opera*, 5 Vols., (Leipzig: B. G. Teubner, 1899, 1903, 1976).

also good in building water clocks and devices for entertainment. He invented toy machines, self-propelling machines, slot machines, omen machines, wine and water vending machines, and a prototype of the steam engine.

In 1851, Bennet Woodcroft, a Victorian professor of engineering at the University College, London, edited one of Heron's most important works, *Pneumatics*. This was the year of the Great Exhibition when the Western world went overboard in extolling machines and celebrating industrialization. Woodcroft, an inventor and specialist on the steam engine, wanted a technological hero from antiquity, someone who had set the roots of the technological craze of the Victorian era. While writing an "Analytical History of the Steam-Engine," he discovered Heron. He was so impressed by Heron's work he dedicated Heron's "Pneumatics" to Prince Albert. In the "Editor's Preface," he said the steam engine Heron described in the "Pneumatics" was the model of the steam engines of the nineteenth century. He reports Heron discussed the "elementary parts of all Steam-Engines, and those also of most other machines."[396]

This assessment by Woodcroft, that the steam engine of Heron was at the heart of the steam engines and the mechanical revolution of the nineteenth century, is extremely important. It rightly undermines modern allegations that the steam engine of Heron, no less than the Antikythera computer, was a toy.

Heron in fact was as modern as one can make him, considering he lived about two millennia ago. Heron went so far in automation that he mechanized theatrical productions. His how-to books made technology easy to understand and put into good use: his geometrical book, *Metrika* (Metrics), explained the measurement of surfaces; *Stereometrika* (Solid Metrics) tackled measuring solid bodies in three dimensions. It also discusses two- and three-dimensional geometry; *Mechanika* (Mechanics, saved in an Arabic translation) explained how to move or lift weights with very little effort; Peri *Automatopoietikes* (About Making Automata) made building robots easy; *Pneumatika* (Pneumatics) showed how to construct machines driven by compressed air, steam, and water; and *Dioptra* was about building surveying instruments measuring horizontal, vertical, and angular distances and heights. *Dioptra* was useful for replacing land property boundaries, digging tunnels through mountains, and measuring

[396] *The Pneumatics of Hero of Alexandria from the Original Greek*, ed. Bennet Woodcroft, tr. Joseph Gouge Greenwood (London: Taylor Walton and Maberly, 1851, facsimile of the Woodcroft edition, London: Macdonald, 1971) vii.

areas under enemy control. *Dioptra* also measured distances on Earth and between the Earth and the stars.[397]

Katoptrika (Catoptrics) was all about constructing and using curved mirrors, especially explaining reflection in mirrors. Like Archimedes, Heron moved smoothly between philosophy and mechanics, and between mechanics and geometry, using one to solve the technical problems of the other.[398] Euclid and Archimedes were Heron's models.

Like Vitruvius, Heron also dispels the illusion that machines among the Greeks were confined to toys. He thought mathematics had a social function, geometry securing equality and justice. In the Introduction of the "Pneumatics," he demystified machines. What you need in order to make machines, he said, is the proper mix of three or four of the fundamental elements: fire, air, Earth, and water. Entwine these elements in ways conducive to the generation of useful power for work. Showing his influence from the atomic theory of Democritus, he envisions invisible atoms and miniscule voids all over matter, including air and water. But the net result of all theory was making machines because he believed that some of the machines supply the most pressing wants of human life and others spark astonishment and admiration.[399]

Proklos agreed with Vitruvius and Heron. Writing more than three centuries after Heron, Proklos had plenty to say about the history of mathematics, especially geometry. Just like the mathematician Pappos of the fourth century, he said geometry was both theoretical and practical. He explained the practical and useful implications of geometry as follows:

"[W]hen... [geometry] touches the material world, it gives birth to a variety of sciences—such as geodesy, mechanics, and optics—by which it benefits the life of mortals. Through these sciences, geometry has devised instruments of war and defenses for our cities, made familiar the succession of the seasons ... taught how to measure distances by land or sea, constructed balances and scales for determining arithmetical equality when a city needs it, invented models for exhibiting the order of the whole heaven [for example, geared devices like the

[397] Heron, *Dioptra* 190.10–21; 238.3–240.27; 268.17–272.15, ed. and tr. into German by H. Schone, in *Heronis Alexandrini Opera*, Vol. 3 (1903).

[398] Karin Tybjerg, "Heron of Alexandria," *The Encyclopedia of Ancient Natural Scientists*, 384–387; Karin Tybjerg, "Hero of Alexandria's Mechanical Geometry," *Apeiron*, Vol. 37, No. 4 (2004) 29–56.

[399] Heron, *Pneumatica* 145.15–20; 146.1–25, ed. and tr. into German by W. Schmidt, in Heronis Alexandrini Opera, Vol. 1 (1899); also: Woodcroft, tr., *The Pneumatics of Hero of Alexandria* Introduction, pp. 1–10.

Antikythera Mechanism]. Geometry has also unveiled many incredible things, making them credible to all."[400]

Finally, there's a medieval Greek intellectual of the twelfth century who had probably read some of the work of Archimedes, Heron of Alexandria, and Pappos. I already cited John Tzetzes. He praised the social role of technology for improving human life. For Tzetzes, mechanics was a boon to national defense and civilization. Like Archimedes, Heron, and Pappos, he credits geometry for being the mother of technology. But unlike Pappos, Tzetzes is superficial and full of self-importance. However, he read books we no longer have. He was a proud polymath who made a living as a scholar and writer. He was evocative and eloquent in his admiration of the achievements of the Greek past.

Geometry, he said, is useful for many mechanical works like the lifting of weights, putting ships to sea, and throwing rocks. Geometry also is necessary for the design of a variety of machines for attacking and defending cities, including building harbors and bridges, and constructing solar weapons like mirrors for setting enemy warships on fire. In addition, geometry helps us build machines that lighten physical labor and give comfort and enjoyment to life. "Fabulous geometry gives birth to myriad machines," he wrote.

Tzetzes talked about a variety of machines: the wedge, pulleys, the lever, the screw, the axle, and the wheel. He mentioned the "baroulkos" (gear machine for lifting heavy weights), "tortoises" (weight-bearing frames with rollers), mining tortoises, armed tortoises, and maneuverable mantlets or tortoises. He also listed siege machines; shooting devices like stone-throwers and catapults for missiles, and stomach-bows; rams that breach city walls; ladders and universal joints; and wheeled towers. Optics together with geometry, said Tzetzes, make possible a variety of mechanical devices. They are also of enormous significance to craftsmen like sculptors and painters.[401]

Serafina Cuomo, expert on ancient technologies, also assures us technologies served a broad social purpose in the ancient world.

"[I]n their written [Greek and Roman] texts," she wrote, "technicians often addressed not just other technicians, but the educated general public, and in particular people of authority ... The unique selling point of technology

[400] Proklos, *A Commentary on the First Book of Euclid's Elements* 63 (I modified the translation somewhat.)

[401] Tzetzae, *Historiae* 11.586–641.

to these people was its usefulness: the capacity, or alleged capacity, to produce results rather than mere words."[402]

True, according to the Danish scholar, A. G. Drachmann, "gear-wheels" developed "by trial and error." They, and all other forms of advanced technology, still develop by trial and error. But it is not true, as Drachmann says, the "rack-and-pinion was confined to toys and smaller instruments, and the worm-and-wheel gear, after a short flowering under Archimedes, was also reduced to toys and instruments, except for the re-setting apparatuses of the surgeons."[403] On the contrary, gears continued to improve both toys and machines, following and sustaining the flowering they displayed in the workshop of Archimedes who was also a social philosopher and a patriot. Indeed, the astronomical models after Archimedes improved. Those models made the mechanical Cosmos of the Antikythera computer possible (chapter 4).

Crafts and mechanics were not trivial but comprehensive, sophisticated, scientific, and fundamental to culture and life. Greeks were technologically literate. They built machines to lessen hard work and improve their lives. The Athenians even had machines for selecting citizens by lot to serve as jurors.[404] According to T. E. Rihll, a British classical scholar: "[Machines] were to be found in every home, many public buildings and spaces, on and under the ground and sea, for work and pleasure. From looms and olive presses, locks and keys, hoists and windlasses, to water-clocks and thunder-makers, machines were a familiar part of life for the ancient Greeks and Romans."[405]

Tassios, member of the Society for the Study of Greek Technology, has been documenting the upsurge of Greek technology during the Alexandrian Age. He describes land-reclamation projects; the building of sophisticated bridges; the construction of the 100-meter-tall Pharos (Lighthouse) of Alexandria; the construction of four aqueducts that brought 2,000 cubic meters of water to Pergamum every day; the flourishing of shipbuilding, especially the construction of the gigantic ship *Syracuse* by King Hieron of Syracuse for King Ptolemy III of Egypt; the deployment of a military mobile siege tower,

---

[402] Serafina Cuomo, "Ancient Written Sources for Engineering and Technology," in *The Oxford Handbook of Engineering and Technology in the Classical World*, 32.

[403] A. G. Drachmann, *The Mechanical Technology of Greek and Roman Antiquity* (Copenhagen: Munksgaard, 1963) 203.

[404] Aristophanes, *The Ecclesiazusae* 681–686; Aristotle, *The Athenian Constitution* 63.2.

[405] T. E. Rihll, *Greek Science* (Oxford: Oxford University Press, 1999) 26.

Helepolis, which was several stories high and armored; the use of huge cranes as offensive weapons by Archimedes; and the invention of the olive screw press and the Archimedes screw-pump that benefited peasants. He assures us metallurgical technologies reached their peak during the Alexandrian Age: the sculpting of excellent statues, and the manufacture of weapons, domestic utensils, medical instruments, and gearwheels and astronomical spheres like the Antikythera computer. Tassios also argues that the Alexandrian Age made progress in the chemistry of "processing metals, precious stones, and dyes of all kinds" by using a great variety of reagents. In addition, scientific measurement of useful things did well in, for instance, the measurement of time, terrestrial and astronomical distances, and weights. Tassios is right. The Alexandrian Era was a golden age of science and technology.[406]

The Greeks used cranes since the late sixth century BCE. Three hundred years later, cranes were "ubiquitous," both among the Greeks and the Romans. The use of machines was widespread among the ancient Greeks.[407] In fact, machines and products of craftsmanship were part of the fabric of Greek culture, both political and religious. In chapter 1, I showed that even gods (Prometheus, Metis, Athena, and Hephaistos) protected the technological civilization of the Greeks. Indeed, the gods created models of craftsmanship that triggered the Greek imagination for a scientific and technological understanding of the Cosmos. Hephaistos was the master of metallurgy and fire, and Athena inspired wisdom and achievements in the arts and crafts. Athena was the daughter of another goddess of intelligence, Metis. It was out of that technological culture, which received a tremendous boost in the Alexandrian Age, that the Greek computer came into being.

François Charette, professor of the history of natural sciences at the Ludwig-Maximilian University in Munich, Germany, studied the Greek computer and he concluded in 2006 that "mind-boggling technological sophistication" must have been available to those who made it.[408]

The makers of the Antikythera computer were the best of scientists and engineers. Their age, the third and second centuries BCE, shone as the climax

[406] T. P. Tassios, "A very brief introduction to Hellenistic Alexandrian technology," in *Hellenistic Alexandria*, 191–198.

[407] Wilson, "Machines in Greek and Roman Technology," in *The Oxford Handbook of Engineering and Technology in the Classical World*, 362.

[408] François Charette, "High tech from Ancient Greece," *Nature*, November 30, 2006, 444: 552.

of innovation and scientific achievement. The final product, the Antikythera Mechanism, was so advanced that, to this day, it astonishes those seeing its surviving fragments in the National Archaeological Museum in Athens. Jonathan Keats, *Forbes* contributor, admits that these fragments elicit "awe and anxiety." Nevertheless, he said on July 30, 2019, that the Antikythera Mechanism "still stands out as one of the most magnificent mechanisms of all time."[409]

According to the Italian engineer and scholar Giovanni Pastore, who also studied the Antikythera Mechanism, those who made the Antikythera celestial computer anticipated by nineteen centuries the law of universal gravitation formulated by Isaac Newton in 1687; they applied the heliocentric theory of Aristarchos; and they anticipated the 1841 work of Robert Willis about epicyclic gearings.[410]

Greeks and Romans

In 2006, the scientists who studied the Antikythera computer remained silent about any connections of the computer with universal gravitation and the heliocentric theory. They admired the ancient device's "great economy and ingenuity of design." The Antikythera Mechanism, they concluded, "stands as a witness to the extraordinary technological potential of ancient Greece, apparently lost within the Roman Empire."[411]

Yes, ancient Greece had extraordinary technological potential but, as the Antikythera computer shows, they also had plenty of technological power.

Political division among the Greeks, a bad habit that survived the unification of Greece by Alexander the Great, enabled Romans to conquer them. This was definitely a serious blow against Greek freedom and, necessarily, an impediment to the continuity of Greek technological development. Living under Roman occupation the Greeks, inevitably, followed the fate of their rulers.

---

[409] Jonathan Keats, "What is the Ultimate Machined Object? A New Exhibit of Ingenious Devices Shows How Innovation Works," *Forbes*, July 30, 2019.

[410] Giovanni Pastore, *Antikythera E I Regoli Calcolatori* (Rome, 2006). On the internet, Pastore reports that, in 2006, Italians discovered in the civic market of Olbia, Sardinia, "a fragment of forty-three-millimeter gear wheel" dated from the end of the third century to the beginnings of the second century BCE. See also the Greek magazine *Davlos*, 314, June 2008, 22245–22248. I am grateful to Nikos Daphnis for this source.

[411] Tony Freeth et al., "Decoding the ancient Greek astronomical calculator known as the Antikythera Mechanism," *Nature*, 30 November 2006, 444: 591.

Strabo was an exemplary Greek scientist who lived the crisis of the political domination of the Greek world by Rome. Strabo was also a polymath and philosopher I have cited repeatedly. He developed scientific geography right when the Roman Empire came into being. Like Poseidonios, whom he knew, he traveled widely in Asia and Europe and left us his "Geography" in seventeen books. This is a rich story of the physical environment and the traditions of the people and countries of the Roman Empire.

Strabo saw his work as philosophy and science. He thought of geometry and astronomy as essential for geographical knowledge. He took for granted the spherical Earth and the spherical geocentric Cosmos. He explained that the curvature of the sea prevents sailors seeing lights at a distance when those lights are at the elevation of the eyes. For the same reason, when sailors are approaching land, the low-lying land grows higher and higher.[412]

Strabo also wrote his "Geography" as a strategic text for military leaders, politicians, and rulers.[413] In fact, he wrote it for Pythodoris, the queen of Pontos who owed her throne to the first Roman Emperor, Augustus. Strabo was also from Amasia, Pontos. He was the son of a wealthy and powerful family that committed treason against Pontos. His family collaborated with the Romans. He heard from his father how the Romans dismembered the countries of their enemies, including Greece and Pontos. We don't know anything of his personal feelings, but he probably interpreted the Roman atrocities and looting of Greece as just punishment for Greek disunity. He saw the Roman Empire as a stabilizing force in the world.

Strabo was also a historian. His "Historical Sketches" was an encyclopedia of forty-seven books, all of which did not make it to our day. But from the surviving "Geography," we see Strabo like a scholar contributing to science, putting geography right in the field of scientific research. His "Geography" was also historical and philosophical. He thought it a waste of time to engage uneducated people with philosophy. These men and women, he assured himself, are like children, necessarily frightened by the monsters of myths. On the other hand, he was certain, it would be a pleasure to have philosophical discussion with the likes of Eratosthenes, Hipparchos, Poseidonios, and Polybios, all outstanding philosophers and scholars.[414]

---

412 Strabo, *Geography* 1.1.20.

413 Strabo, *Geography* 1.1.16.

414 Strabo, *Geography* 1.2.1, 1.2.8.

I already mentioned Polybios because he admired Archimedes. Polybios explained to the Greeks and Romans why Rome became a superpower in the Mediterranean. Strabo liked Polybios because, like him, he worked in a world run by Rome. Polybios tried to be dispassionate in his account but the Romans loomed large, too large to be taken lightly or to be treated like trusted friends. After all, in the mid-second century BCE, the Romans kidnapped him, along with the entire leadership of Greece. Polybios had to stay in Rome for twenty years before the Romans allowed him to return home. Polybios had to be pro-Roman, and he was. Nevertheless, his history is extremely useful and philosophical; it is almost a letter to his fellow Greeks.

For example, he reported that in an effort to cease hostilities among Greeks, Greek military leaders and the Macedonian Greek King Philip V, 221–179 BCE, met in northern Peloponnesos in August 217 BCE. Polybios recorded the speech of a Greek leader, Agelaos of Naupaktos, because this man captured the unspoken anguish of the Greek people.

Agelaos said: "It would be best if Greeks never went to war against each other. They should regard such a blessing the greatest gift of the gods, speaking with one voice and joining hands like men crossing a river. That way they could repulse the incursions of the barbarians and preserve themselves and their cities." Agelaos warned the assembled Greeks that at a time of huge Roman and Carthaginian armies fighting in the West, the Greeks ought to know that the victor of those intense battles would eventually turn against them. So Agelaos pleaded with King Philip to end his quarrels with the Greeks before "the clouds which are now gathering in the west settle upon Greece."[415]

Unfortunately, Agelaos spoke to the wind. Rome never let Greece out of its strategic vision. The clouds of Rome's African wars settled in Greece sooner than the Greeks thought possible. By the early second century BCE, Rome had made Greece a dependency, always proclaiming to be a defender of Greek freedom while preparing to inflict a fatal blow. Rome was also apprehensive about Greek Macedonia. And when Perseus, the older son of King Philip V of Macedonia, took over, Rome decided to intervene decisively. After lengthy but futile negotiations in 166 BCE, Rome sent general Aemilius Paulus against Perseus. Paulus defeated Perseus and dragged him and his family, in chains, to Rome for a spectacular if barbaric display of Roman power. Paulus looted

[415] Polybios, *The Rise of the Roman Empire* 5.104.

Macedonia so thoroughly that it took three days to parade the looted wealth, precious treasures, military weapons, and slaves in Rome.[416] Polybios says the Romans destroyed the Macedonian kingdom, "root and branch."[417]

The extraordinary thing was that despite this and other political misfortunes, the Greeks did not cease their work with science. In less than twenty years after the Romans dismantled Macedonia, they wiped out Corinth and annexed Greece. This took place in 146 BCE. Six years later, in 140 BCE, Hipparchos started working in Rhodes where he created all those astronomical marvels that contributed to the making of the Antikythera computer.

Strabo, writing about a century after the Roman occupation of Greece, illustrates the insight and originality of those Greeks caught in the spirit of the scientific revolution of the Alexandrian Era, even though the Romans cared very little about science.

After Strabo, another Asian Greek named Dio Kokeianos experienced the unpleasant reality of Roman domination: He was loved and hated by Roman officials and emperors. He tasted exile and other humiliations.

Dio was born in the mid-first century in Prusa, capital of Bithynia in northwestern Asia Minor. He died after the first decade of the second century. He was the son of a rich landowner who gave him an excellent education. He used his wealth to travel and write. He visited several poleis all over the Greek world. He also visited Rome. His orations were so eloquent, thoughtful, critical (sometimes too critical), and timely he was known as the "goldened mouthed" Chrysostom (*Chrysostomos*). He had lots of advice for Greek city councils, often telling them things they did not wish to hear: that their ancestors were so much superior to them that, with the Romans commanding Hellas, the only safe thing to do was to, at least, keep alive traditions in limited democracy and Hellenic way of living.

However, Dio had harsh—nay, insulting—things to say about his contemporary Greeks. He blamed them for their historic failure to defend their independence and freedom as one united country. Their jealousies and antagonisms brought about their ruin. In fact, he said, in some poleis, it would have been better if there were no surviving Greeks. That way, their past glories would not be compromised by their degraded living under Roman rule. He praised the

[416] Ploutarchos, *Aemilius Paulus* 32–34.

[417] Polybios, *The Rise of the Roman Empire* 36.9.

Greeks of Rhodes for maintaining some kind of a semblance of Greek traditions. But, in general, Dio was very disappointed with his fellow Greeks. He went so far in his disapproval and, probably, immense personal sorrow of how low the Greeks of his time had fallen, that he suggested that the stones and ruins of ancient Greek poleis were by far the most important evidence of the glorious past and greatness of Hellas.[418]

There's no doubt that, in some of his denunciations of his contemporary Greeks, Dio hit the nail on the head. Occupation is never pleasant. Yes, the political independence of the small city-states and the freedom of their citizens probably explains the Greek genius. But this independence went too far and precipitated civil wars like the Peloponnesian War and, indirectly, led the Romans and barbarians to Greece. Like other prominent Greeks, Dio lived in the Roman Empire. To what degree did the burdens of that empire muffle his speech? Or make his orations so controversial? We don't know. Then, like most surviving texts, the orations of Dio may not be, even in a few paragraphs, what they appear to be. There's always that nagging suspicion the fingers of an enemy might be hiding behind Dio's offensive language.

Like Strabo and Dio, Athenaios shed light on the uneasy relations between Greeks and Romans. He was a Greek intellectual of the second century of our era. He knew the Romans very well. In fact, a rich Roman funded him. He reported that the Romans sucked the life out of their subjects, leaving them to rot. He explained this colonial policy as follows:

"[The Romans] took over from the conquered anything beautiful and useful they had ... leaving to them only their bad and useless practices, so that they would never be able to regain the freedom and culture they had lost. For example, the Romans learned from the Greeks how to use machines and siege-engines. Then they used those machines and knowledge to conquer the Greeks. While the Phoenicians discovered seafaring, the Romans defeated them at sea. From the Etruscans, who fought in close phalanx formation, they took close-fighting tactics, while from the Samnites [of Italy] they took the use of the rectangular shield, and from the Iberians they took the javelin and from

---

[418] Dio Chrysostom, *The Rhodian Oration* 31.160.1–3: "ἀλλ' οἱ λίθοι μᾶλλον ἐμφαίνουσι τὴν σεμνότητα καὶ τὸ μέγεθος τῆς Ἑλλάδος καὶ τὰ ἐρείπια τῶν οἰκοδομημάτων." ("Certainly, the stones and the ruins of ancient Greek buildings reveal the majesty and greatness of Hellas.") Read also the following sections of the thirty-first oration: 18–20, 157–165. Unfortunately, we have no means to judge the integrity of Dio's text. Did he really say what he "says" in his orations?

others they took other practices and perfected them. The Romans also copied the constitution of the Spartans and maintained it better than the Spartans themselves. However, nowadays [late second century] in selecting useful traditions from their enemies, the Romans also adopt pernicious practices."[419]

So, the Romans befriended the Greeks until they had mastered some of the things that made the Greeks superior to them: technology, especially engines of war. Then, at the opportune time, the Romans turned around and smashed the Greeks using the weapons they had borrowed from them. One reads Cicero and sees the educated Roman's tortured thoughts about the Greeks.

In about 60 BCE, Cicero advised his brother Quintius to be wary of Greeks bearing gifts. He said to his brother in Asia, where he served as Roman governor, many fickle and deceitful Greeks had "trained by a long period of servitude to be excessively flattering." Of course, Cicero admitted, some of those Greeks may be "worthy of ancient Greece." But Cicero was also honest enough to admit he was who he was because of the educational and cultural gifts of the Greeks. He said to his brother that he was governing not Africans, Spaniards, or Gauls, all "uncouth and barbarous nations," but Greeks who have had civilization. Greeks spread that civilization for the benefit of others like Romans. Quintius should at least give the Greeks "what we have received from them. For I am not embarrassed to say now... that what I have accomplished has been achieved because of the studies and arts handed down to us in the records and teachings of Greece."[420]

Vitruvius, a Roman engineer and architect I cite extensively, had only good things to say about the Greeks. He truly admired them. He lived in the late first century BCE.

Sure, the Romans robbed the Greeks and, in general, they crippled and wrecked Greece, making the country an anthropological museum for the fun of rich Roman tourists. Their emulation and admiration of Hellenic culture was superficial. Some Romans wanted to believe they were Greeks. Others tried to prove that Latin was an offshoot of Greek, or that they were descendants of

[419] Athenaios, *Deipnosophistai (The Learned Banqueters)* 6.273e–f, tr. S. Douglas Olson (Loeb, 2006), tr. mine.

[420] Cicero, *Letters to his Brother Quintius* 1.1.8–35 in *The Roman World: A Sourcebook*, ed. David Cherry (Oxford: Blackwell, 2001) 156–162.

Herakles. In general, most Romans had an ambiguous, nay, hostile, attitude towards the Greeks.[421]

It is worth repeating that the Romans assassinated Archimedes in Syracuse in 212 BCE. Sixty-six years later, in 146 BCE, a Roman army under Lucius Mummius wiped out Corinth, the mother-polis of Syracuse that gave birth to Archimedes. After the atrocity, Mummius shipped to Rome lots of treasures from Corinth and the rest of Greece. Pliny the Elder says as much: "Mummius filled Rome with statues after his conquest of Achaia."[422]

In the times of Homer, Achaia meant most of Greece, especially southeast Thessaly and northeast Peloponnesos or Argolis. The Romans did not discriminate. Achaia was Greece to them. They honored the killer and plunderer of Corinth, Mummius, with the title Achaicus, the conqueror of Greece.

In 86 BCE, another Roman general, Cornelius Sulla, repeated the atrocity of Mummius. Sulla captured Athens and slaughtered thousands of Greeks, sparing neither children nor women. The bloodbath ended only when Roman senators on Sulla's staff and some Athenian allies interceded with Sulla. Ploutarchos reports that the Athenian agora was flooded with blood.[423] Sulla's atrocity in Athens confirmed the ambiguity of Romans towards the Greeks.

With Sulla in command, the Romans abandoned the republic and embarked on several decades of civil wars—before and after Sulla. They fought those bloody wars in Greece, further devastating the country. The Romans thought of themselves primarily as soldiers whose destiny was to rule other people. This was the ideology of Sulla, the educated class, and the leaders of the civil wars who sowed the ground for the Roman Empire.

Virgil, a Roman poet of the first century BCE, witnessed the violent birth of the Roman Empire. Yet he learned to love it. Virgil modeled his epic poem, *Aeneid*, after the epic poems of Homer, the *Iliad* and the *Odyssey*. The *Aeneid* invented a heroic origin for Rome. He dedicated his poem to the first Roman emperor, Augustus. Virgil said let other people (the Greeks) sculpt statues so perfect as if ready to talk, pleading their legal cases better than the Romans, even examining the heavens and predicting the rising of the constellations. But you Roman, he said, remember to govern the world with your

[421] Elizabeth Rawson, "The Expansion of Rome," in *The Roman World*, ed. John Boardman et al. (New York: Oxford University Press, 1995) 55.

[422] Pliny the Elder, *Natural History* 34.36.

[423] Plutarch, *Sulla* 14.3–5.

power and enforce peace. "Mercy for those cast down and relentless war upon proud men."[424] So the Romans stepped into Greece with hubris and all the violence that accompanies war. They "spared" the conquered Greeks but did not try to change them.

In chapter 9, I will show that the Christians were worse than the Romans—by far. They burned most of what had survived of the great Greek Library of Alexandria in 391; they brought the Olympics to an end in 393; they butchered the Greek philosopher and mathematician Hypatia in 415; and, in 529, they shut down the Platonic Academy that had been the greatest university of Greece for about 900 years. These barbaric acts were part and parcel of an extremely effective and sustained attack against the Greeks, almost making them and their civilization disappear.

Neither Price in 1974 nor later investigators of the Antikythera Mechanism made any connection between Christianity and the elimination of the Antikythera Mechanism–like machines. They ignored the Christianity effect that almost obliterated technology and mechanical devices throughout the Greek world. Christianity explains the rarity of the Antikythera computer. Why do we have the fragments of only one? Even that single mechanism was discovered quite by accident in the deep sea.

The Fate of the Tablet

The technology of the Antikythera computer was very much in use during the time of Cicero and Vitruvius—the first century BCE. This was a time when the Greeks in Alexandria, Rhodes, and Asia (more so than the Greeks of mainland Greece) were free, deciphering the heavens and the world. The Romans first put the pressure on the Greeks in mainland Greece and Macedonia. Athenaios is right. The Romans smashed the Greek world. That is the reason Cicero was urging his brother to return the cultural favor. By the time the Romans turned to the kingdoms of Alexander in the second century BCE, the Antikythera machine was already at work.

Indeed, Antikythera-like machines go back in Greek history. One such spectacular astronomical device, a less sophisticated version of the Antikythera

[424] Virgil, *Aeneid* 6.847–853, tr. Frederick Ahl, introduction by Elaine Fantham (New York: Oxford University Press, 2007).

computer, became part of the popular imagination in *The Alexander Romance*, better known as *The Life and Deeds of Alexander of Macedon*. This was the most popular novel of the ancient and medieval world. Pseudo-Kallisthenes wrote *The Alexander Romance* sometime between the third century BCE and the third century of our era. The first version of this story probably circulated sometime after the time of Alexander's death in 323 BCE. It is possible the Greek author of *The Alexander Romance* knew some version of an astronomical and calendrical device that resembled the device we name the Antikythera Mechanism. In his story, Pseudo-Kallisthenes has an Egyptian king, Nektanebos, talking to Olympias, mother of Alexander the Great. In an effort to build up his credentials as one who could read the heavens, Nektanebos showed Olympias a little astronomical tablet. Pseudo-Kallisthenes says mere words could not describe that dazzling tablet made up of gold, ivory, ebony, and silver. He wrote:

"The tablet had three engraved zones. There were thirty-six decans on the first circle. The second circle had the twelve signs of the Zodiac. The third or inner circle had the Sun and the Moon. Nektanebos put the tablet on a chair. He then opened a small ivory box and emptied out the seven stars and the horoscope made up of eight precious stones. These stones illuminated the miniature heaven in a small circle. The Sun was of crystal, the Moon of diamond, Ares of hematite, Hermes of emerald, Zeus of air-stone, Aphrodite of sapphire, Kronos of serpentine, and the horoscope of white marble."[425]

A decan comes from *deka* (δέκα), the Greek number ten. This means the first circle had 360 marks representing a calendar for 360 days. The second engraved zone of the precious tablet had the twelve signs of the Zodiac. These signs were the twelve constellations in the path of the Sun known as the ecliptic. Geminos names the signs: Aries, Taurus, Gemini, Cancer, Leo, Virgo, Libra,

[425] *Historia Alexandri Magni*, Recensio α sive Recensio vetusta, 1.4.5–6. Greek text from Thesaurus Linguae Graecae, 1386.001 (University of California-Irvine). Another version of the Greek text has been edited and translated into German by Helmut van Thiel, *Leben und Taten Alexanders von Makedonien* 4.22–28 (Darmstadt: Wissenschaftliche Buchegesellshaft, 1974). For translations of "The Alexander Romance": *The Greek Alexander Romance* 1.4, tr. Richard Stoneman (London: Penguin Books, 1991). Stoneman incorporated the major versions of "The Alexander Romance" in his translation, including the Greek, hence "The Greek Alexander Romance." For translations without the Greek version see *The Alexander Romance*, tr. Ken Dowden in *Anthology of Ancient Greek Popular Literature*, ed. William Hansen (Bloomington, IN: Indiana University Press, 1998); *The Romance of Alexander the Great* by Pseudo-Callisthenes, tr. from the Armenian version by Albert Mugrdich Wolohojian (New York: Columbia University Press, 1969).

Scorpio, Sagittarius, Capricorn, Aquarius, and Pisces.[426] The middle zone of the tablet had the Sun and the Moon.[427]

Nektanebos, wearing the robes of a prophet, impressed Olympias with his pocket-sized astronomical tablet. She asked him to cast her horoscope.

This fragment from popular literature does say that astronomical devices lasted for a very long time in late antiquity. *The Alexander Romance*, a mixture of history and fiction about the ancient world's greatest celebrity and beloved hero, Alexander the Great, survived in more than eighty versions in twenty-four languages. Its brief description of the astronomical tablet is exceedingly limited but leaves no doubt it mirrors the geocentric cosmological system of the Greeks. The small tablet is most probably related to the Antikythera Mechanism or Antikythera Mechanism–like devices. Its engraved zones of a calendar and the Zodiac existed in the Antikythera Mechanism. Indeed, the outer and inner circles on the front dial of the Antikythera Mechanism have short marks at degree intervals. There are longer marks across both rings every thirty degrees. The inner circle is the Zodiac; the outer is a calendar of 365 days. When the Sun lights the marked degrees of the Zodiac, the *Parapegma* calendar of the Antikythera computer would tell the heliacal risings and settings of the most important bright stars. This was a function of Greek astronomical calendars.

The name of a geared device dedicated to deciphering the heavens in ancient Greece would most likely be *Pinax*, *Pinakion*, or *Pinakidion*, tablet or small tablet. That's what Greeks called astronomical instruments that modeled the heavens. They also called them spheres and astrolabes. *Meteoroskopeion* (telescope-like device) also meant a geared mechanical universe. And *Oroskopeion* (sundial or clock) could also mean a mechanical device dedicated to reading the heavens.

As far as we know, the Antikythera Mechanism–like mechanical universe made it under the Romans relatively intact for a considerable amount of time, for sure down to the time of the great astronomer Ptolemaios in the second century, and probably for a few more hundred years. In the previous chapter, I documented that Archimedes and Poseidonios constructed models of the mechanical universe, probably very much like the Antikythera Tablet. But I don't have direct evidence showing that Hipparchos, Geminos, and Ptolemy owned a celestial computer. Yet, I cannot imagine they did not. In fact, this

[426] Geminos, *Introduction to the Phenomena* 1.1 in *Geminos's Introduction to the Phenomena*, tr. and ed. James Evans and J. Lennart Berggren (Princeton, NJ: Princeton University Press, 2006).

[427] Pseudo-Callisthenes, Historia Alexandri Magni 1.4.

book shows that, in the second century–first century BCE, the fingerprints of Hipparchos and Geminos were all over the Antikythera machine. This means their science was the software of the computer. The technology of this astronomical and calendrical device was not lost.

I repeat: the Antikythera computer-tablet was a very small and portable device of bronze gears with front and back plates. It operated with a hand crank. It had considerable scientific, astronomical, practical, and political usefulness, which guaranteed its widespread ownership. There must have been hundreds if not thousands of such computers all over the Greek world and more than one place manufacturing them.

The paradox of finding very little precise information in the literature about a fabulous device like the Antikythera Tablet is not difficult to explain. Ktesibios, who made gears possible in the early third century BCE, was a barber-engineering genius whose written work disappeared. Archimedes wrote an entire book on such devices and the book never made it to our times. We cited *The Alexander Romance*, mentioning an early antecedent of the Antikythera mechanical universe. We also cited Cicero (The *Republic* 1.14.21–22; *Tusculan Disputations* 1.63; and *The Nature of the Gods* 2.87–89) describing the planetariums of Archimedes and Poseidonios: the closest literary evidence we have of the possible prototype or, in the case of the sphere of Poseidonios, a working model of the Antikythera computer.

Second, in the *Introduction to the Phenomena*, Geminos, student of Poseidonios, is describing the science of the calendars of his time. I discuss his contribution in chapter 7. Suffice it to say, Geminos knew that in a Metonic calendar of 235 months things get complicated. First of all, the Metonic 19-year period has 6940 days (from counting 12 years of 12 months and 7 years of 13 months).[428] Some months had 29 days and some others had 30 days. The 29-day month Metonic cycle equaled to 6815 days (29 × 235 = 6815) and the 30-day cycle was 7050 days (30 × 235 = 7050). Second, Geminos harmonized the lunar days and months with the solar year by subtracting a day every 63 days in the 19-year Metonic calendar of 235 months.[429] Sixty-three comes from dividing 6940 days by 110 days; the number 110 is the difference between 6940 and 7050. The astonishing thing about these calculations of Geminos is that

[428] Geminos, *Introduction to the Phenomena* VIII.50–54.

[429] Geminos, *Introduction to the Phenomena* VIII.50–57.

we find the same math inscribed in the back cover of the Antikythera computer. That is not accidental. Geminos was an astronomer, a mathematician, and a philosopher who wrote for non-experts. He taught astronomy and he probably designed and constructed astronomical calendars and instruments like the Antikythera computer. That's why his *Introduction to the Phenomena* captured the Greek science built onto the Antikythera Tablet.

The calendar Geminos talked about happened to be the calendar on the Metonic spiral of the Antikythera computer. This calendar fits the calendar Corinth used in its colonies in northwest Greece, especially Dodona and Ambrakia in Epiros. This was also true for the calendar of Tauromenion (now Taormina) in Sicily.[430] Geminos also writes about the celestial spheres/planetariums he employed in his research: one was a solid sphere (*sphaira steraia,* XVI.12) and the other was a ringed sphere (*sphaira krikote,* XVI.10, 12). Probably, these spheres were models of the circulating Antikythera computer.

Third, Heron of Alexandria is probably a key witness to the survival of the technology of the Antikythera computer. I discussed his contributions to the technological legacy of ancient Greeks earlier in this chapter. Heron, I repeat, worked in the first century. He probably left us a detailed description of the celestial computer and how to build it, but the text did not survive. Yet in his *Pneumatics*, he describes the construction of mechanical devices in geometrical terms, in a sense, informing us how to use simple machines to build more complex machines.[431] Also, immersing myself in his texts, as I did, I had the feeling Heron built and used Antikythera-like devices. In his *Pneumatics*, he says he wrote extensively on water-powered sundials he calls *Oroskopia*, which might have been geared clocks. However, only two fragments on *Oroskopia* survived. Nevertheless, I am convinced Heron's knowledge of mathematics, physics, astronomy, and engineering was profound and modern. His robots are of extraordinary ingenuity and mechanical sophistication.

Fourth, the astronomer Ptolemaios, who flourished after Heron, in the second century, described his construction of a *Meteoroskopeion*, one of the names Greeks used for celestial devices or astronomical computers. Such an instrument for the observation of the heavens would be the equivalent to an

[430] Iversen, "The Calendar on the Antikythera Mechanism and the Corinthian Family of Calendars," *Hesperia* 86 (2017) 129–203.

[431] Heron, *Pneumatics* 2.14–16; 56.12–16; 70.13; 120.4–5, ed. and tr. into German by W. Schmidt, in *Heronis Alexandrini Opera*, Vol. 1 (1899).

Antikythera-like mechanism. Perhaps, it was an advanced version of an armillary sphere for his astronomical studies.[432] Ptolemaios also spoke eloquently and with knowledge about the science of prediction, especially human and natural, and phenomena and the movements and positions of the Sun-Helios, the Moon and the planets (*Tetrabiblos* 3.4–3.20; 2.7.1–7.4; 2.10.1–10.4; 3.1.1–1.3). And like Geminos, he routinely predicted the eclipses of the Sun and the Moon.

Fifth, the great physician Galen of the second century was well aware of Greek astronomical achievements. He admired the beauty and order of the Sun-Helios, the Moon, and the "whole chorus of stars," their size, and ceaseless motion. He spoke of mechanical models built in imitation of the "revolutions of the wondering stars [the planets]."[433]

Sixth, in the fourth chapter, I cited the Greco-Roman poet Claudian of the fourth century who praised Archimedes's glass sphere for enclosing a human-made Cosmos challenging Zeus.

Seventh, the fifth-century Platonist and historian of science, Proklos, wrote about "the art of making spheres imitating the revolutions of the heavens."[434] Clearly, those spheres were mechanical universes that, like the Antikythera Tablet, brought the heavens down to Earth.

Eighth, another witness of the continuing influence of Antikythera Mechanism–like devices was Flavius Magnus Aurelius Cassiodorus Senator, c. 485–585. Cassiodorus was a politician, writer, and monk. He was the publicist of the Ostrogothic king Theoderic of Italy for several years. He was educated in Greek culture and had a grasp of the scientific and technological achievements of the Greeks. He corresponded with another outstanding scholar also versed in Greek culture and science. This was Anicius Manlius Severinus Boethius, c. 480–c. 525. In a letter to Boethius, Cassiodorus wants him to send him a sundial and a water clock. He marvels at the machines and automata from the legacy of Heron of Alexandria. He describes a "tiny device" with great astonishment and enthusiasm. He is so taken by the beauty, complexity, and purpose of what he calls "Archimedes' sphere," he becomes poetic in his understanding of what the device is and what it represents. He says this

---

[432] Ptolemaios, *Almagest* V.1.

[433] Galen, *On the Usefulness of the Parts of the Body* 3.10; 14.5, tr. Margaret Tallmadge May (Ithaca: Cornell University Press, 1968), 2 vols.

[434] Proklos, *A Commentary on the First Book of Euclid's Elements* 41–42.

"tiny device [is] pregnant with the world, [being] a portable sky, a compendium of the universe, a mirror of nature which reflects the heavens."[435] This is a marvelous metaphor for the Antikythera Mechanism, capturing its scientific and philosophical design and purpose.

Finally, I read more than Geminos's *Introduction to the Phenomena*. I read as much as I could from the surviving books and literary fragments of Eudoxos, Aristotle, Euclid, Ktesibios, Archimedes, Herophilos, Erasistratos, Philon of Byzantium, Eratosthenes, Aratos, Aristarchos of Samos, Hipparchos, Poseidonios, Geminos, Heron, Strabo, Vetruvius, Ptolemaios, Athenaios, Galen, Pappos, Proklos, and Eutochios (sixth century). This immersion almost made me a participant in the evolution of science and technology that assured the building of a mechanism of advanced technology like the Antikythera computer.

It is possible, of course, we have not searched hard enough for more versions or copies of this ancient device. After all, the Antikythera machine did not vanish.

In 1983, a Lebanese person sold to the Science Museum of London a sundial constructed in Constantinople, in all probability around the early sixth century. This device had four-toothed wheels and a ratchet carried on two small axles. One of the wheels had fifty-nine teeth. This wheel measured the movement of the Moon. Like ancient lunar calendars, this sundial represented a thirty-day month followed by a twenty-nine-day month, the gearwheel turning a tooth a day. This sundial is the oldest surviving Greek clockwork device after the Antikythera device.[436] To find anything resembling this technology, one has to wait until the eleventh century when Abu Rayhan al-Biruni, 973–1050, an Arab natural philosopher, described a geared calendar.[437]

Michael Wright, who has a talent for building machines, made a working model of this medieval Greek sundial. Like Price, Wright saw the gearing of the old clocks—gearing that used to model the movements of the Sun, Moon, and planets—into the big engines of the Industrial Revolution.

---

[435] "King Theoderic to the Illustrious Patrician Boethius" in *Selected Variae of Magnus Aurelius Cassiodorus Senator* 1.45.3–11, tr. S. J. B. Barnish (Liverpool: Liverpool University Press, 2010).

[436] J. V. Field and M. T. Wright, "Gears from the Byzantines: A Portable Sundial with Calendrical Gearing," *Annals of Science*, 42 (1985) 87–138.

[437] John Freely, *Aladdin's Lamp: How Greek Science Came to Europe Through the Islamic World* (New York: Alfred A. Knopf, 2009) 88–89.

In the case of the Greek sundial, its technology of gearwheels was pretty similar to the description of al-Biruni. This was evidence that Price was right that the Greek knowledge of gears passed on to the Arabs who made their own calendars. In addition, the construction of the Greek sundial in Constantinople showed that the instrument, clearly descended from the Antikythera model of Greek computers, was not a luxury toy but an everyday calendar. There could have been thousands of these sundial calendars all over the Greek medieval world.

The Arabs copied the Greek geared machine and, eventually, they passed it on to the Europeans in thirteenth-century Muslim Spain. It then became the model for Europe's sophisticated clockwork technologies that underpinned the scientific and technological revolutions of the sixteenth and seventeenth centuries.

Chapter 7

The Science and Civilization of Astronomical Prediction

Among the stars, the Sun-Helios and the Moon were life and death to the Greeks. They studied them and hoped their understanding of their movements and those of the heavens might crown harmony between the Cosmos and their society. The Cosmos included the natural world of the Earth and the heavens. For this reason, the prediction of the lunar and solar eclipses was of paramount importance to them. These cyclic events held the key to both the scientific and religious understanding of the universe. The eclipses of the Sun and the Moon brought the Greeks face to face with their gods—and intense fear. Why did the Sun and the Moon occasionally disappear? Were these gods angry with them? This chapter explains how the Antikythera Mechanism could make accurate predictions based on its scientific technology and, in particular, on the complex Sarosian and Metonic cycles of the Moon. These cycles in turn came out of the understanding of the heavens and the evolution of astronomy in Greek thought.

Landmarks in the Sky

Astronomers relied on the cosmological ideas of Thales, Anaximander, Pythagoras, Herakleitos, Parmenides, Empedocles, Anaxagoras, and other philosopher-scientists like Democritus, Plato, and Aristotle for further advancements of their science. For example, designating stars in constellations was the consequence of this theological, astronomical, and natural evolution of Greek science and society. But constellations had also a practical utility. They were the guideposts of the heavens. These guideposts helped the Greeks to explore the secrets of the sky as well as facilitate their journeys on Earth. The "great purpose" of the constellations was "mnemonic: to define patterns and 'landmarks' in the sky, with which to measure time and direction."[438]

Indeed, that was the purpose of the constellations among ancient Greeks. Astronomers like Aratos (third century BCE), Hipparchos (second century BCE), and Geminos (first century BCE) demonstrated the intimate connection between the heavens and the world of the Greeks. They adopted the constellations and stars of Eudoxos.

Two of the constellations invented after Eudoxos were of enormous historical significance. They represented the beloved Ptolemaic queens Arsinoe II (Parthenos or Virgo) and Berenice II (Coma or Hair of Berenices). Arsinoe and Berenice had a passion for prosperous rural life and civilization. They had special honors and celebrations for the gods Dioscuri-Kabeiroi and the goddess of wheat and agrarian civilization, Demeter. Arsinoe II was the daughter of King Ptolemy I Soter and sister-wife of King Ptolemy II Philadelphos. She died about 270 or 268 BCE. Berenice II was the daughter of King Magas of Cyrene and wife of King Ptolemy III Euergetes. She flourished from about 264 to 221 BCE. Like Arsinoe, she was at the center of intellectual life in Alexandria, especially encouraging and lavishing support on the poets and scholars working at the Mouseion and the library. In fact, Berenice and her husband, Ptolemy III, brought a golden age to Egypt. For a magic moment, "The forces of politics, economics, religion and culture, Greek and Egyptian, were all in balance."[439]

Poets praised Arsinoe and Berenice for their beauty, enlightenment, piety, and Hellenism. This and their royal connections opened the highway to becoming constellations near the constellations Bootes, the cattle herdsman and ploughman, and the Wagon, namely Ursa Major.

Dee Clayman, professor of classics at the City University of New York, praises Arsinoe and Berenice for their power and policies in fostering justice and prosperity in rural Egypt. She is right in associating Virgo (Arsinoe) and Berenice constellations with the core values of Greek civilization. The two queens, says Clayman, are always guiding sailors safely through the sea, giving clues to the peasants about the seasons and the weather, including "inspiring

---

[438] Peter Whitfield, *The Mapping of the Heavens* (San Francisco: Pomegranate Artbooks, 1995) 2.

[439] Dee Clayman, *Berenice II and the Golden Age of Ptolemaic Egypt* (New York: Oxford University Press, 2014) 178.

mankind to live in justice and peace." In other words, people seeing them full of light in the darkness of the night sky think of them as "guarantors of universal order, like [the Egyptian goddess] Isis, rising and setting with the sun, the moon, and the stars." And who "could look at the night sky," asks Clayman, "and doubt their power? Or doubt the power of the poets and poetry that put them there?"[440]

Exactly right. Bright constellations and beautiful poetry mattered. Aratos and his contemporary Alexandrian poet, Callimachus, about 305–240 BCE, were both poets and natural philosophers. They were partly responsible for raising these Ptolemaic queens to the heavens. In his astronomical-poetic book, *Phenomena*, Aratos described the constellations of his day, including: the Bears, Herakles, Corona Borealis, Ophiochos or Asklepios holding a snake, Scorpio, Zygos or Libra, Arktophylax or Bootes with its bright star Arktouros, Virgo (Arsinoe), Hair of Berenice, Gemini, Cancer, Leo, Auriga or the Charioteer, Taurus, Cepheus, Kassiopeia, Andromeda, Equus or the flying horse Pegasos, Pisces or Fishes, the Swan, Aquarius, Capricorn or the Goat, Sagittarius or the Archer, Sagita or the Arrow, Aquila or the Eagle and Delphinus or the Dolphin, Orion, Canis Major or the Dog and Sirius, Lepus or the Hare, Argo, Cetus or the Whale, River Eridanos, Piscis Australis or the Southern Fish, Hydor or Water, Ara or the Altar, Centaurus or the Centaur, Therium or the Wolf, Hydra, Crater or the Cup, Corvis or the Raven and Procyon.[441]

The astronomical poem of Aratos was popular for centuries. Its attraction mirrored more than the citing of the names of constellations and stars. The beauty of the poem was in its poetry of singing the truth and the science of the celestial and terrestrial Cosmos of Homer, Hesiod, and Eudoxos. This was the natural and social world of the Greek countryside. Now, thanks to the astronomical knowledge of Eudoxos, put in Homeric poetry by Aratos, the heavens spoke to the Greeks quite clearly. Leonidas of Tarentum left us an epigram in which he praises Aratos, through literature, to the heavens. He says Aratos was second to Zeus because he studied the "long-lived stars" and made them "brighter."[442]

---

[440] Dee Clayman, "Royal catasterisms: Arsinoe II and Berenice II translated to the stars," in *Hellenistic Alexandria*, 201–208.

[441] Aratos, *Phenomena* 26–732, tr. G. R. Mair (Loeb, 1921, reprint 2006). The translation of the Greek text is awful—and almost a century old.

[442] *The Greek Anthology* 9.25.

Eratosthenes borrowed his constellations from Aratos. Eratosthenes included the following constellations on the Zodiac: Scorpios (Scorpio, "two-twelfth parts" occupying two zones of the Zodiac), Libra (Zygos or Scales, was a Roman invention), Parthenos (Virgo the Virgin, Arsinoe), Didymoi (Gemini the Twins), Karkinos (Cancer the Crab), Leon (Leo the Lion), Tauros (Taurus the Bull), Krios (Aries the Ram), Ichthyes (Pisces the Fish), Hydrochoos (Aquarius the Water Pourer), Aigokeros (Capricorn the Goat-Horned), and Toxotes (Sagittarius the Archer).[443] Geminos added Libra and changed the order but he gave us the same constellations.[444] However, the astronomer Claudius Ptolemaios made meticulous lists of constellations like Ursa Minor, Ursa Major, Cepheus, Corona Borealis, Lyra, Cassiopeia, Pegasus, Aries, Taurus, Gemini, Leo, Virgo, Scorpius, Capricornus, Aquarius, Pisces, Eridanus, Argo, and Centaurus.[445]

The most well-known stars among the Greeks included the following: Arcturus; Sirius or "Kyon" (dog); Antares; Procyon; Canopus; Aix (goat), which is known with its Latin name Capella from the Greek goat; Lyra, the Lyre, but named Vega from a medieval mistranslation of an Arabic version of the Greek Lyra; Aetos (eagle) or Altair from the Arabic; Stachys (ear of wheat) or Spica, the Latin translation of the Greek Stachys; Basiliskos or Regulus, the Latin translation of the Greek "little king"; Protrygeter (harbinger of the grape harvest) or Vindemiatrix, the Latin version of the Greek name; Onoi (asses) or γ (gamma) and δ (delta) Cancri; and Eriphoi (the kids) or the dim stars η (eta) and ζ (zeta) Aurigae.

I already mentioned Aratos. He wrote "Phenomena" in which he summarized the astronomical knowledge of the Greeks and, especially, the contributions of Eudoxos. This was a popular poem that, eventually, a millennium or so after his time, circulated in translation among Arabs and Latin Westerners.

Aratos started his poem with a paean to Zeus: the god that filled the streets and markets, the seas and heavens. Zeus was the immortal father of the gods whom the Greeks considered their father: he gave them life and helped them to earn a livelihood. Aratos says Zeus put stars and constellations in heaven as sure signs for the seasons, guiding human beings to the best time for the cultivation of the land and for sowing all kinds of plants and seeds and assuring

---

443 Eratosthenes, *Katasterismoi (Constellations)* in *Anthology of Classical Myth*, ed. and tr. by Stephen M. Trzaskoma et al. (Indianapolis/Cambridge, 2004) 102–106.

444 Geminos, *Introduction to the Phenomena* III.

445 Ptolemy, *Almagest* VII. 5–VIII.1 (ed. and tr. G. J. Toomer, Princeton, NJ: Princeton University Press, 1998).

an unfailing harvest and agrarian prosperity. "Hail, O father," says Aratos to Zeus, "great god, mighty blessing onto human beings."[446]

Aratus described the stars and constellations that made life possible. He said an unchanging axis is holding the Earth at the center of the Cosmos. The axis wheels around the Earth. He talks about the imaginary circles of the celestial sphere: Gala (milk), Galaxy or Milky Way; the Equator; the Tropic of Cancer: north of the Equator, marking the extreme northern limit in the annual path of the Sun; the Tropic of Capricorn: south of the Equator, marking the extreme southern limit in the annual path of the Sun; and the Zodiac: like the ecliptic or annual path of the Sun in the celestial circle, the Zodiac represents the annual movement of the Sun among the stars.

But Aratus, much like Hesiod, uses the astronomy of his day to advise the Greeks, not merely how to interpret the movement of the constellations but how to read the movement and behavior of wild and domesticated animals to predict the weather. This was no small task. It would be the equivalent skill of a present-day astronomer, climatologist, agronomist, ornithologist, and biologist together. So, the careful Greek farmer was a polymath. In fact, Aratus told him to double-check every sign of nature for accuracy. "Study all the signs of nature together the entire year," Aratus says to the farmer, "and your forecast of the weather will never be a random guess."[447]

This demand of acting according to nature opened the Greek mind to investigation and the use of reliable evidence for reaching decisions; in other words, Greeks looked at the natural world as their ally, not their enemy, which they had to subdue. This practice also taught the Greeks how nature works and brought them even closer to their gods who operated in nature or as natural phenomena. Zeus, for example, was the "black cloud" or the god of the weather. That's why Aratus, a popularizer of astronomical knowledge, paid such homage to Zeus.

Was the Earth or the Sun at the Center of the Cosmos?

Nevertheless, the poem of Aratus was extolling the Greek geocentric system. This view of the Cosmos with the Earth at its center was complicated. One used eccentric and epicycle theories in explaining the movements of the stars. The eccentric model has the planet rotating in a circle, which does not have

---

[446] Aratos, *Phainomena* 1–15.

[447] Aratos, *Phainomena* 553–461, 963–1154.

the same center as the Earth. The epicyclical model, represented primarily by Apollonios of Perga and Hipparchos, makes the planet a point on a small circle (the epicycle). The center of the planet is on the circumference of a larger circle with its center at the center of the universe: the Earth. Greek mathematical astronomers realized that these two hypotheses were not that much different.

In the third century BCE, Aristarchos of Samos proposed his pioneering and famous heliocentric theory that placed the Sun at the center of the universe.

The Earth, like the rest of the planets, rotates around the Sun while it also rotates daily on its own axis.[448]

Aristarchos wrote a book in which he measured the sizes of the Sun-Helios and the planets, including figuring out the distances of the Sun-Helios and planets from the Earth.

Philolaos was one of the philosophers who contributed to the ferment that made Aristarchos possible. His primary contribution was to bring his Pythagoreanism into the explanation of the universe. He was born about 470 BCE in Kroton, Italy. He never met Pythagoras, who died in 494 BCE. The Pythagoreans thought the number divine and living, a universal phenomenon like light or sound: the essence for understanding the universe, which is harmonious. That is why the Greeks followed the example of Pythagoras and called the universe Cosmos (Κόσμος), an ordered world, a perfect and just universe and ornament. This Cosmos, full of order, harmonies, and beauty, was known through numbers.

Aristotle reports that the followers of the philosopher Pythagoras advanced mathematics by suggesting that mathematical principles and numbers are the keys to studying and understanding nature and the foundations of all existing things.[449] Aristotle also reports that the Pythagoreans did not put the Earth at the center of the universe. They put fire at the center of the Cosmos, calling that fire the guardhouse of Zeus. The Earth, they said, is just one of the stars, circling around the cosmic center, thus creating day and night.[450]

Herakleides of Pontos operated in this Pythagorean Cosmos. He had already spoken about the Earth rotating on its own axis in the fourth century BCE. Herakleides also taught that Aphrodite (Venus) and Hermes (Mercury)

---

[448] Thomas Heath, *Aristarchus of Samos: The Ancient Copernicus* (First published by Oxford in 1913: Elibron Classics, 2007) 301–316.

[449] Aristotle, *Metaphysics* 985b23–986a2; 986a17–21; 987a13–19.

[450] Aristotle, *On the Heavens* 293a20–293b6.

moved around the Sun. Herakleides of Pontos, student of Plato, was one of the scientists and philosophers who, like Philolaos, prepared the road for Aristarchos.

Another important precursor of Aristarchos of Samos was Archytas of Tarentum. He knew Plato; indeed, he saved his life in 361 BCE from potential death at the hands of the tyrant Dionysios II of Syracuse. Archytas was a Pythagorean philosopher who also served his country as a chief political leader. He was elected seven times in the office of *strategos*, general, in the first half of the fourth century BCE. He is credited with achievements in mechanics and mathematics, especially in the duplication of the cube.[451] Horace, the Roman poet of the first century BCE, says Archytas measured the sea and the Earth and investigated the heavens, "the house of the gods in the sky."[452] Archytas was important. He invented a flying wooden dove. He loved the sciences and led one of the most powerful Greek states. He was a philosopher-mathematician-inventor king.[453]

Straton of Lampsakos also influenced Aristarchos of Samos. Straton was his teacher. Straton, c. 295–268 BCE, studied at the school of Aristotle in Athens, the Lyceum or Peripatos. After his studies, he tutored Ptolemy II Philadelphos, son of king Ptolemy I of Egypt. He then returned to Athens, where he directed the Lyceum for close to twenty years until his death. Straton studied falling bodies and noticed their acceleration. Rainwater falling from a roof moves faster when it hits the ground than when it leaves the roof. Second, a falling object from a three-foot height compared to a one-foot drop makes more of a thud when it strikes the Earth. Thus, in contrast to Aristotle, Straton said that falling bodies do not reach final speed in short time and then fall steadily. Straton did not allow theological issues to intrude into his study of the natural world; probably this is why he was known as the natural philosopher (*ho physikos*). He spoke about void as a theoretical possibility of being intimately related to space. He studied vacuum and the possibilities for power from air, water pressure, and steam.[454]

---

[451] Laertios, *Lives of the Philosophers* 8.79–83.

[452] *The Odes of Horace* i28, tr. David Ferry (New York: Farrar, Straus and Giroux, 1997).

[453] Carl A. Huffman, ed. and tr., *Archytas of Tarentum: Pythagorean, Philosopher and Mathematician King* (New York: Cambridge University King, 2005). Flying dove: 82–83, 570–577.

[454] Wikander, "Gadgets and Scientific Instruments," in *The Oxford Handbook of Engineering and Technology in the Classical World*, 787.

Despite the new cosmological ideas of the Pythagoreans and especially those of Anaximander, Philolaos, Herakleides of Pontos, and Aristarchos of Samos, who articulated a new vision of how the universe works, the old geocentric view survived and triumphed.

The science and simplicity of Aristarchos's heliocentric hypothesis failed to convince the majority of Greek astronomers, who supported the Earth-centered universe. Even Archimedes who reported the heliocentric hypothesis of Aristarchos in his *Sand-Reckoner* remained unconvinced it could replace the geocentric model of the heavens.[455] Archimedes was wrong. Heliocentrism triumphed—nearly two millennia after his time. Nevertheless, in the very beginnings of the heliocentric idea, the followers and advocates of this new view of the Cosmos advanced mathematical physics, not a small contribution to the culmination of the golden age of Greek science in the third century BCE.

The Greek physicist Antonios Pinotsis argues that the ancient Greeks' use of mathematics and, especially, geometry, made significant contributions in the development of cosmological ideas and theories. "The models of epicycles and of eccentric circles," he says, "were complex geometrical models," which created new horizons in mathematical interpretation of the geocentric system. These geometrical models mirrored a deeply rooted belief that, first, the Earth was at the center of the Cosmos, and, second, the orbits of the celestial bodies were normal motions.[456]

It is also possible that the ancient religious connection between the Greeks and the Earth made a difference in the victory of the old geocentric cosmological model. The Greeks worshipped the Earth as their mother and mother of the gods. In fact, Plato thought of the Earth as the oldest of the gods.[457] It was not easy to dethrone such a mighty and senior goddess. Besides, the science of the geocentric universe was the best for a very long time.

The heliocentric idea rose and declined rapidly. About four hundred years after Aristarchos, the famous astronomer Ptolemaios knew of the heliocentric system but he rejected it on the grounds of physics. If the Earth moved, he said,

---

455 Archimedes, *The Sand-Reckoner* 135.8–23.

456 Antonios D. Pinotsis, "Comparison and historical evolution of ancient Greek cosmological ideas and mathematical models," *Astronomical and Astrophysical Transactions*, Vol. 24, No. 6, December 2005, 481.

457 Plato, *Timaios* 40c.

"living things and individual heavy objects would be left behind, riding on the air, and the Earth itself would very soon have fallen completely out of the heavens. But such things are utterly ridiculous merely to think of."[458]

This brings us back to Hipparchos and the realization that his legacy made a difference in the building of the Antikythera machine.

Harmonizing Heavens and Earth

The Antikythera Mechanism was not that much different from a laptop computer. It was a Greek Apple BCE. It was a reliable calendar for the Sun and the Moon as well as a calendar for the most important agricultural, religious, and cultural events of the Greeks: it calculated the dates of the eclipses of the Moon and the Sun; it did other complicated astronomical time calculations for balancing the lunar month with the solar year; and both of those with the phases of the Moon.

The small group of scientists who reported on the Antikythera Mechanism in July 2008 were right that this geared machine was more than a device of pure astronomy. It exhibited longitudes of planets on the front dial and predicted eclipses of the Sun and the Moon on the lower back display. The Olympics dial of the Antikythera Mechanism provided the names and schedule for the Panhellenic games: OLYMPIA for the Olympics; NEMEA for the Nemean; ISTHMIA for the Isthmian in the Isthmus of Corinth; PYTHIA for the Pythian celebrations at Delphi, and NAA games in Dodona in Epiros. NAA stands for Naia in honor of Naiou Zeus. The Games dial also included HALIEIA (ΑΛΙΕΙΑ) in year 4 for the games in Rhodes.

Since the Greeks did not need a computer to tell them the time of their Panhellenic games, which for more than a millennium were the most important political, social, athletic, and religious celebrations in their civilization, the Olympics dial of the Antikythera device had another purpose. It made the most important part of the lives of the Greeks—that which they devoted to the gods—an integral part of the machine decoding the heavens. Both the Greeks and the computer represented a Cosmos within a Cosmos. The Olympiad dial also highlighted their conception of the divine nature of the Cosmos. The gods

[458] Ptolemaios, *Almagest* I. 7.

were both in Earth and in the heavens. The Greeks honored the gods with their athletic celebrations. The Olympic games were primarily for honoring Zeus, for example. Observing and studying the heavens was also natural. The heavens were the real home of the gods.

"Is it perhaps not extravagant to see the Mechanism as a microcosm illustrating the temporal harmonization of human and divine order,"[459] wrote the scientists in their 2008 report. That's exactly the nature of the Antikythera Mechanism. Price would have agreed with this conclusion. The Antikythera computer expanded the human view of the heavens. Now it was like opening a window to a vast decoded sky, bright and inviting with light, knowledge, and beauty.

The Almagest

The advancement of science among the Greeks went through the shock of the Roman conquest. With the death of Kleopatra, the last Greek monarch of Egypt, in 30 BCE, Rome completed its occupation of the empire of Alexander the Great. The new political bosses of the Mediterranean started the new millennium with a different vision than that of the Greeks. The Romans did not see much import in the support of libraries, research institutes, and science. This started a slow but steady decline in both science and technology and, in broader terms, Greek culture.

But the greatest blow of the Roman Empire against learning was at the heartland, Rome. Writing during the first century of the empire, Pliny the Elder could look back nostalgically at the time when the Greeks were free. He had the writings of more than twenty Greeks in front of him. Those writings gave him facts and delight. He wondered how those Greeks did so well in matters of thought at times, which to him were full of wars and upheavals. And now that Rome was in charge of the world, when the Roman emperor, according to Pliny, was promoting peace, literature, and science, something was amiss. Pliny was angry: "[A]bsolutely nothing," he said, "is being added to the sum of knowledge as a result of original research; indeed, not even the discoveries made by people long ago [Greeks] are thoroughly assimilated."[460]

---

[459] "Supplementary Information," 21 in Tony Freeth et al., "Calendars with Olympiad display and eclipse prediction on the Antikythera Mechanism," *Nature*, July 31, 2008, 454: 614–617.

[460] Pliny the Elder, *Natural History* 2.117.

The Greek East, however, was more resilient. The last glow of Greek science in the Roman age came out of the astronomical studies of Ptolemaios—in the second century.

Aside from the impact Hipparchos had on Poseidonios and Geminos, his immediate influence was in the growth of horoscopic astronomy that relies on the computation of the position of certain stars at birth or other vital stages of human life. The real influence of Hipparchos, however, had to wait about three centuries to find a lasting reception. This happened with the astronomer Claudius Ptolemaios.

After about three centuries of the Romans ruling Greece and the Mediterranean, Alexandria in Egypt had the greatest Greek library in the Mediterranean. Ptolemaios did his research in that library. He was a prolific writer and a genius. His *Geography* was the greatest geography book of the ancient world. It served the Arabs and the West until the Renaissance.

The geographical knowledge and wisdom of Ptolemaios spilled into cartography. Ptolemaios filled books 2 to 7 of his *Geography* with more than 8,000 place names. For each of these place names, he calculated the longitude and latitude. His hope was that with these coordinates, cartographers anywhere in the world would be able to draw maps of the world. His own remarkably accurate maps served humans for several centuries.

Astronomy sharpened and expanded the science of Ptolemaios. It helped him write his *Geography* and produce those great maps. It even led Ptolemaios to the uncharted field of anthropological astronomy, which we dub astrology in order to downgrade it to superstition. Ptolemaios and his predecessor Greek scientists did not divide the study of the stars into astronomy and astrology. Some of them, including Ptolemaios, used the word astrology as another version of astronomy.

Ptolemaios used his formidable astronomical knowledge in the writing of the *Tetrabiblos* (Four Books). This is an extraordinary document. It allows one to enter a culture that was spectacularly close to the stars. These heavenly bodies were far from the Earth but, in some ways, very close to the inhabitants of the Earth. Ptolemaios is telling his Greek and Roman readers that the stars in the sky were intimate parts of their lives and civilization. In fact, like the rest of the gods, Sun-Helios, the Moon, and the planets were all powerful in shaping their lives. They exercised a dramatic influence on their behavior and policies. Humans were creatures of the heavens. Stars formed their character. Stars also shaped the natural world.[461]

[461] Ptolemaios, *Tetrabiblos* 1.1.1–2–20, ed. Kaktos Philological Team (Athens: Kaktos, 2001).

Reading *Tetrabiblos* is shocking. Stars are masculine and feminine. They love and hate each other. They dominate the sky and humans like monarchs do. Alas to humans conceived or born at a time nasty stars are the big forces in the heavens. They can expect diseases, poverty, and death. Yet good planets and stars are the carriers of health, intelligence, and civilization.

Tetrabiblos is all about a magical future. It's not superstition, but the other side of reason. As I will explain below, confronting the gods in the sky is neither a rational nor an irrational experience. It is an earthquake. *Tetrabiblos* is such an earthquake. In the second century, for instance, it mirrored some of the prevailing culture of the fourth century BCE, when a Moon eclipse terrified a large Athenian army in Syracuse. That fright—seeing the Moon disappear at night—triggered the annihilation of the army.

Yet the four books of the *Tetrabiblos* percolate with philosophical and astronomical ideas. *Tetrabiblos* reminds me of reading *Works and Days* of Hesiod, the *Phenomena* of Aratos and an *Introduction to the Phenomena* of Geminos: the same convictions about the role of the Sun-Helios in giving life to the crops and natural world; the same anxieties about the influence of the Moon, rains, and the winds on the growing of food and the behavior of farm and wild animals; and the same veneration of the gods.

Nine hundred years separate Ptolemaios from Hesiod; about five hundred from Aratos; and about two hundred from Geminos. Ptolemaios has the knowledge of those centuries at his fingertips. However, he is also deeply immersed in Hellenic culture. He speaks with the authority of Homer, natural philosophers, and scientists. He argues convincingly that knowledge of astronomy, mathematics, and appropriate instruments open the door to the prediction of natural phenomena and the eclipses of the Moon and the Sun-Helios.

He explains the extremely difficult and fascinating anthropological and social reality of the world's people strictly on the basis of the influence of the Sun-Helios, the Moon, and the planets and stars. No doubt, the autocracy of the Roman Empire did not encourage freedom of thought. So, Ptolemaios did the best he could with the stars. Given where you are born and live, he said, you are bound to be white, black, civilized, or barbarian (uncivilized). Astronomy for Ptolemaios is a civilization-shaping universal force.

Despite the vast chronological distance between Hesiod, other natural philosophers, and Ptolemaios, natural philosophy and astronomy remain attached to the same metaphysics: the heavens and its stars are the divine teachers and saviors of the natural world and humanity.

Ptolemaios often spoke of the ecliptic as that great central circle paralleling the belt of signs of animals and heroes making up the Zodiac. The ecliptic is divided into twelve 30-degree sections named after the animals representing constellations like Aries, Taurus, and so on. However, Ptolemaios does not use animal signs for the ecliptic. He says each section of the ecliptic represents a twelfth. He was right. The Zodiac was the traditional band of animals and heroes. The Zodiac, however, was not simply a pretty celestial zone decorated with the images of animals and heroes, which mirror and receive energy from the planets and the fixed stars; it is the path of those planets and the Sun-Helios in the heavens. Each constellation on the Zodiac is more than a landmark in the sky. It has an oversight of a region of the Earth and the people living in that region. These constellations work with and against each other, all the time making, inspiring, and unmaking humans.

Ptolemaios wrote his books to advance the science of astronomy. But his vision was enlightenment. A very old civilization had accumulated lore, mythology, and facts about the heavens. Ptolemaios drew from that culture and knowledge, smoothing the path humans take to reach the stars and, especially, improving their vision of the heavens.

In addition to *Geography* and *Tetrabiblos*, the books of Ptolemaios include the *Handy Tables*, the *Harmonics*, and the crowning glory of his life: a book on mathematical astronomy, *Mathematike Syntaxis,* Systematic Mathematical Treatise (Μαθηματικὴ Σύνταξις). In short time, *Syntaxis* became *Megale Syntaxis* or Great Systematic Treatise. Soon *Megale* became *Megiste* (The Greatest). When the Arabs translated *Megiste* into Arabic in the ninth century, they added "al" to *Megiste* the Arab title, becoming (*Al-Megiste*).

When Gerald of Cremona translated *Al-Megiste* from Arabic into Latin in 1175 in Toledo, Spain, *Al-Megiste* became *Almagest*, which remains the title of the treatise to this very day: The Greatest. Ptolemaios probably wrote the Almagest in the mid-second century during the reign of Emperor Antoninus, 138–161. The *Almagest* is one of the most seminal books of all time, remaining a textbook and masterpiece of mathematical astronomy for 1,400 years. Said al-Andalusi, an eleventh-century Arab scholar working in Spain, almost worshipped Ptolemaios and the *Almagest*. He said that with Ptolemaios, "the science of astronomy and the knowledge of the secrets of the skies reached perfection."

Al-Andalusi placed Aristotle next to Ptolemaios and an Arab scholar, Sibawayh al-Basri [from Basra]. Their works, he said, especially the *Almagest* of Ptolemaios, the logical studies of Aristotle, and the Arabic Grammar and

Language of al-Basri were comprehensive in "all the essentials and the secondaries of the subjects they treat. What is omitted has no importance whatsoever."

Al-Andalusi admired the Greeks for "cultivating all the branches of knowledge, including mathematical sciences, logic, natural philosophy, and theology, as well as the political sciences that deal with the home and family and the society as a whole." He either knew Greek or, more likely, he relied on Greek texts translated into Arabic. Like the Greeks, he equated philosophy to science. In his summary of Greek philosophy, he praised those Greeks who contributed to philosophy or science. He was especially effusive of Pythagoras, Empedocles, Socrates, Plato, and Aristotle. These philosophers, he said, were "the greatest and the most famous among the Greeks." But he had no doubt the Greeks "served humanity with their labor and guided it with their lights."

Al-Andalusi's insight about Greek philosophers/scientists (they spread light and enlightenment among humans) captures the essence of the contribution Greeks made to civilization.

In the twentieth century, the Danish scientist and author of a comprehensive study of the *Almagest*, Olaf Pedersen, updated al-Andalusi's assessment of the work of Ptolemaios. He put it like this:

"The Almagest was the culmination of Greek astronomy, and unrivalled in Antiquity as an example of how a large and important class of natural phenomena could be described in mathematical terms in such a way that their future course could be predicted with reasonable precision. It taught scientists of many ages how geometrical and kinematical models could be constructed and, by means of empirical data stemming from careful observations, made to simulate nature in a way which came to influence scientific method until the present day."[462]

Otto Neugebauer, a historian of ancient astronomy, also of the twentieth century, described the *Almagest* as the "keystone" necessary for "the understanding of all ancient and medieval astronomy." Medieval astronomy, however, was very rudimentary. Medieval Cosmos resembled this disorder: the Earth was at the bottom, fire was above the Earth, and then there were colorful zones going upwards towards the Moon, Mercury, Venus, Sun, Mars, Jupiter, and Saturn. The fixed stars and, finally, the angels filled the heavens. There were bands of color hosting the planets, which did not interact.

[462] Pedersen, *A Survey of the Almagest*, 11.

That's the impression one gets from the *Buch der Natur*, *Book of Nature*, published in 1481.

The *Almagest*—even in its summary Latin version—must have unsettled the minds of those dividing the space between the Earth and the sky into impenetrable zones. This was no astronomy but theology.

Only the *Elements* of Euclid outlasted the *Almagest* in influence. Yet the *Almagest* educated Greeks, Arabs, and Westerners until the 1600s. Neugebauer says that Ptolemaios was "one of the greatest scientific minds of all time."[463] He preserved Hipparchos and, in the process, we don't know how much of Ptolemaios is Ptolemaios or Hipparchos in Ptolemaios's clothing. Ptolemaios had no problem in assigning originality to Hipparchos: an "industrious [scientist] and a lover of truth."[464]

In the early 1990s, Owen Gingerich, American historian of science, said the *Almagest* was an "epoch-making book" that deserved its reputation of being the "greatest" astronomical book that has come down to us from the ancient world. In an incomparable way, the *Almagest* showed that complex natural phenomena "could be described by relatively simple underlying regularities in a mathematical fashion that allowed for specific quantitative predictions."[465]

Carlo Rovelli, an Italian theoretical physicist, writing in 2017, praises the *Almagest* for being "a major work of science. Rigorous, precise, and complex, it presents a mathematical system of astronomy capable of predicting the seemingly random movements of the planets in the sky with almost complete precision." Rovelli then adds that even today, about 1,800 years after Ptolemaios wrote the *Almagest*, one can go to the *Almagest*, learn Ptolemaios's techniques, and "calculate" the position of a planet in a "future" sky. "The realization," Rovelli says, "that working this magic is really possible is the basis of modern science, and owes not a little to Pythagoras and Plato."[466]

True, Plato had a significant influence in the spread of modern scientific ideas like the extraordinary insight of Pythagoras that mathematics was a language for understanding the Cosmos. Plato and Pythagoras are behind the genius in the *Almagest* of Ptolemaios. But the *Almagest* also connects us to

[463] O. Neugebauer, *A History of Ancient Mathematical Astronomy*, 3 vols. (New York: Springer-Verlag, 1975) I, vii, 5.

[464] Ptolemaios, *The Almagest* III.1.

[465] Owen Gingerich, *The Eye of Heaven: Ptolemy, Copernicus, Kepler* (New York: American Institute of Physics, 1993) 4–5.

[466] Carlo Rovelli, *Reality Is Not What It Seems* (New York: Riverhead Books, 2017) 44–45.

the magic of the Antikythera computer. In fact, the *Almagest* is an elaborate mathematical and astronomical explanation of the science and technology of the Antikythera computer. Of course, Ptolemaios does not name this marvel of heavens and Earth. There was no such mechanism with the Antikythera name in his time. But what existed in the second century was advanced technology of gear mechanisms for predicting human and celestial phenomena. Ptolemaios translated the advanced technology of eclipse prediction and sophisticated calendars into the mathematical astronomy of his magnificent *Almagest*. In other words, the *Almagest* was the theory of the Antikythera computer.

The *Almagest* provides additional evidence of why the Greeks had advanced technology three centuries before Ptolemaios. The *Almagest*, after all, is but a glimpse of the Antikythera computer and, more broadly, the golden age of Greek science. It is a textbook of the achievements of that era, a strategy for studying the heavens. Ptolemaios studied the heavens with the plinthis (Πλίνθις) of Hipparchos—a quadrant (a 90-degree graduated arc for calculating angles in order to measure the distances of stars from the Earth).

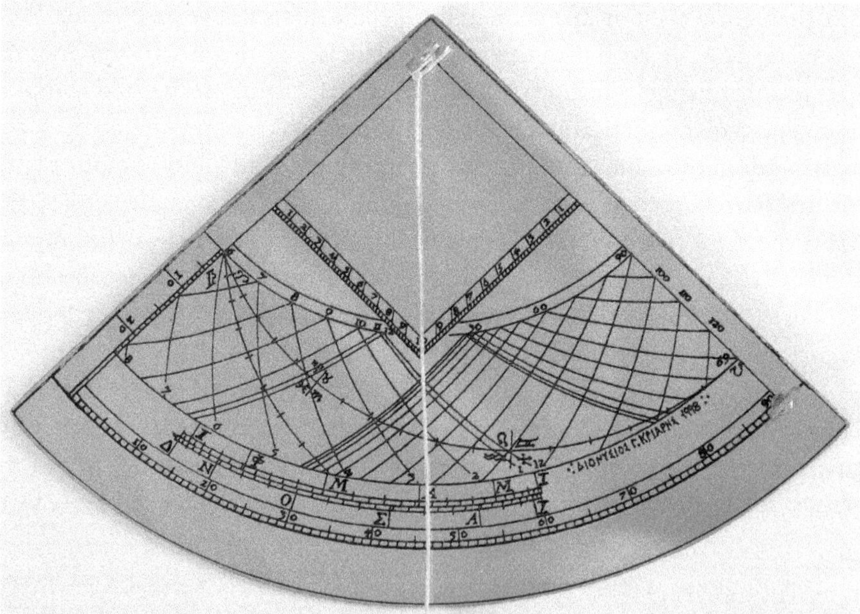

Plinthis (quadrant) of Hipparchos, Courtesy Dionysios Kriaris.

Ptolemaios illustrates how observations can be translated into mathematical tables for the calculation of eclipses and positions of the Sun, the Moon, and the stars. The Antikythera computer did the same thing with interlocking trains of toothed gears. Hipparchos guided the design of the Antikythera device as much as he influenced Ptolemaios.

When the Pleiades Rise, Begin the Harvest

The Greeks' mathematical astronomy had been a labor of observing the sky for several centuries. This and the clear insight of trigonometry in its applications to the problems of the heavens established the data for measuring the phenomena of the stars. But the phenomena of the heavens had something to say on natural phenomena, even regulating the most basic of human activities like raising food. So, it was critical for the farmer to be able to read the heavens. And this is where the Antikythera computer and similar devices came in handy. Price was right in calling the Antikythera computer a calendrical device. It was. It served farmers and others living and working in specific geographic locations governed by specific calendars.

The calendar etched right on the back of the Antikythera Mechanism is telling evidence of its origins. The names of the months inscribed in the computer are names of months one finds in the calendar of Corinth and its colonies, including Syracuse, home of Archimedes. Seven of those names are identical to the names of the months in the calendar of Tauromenion in Sicily.[467] Greeks from Syracuse founded Tauromenion in the fourth century BCE.

The names of months appearing on a calendar reconciled the lunar month to the solar year. Why should this be important?

The Greeks, we repeat, were primarily very small family farmers, peasants who were intimately connected to the land and the growing of food. Agriculture was food, economics, natural history, science, religion, and politics. In other words, agriculture was civilization. Homer reports that when *Odysseus* came into contact with the Cyclops Polyphemos, he faced an uncivilized barbarian. The Cyclopes, Homer writes, were lawless savages who never planted crops or plowed the land. They never sowed wheat, barley, and grapevines.

[467] M. Anastasiou et al., "The Back Dial and Back Plate Inscriptions," *Almagest*, 7-1, May 2016, pp. 139–215.

"Rain from Zeus waters these crops. However, the Cyclopes have neither assemblies for law making nor any established customs."[468]

Indeed, according to Euripides, Polyphemos was also full of hubris. The one-eyed monster Cyclops said to Odysseus that he could have cared less about the gods. As far as he was concerned, the greatest god was his own stomach. The Cyclops bragged that eating and drinking for a day and causing no pain to himself was his greatest happiness. Polyphemos had contempt for those who established laws, which complicate people's lives.[469]

Hesiod, the epic poet only second to Homer, focused his masterful poem, *Works and Days*, on farming, urging the Greeks to work the land for food, security, community, and for the favor and pleasure of the gods. In addition, he connects farming to the Cosmos; the peasant farmer labors in accord with the seasons and the weather, the position of the stars and, especially, the phases of the Moon. This knowledge of the heavens was part of farming and, as we will see later on, the farmer could not afford to neglect that knowledge.

Herodotus, the father of history who lived in the fifth century BCE, also shares the prevailing idea that agriculture was a mark of civilization. He says that in Scythia, present-day Ukraine, there was this community, the Kallippidai, in the middle of the coastline of Scythia. These Kallippidai were Greek Scythians who in all manner of customs were like the rest of Scythians, save for their food. The Greek Scythians "cultivate the land and eat grain, onions, garlic, lentils, and millet." Herodotus said there were other Scythians that cultivated grain only to export it while others made a living by farming. There were also nomadic Scythians who "do not grow food or cultivate the land."[470] In the eyes of Herodotus, those Scythians who, like the Greeks, cultivated the land and ate grains, were preferable to the nomads. They were civilized.

Xenophon, much like Herodotus, loved farming. He believed that agriculture was the foundation of Greek society. Farmers were, by necessity, the best of citizens because they worked the land and learned justice from goddess Earth. They had the skills and courage to defend their own freedom and the freedom of the polis. They learned the advantages of cooperation and independence. They invented and defended civilization. Besides, owning a farm was equivalent to immersion into the beauties and pleasures of the natural world.

[468] Homer, *Odyssey* 9.106–112.

[469] Euripides, *Cyclops* 332–340, tr. Arthur S. Way, Loeb, 1988.

[470] Herodotus, *The Histories* 4.17–19.

The farm was the ideal place for spending the seasons, taking hot baths in winter, and taking advantage of the water of the streams and shade of the trees in the summer. "What other occupation," he asks, "provides more appropriate first fruits for the gods or produces festivals with a greater abundance of offerings?" Xenophon connected farming to the gods, saying that "the gods control farming matters no less than matters of war." Soldiers offer sacrifices to the gods before combat in order to discover what they should or should not do. In the same way, farmers "pay respect to the gods" for the protection of their horses, cattle, and sheep as well as fruits and crops and other possessions. Xenophon also considered agriculture the mother and nurse of the arts of civilization.[471] Harvesting of olives was one of the bedrocks of that civilization. Everything they did was in the spherical orbit of agriculture. Their agrarian civilization combined polytheism, theater, athletics, beautiful temples, democracy, and science. They fished and traded and manufactured things. But they depended on the land. They had to know the best time to plant and to harvest their food. They had a star calendar they called *Parapegma*, which listed the risings and settings of the stars and constellations in relation to the Sun. The Pleiades, for example, emerged just before dawn in the east in late May, as good an astronomical message of impending wheat harvest as one could get. This sighting also meant the beginning of summer. When Arcturus, the brightest star in the constellation Bootes, rose, they knew that the autumn was with them. The sailors also knew they should not venture sailing.[472]

Conventional scholars date Homer and Hesiod, the greatest Greek epic poets, in the eighth century BCE, Homer being the more senior of the two. These poets incorporated in their writings the astronomical knowledge of their time, converting it into practical advice for travelers and farmers. In the eighteenth book of the *Iliad*, the god of metallurgy Hephaistos makes a shield for Achilles on which he imprints images of the heavens and Earth. Prominent among the inhabitants of the sky are the "tireless Sun, and the Moon waxing into her fullness." Then the constellations filling the heavens include the Pleiades, the Hyades, mighty Orion, and the Bear, which the Greeks also call the Wagon. "The Bear," Homer says, "turns about in place, looking back at Orion. The Bear alone remains aloof from the wash of the Ocean."

[471] Xenophon, *Oeconomicus* 5.7–14; 17, 20, ed. and tr. by Sarah B. Pomeroy (New York: Oxford University Press, 2001).

[472] *The Greek Anthology* 7.495.

Finally, Hephaistos added the great Ocean River to the outermost rim of Achilles's Shield.[473]

Homer's Bear or Wagon is Ursa Major. The Bear is a circumpolar constellation, turning around the celestial pole all night long without rising or setting. Homer knew that the Bear was also a guide to sailors. Odysseus, the god-like hero of the *Odyssey*, sails east by keeping the Bear to his left. Odysseus also sees the Dog Star (Sirius) and the constellation Bootes.[474]

In book five of the *Iliad*, Homer calls Sirius "aster oporino" (ἀστέρ ὀπωρινῷ), autumn star. He knew that Sirius was extremely bright in late summer. He has Pallas Athena make Diomedes, one of the Greek heroes in the Trojan War, fight with the courage and strength that could only bring him glory. He says: "Starlight flowed from Diomedes' helmet and shield like Sirius had just risen from the sea before dawn in the fall. That brightest of stars was blazing from his torso and face rather than from the sky."[475]

The poems of Homer are mirrors of early Greek history and culture, including scientific culture. The scenes of agrarian life in the Shield of Achilles complemented the technological works of Hephaistos and the reliance of farmers and sailors on constellations and stars. Homer was versed with astronomical timekeeping and the heavens, including the eclipses of the Moon and the Sun.

A Terrible Darkness Swallowed the World

In book 20 of the *Odyssey*, we see a herdsman named Philoitios bringing cattle for the feast of suitors. The cattle are from the herds of Odysseus in Kephalonia. Philoitios had served Odysseus since he was a small boy. He was angry with the suitors plundering his master's property. The seer Theoklymenos, who caught a ride to Ithaca with Telemachos, was also angry. He warned the suitors they were doomed: shrouded in night and death from head to toe. He says: "The Sun disappeared completely from the sky and a terrible darkness attacked the world."[476] When Theoklymenos spoke about the obliteration of the Sun from the sky, he had probably witnessed a total solar eclipse,

---

[473] Homer, *Iliad* 18. 478, 483–489, 606–607, ed. M. M. Willcock (London: Bristol Classical Press, 2001).

[474] Homer, *Odyssey* 5. 272–277.

[475] Homer, *Iliad* 5. 4–7.

[476] Homer, *Odyssey* 20. 356–357 (ἠέλιος δὲ οὐρανοῦ ἐξεπόλωλε, κακὴ δ' ἐπιδέδρομεν ἀχλύς).

a rare and dramatic phenomenon. Seconds before the Moon hides the Sun, the Sun illuminates the Earth in great brightness. Then, suddenly, the Sun is gone and the horizon turns ink-blue and the stars become visible. The temperature drops a few degrees; winds pick up and animals and humans are caught in an unprecedented and frightful condition as if the world is being swallowed by the darkness of Erebos. That is why eclipses, and this kind of eclipse of Helios in particular, were bad omens.

Archilochos, a lyric poet of the seventh century BCE from the Aegean island of Paros, gives us a taste of the terror accompanying the eclipses of the Sun in 648 BCE. In a fragment of his poetry, he says Zeus, father of the Olympian gods, "made night out of noontime, keeping back the light of the beaming sun; and upon mankind came fear."[477]

In a similar reaction, another great Greek lyric poet, Pindar, saw the end of the world in the temporary disappearance of Helios. He witnessed a solar eclipse on February 17, 478 BCE, and another on April 30, 463 BCE. He wrote an ode based probably on the second eclipse of the Sun. This poem was popular for centuries, a clear mirror of what such a spectacular phenomenon meant to people all over the Greek and Roman worlds: Pindar said the Sun-Helios "was mother of the eyes and brightest of the stars, so, hiding itself, running down the highways of darkness during the day was making men and wisdom powerless."

"By Zeus, Helios, driver of swift horses, don't turn your disappearance to harm like war, blight of crops, too much snow, flood, summer heat. Will you overwhelm the Earth with flood and create a new race of men from the beginning?"[478]

Fifty years after Pindar immortalized a universal fear of solar eclipse, an eclipse of the Moon had unforeseen consequences for an Athenian army in Sicily. The Athenians paid a tremendous political price for misreading a heavenly omen during the Peloponnesian War. In 415 BCE, they sent an armada against Syracuse. But things turned bad and the Athenian generals were ready to order a retreat and return to Athens. Unfortunately, a celestial phenomenon intervened and the Athenian army in Syracuse suffered a catastrophic defeat.

According to Thucydides, the Athenian general and historian who fought and lived through the Peloponnesian War, all started going haywire when the

[477] "Archilochos," in *Anthology of Classical Myth*, 58.

[478] Pindar, *The Odes and Selected Fragments* Paean 9.1–21, tr. G. S. Conway and Richard Stoneman, ed. Richard Stoneman (Everyman, 1997).

departure of the Athenian army from Syracuse coincided with the disappearance of the Moon. A full eclipse of the Moon made most Athenian soldiers feel uneasy. They urged their generals to stay. Nikias, the commander-in-chief, shared the soldiers' fear and ordered that before departure the priests should do what was necessary to appease the Moon. This required "three times nine days."[479]

The lunar eclipse of August 27, 413 BCE, unsettled and frightened the Athenian commanders. Nikias and his troops had witnessed a solar eclipse before the Moon disappeared in the darkness of the night. They understood the Moon was between themselves and the Sun so the Sun went dark. But in the case of the Moon becoming invisible late in the evening of August 27, 413 BCE, something strange happened. The Athenian soldiers and Nikias saw a full Moon suddenly becoming dark, the sky filling with different colors coming out of the dying Moon. They found the night spectacle incomprehensive. They acted with dread. They probably saw the eclipses of the Moon as a sure thing of divine calamities to come.[480]

The consequences of this spectacular natural phenomenon were deadly. The decision of Nikias to stay because of the eclipse of the Moon resulted in the near obliteration of thousands of Athenian soldiers in Syracuse. The expedition of Athens to Sicily was the largest in Greek history. In 413 BCE, Athens lost everything. The huge armada was destroyed and the 40,000 soldiers, both Athenian and other Greeks fighting on the side of Athens, were slaughtered or sold into slavery.

This tragic incident connected the heavens, religious practice, and politics in the field of battle. It was clearly unusual. But what was not unusual was the respect and often fear the Greeks had for natural and celestial phenomena. According to Pausanias, the second-century Greek historian and traveler, the Spartans would never "march to battle until the circle of the Moon was full."[481] Seers accompanied armies to decipher the will of the gods written in nature: in the liver of birds and sacrificial animals, eclipses of the Sun and the Moon and the interpretation of dreams. Poleis took no major decision without consulting the oracle of Apollo at Delphi.

Herodotus acknowledged the importance of omens among the Greeks. He warned how difficult it was to interpret them and other terrestrial and celestial phenomena, and how deadly to misread them.[482] This was the reason why the poleis and countryside of Greece were dotted with altars and temples and oracles

[479] Thucydides, *The Peloponnesian War* 7.50.
[480] Ploutarchos, *Nikias* 23.
[481] Pausanias, *Guide to Greece* 1.28.4, tr. Peter Levi (Penguin Books, 1984).
[482] Herodotos, *The Histories* 1.34–45, 3.124, 5.55–56, 6.107.

of the gods. These sacred places connected the polis of people to the polis of the gods. Greeks prayed and offered sacrifices to the gods; they lavishly celebrated their birthdays and honored them with elaborate athletic and cultural competitions. They sought the guidance and insight of the gods for leading their lives and reading the heavens. In addition, astronomers and Antikythera-like mechanisms tried to predict eclipses of the Moon and the Sun, by far the most telling of the celestial messages the Greeks wanted to understand in the workings of the heavens. For example, Herodotus reports that Thales of Miletus predicted an eclipse of the Sun that took place on May 28, 585 BCE. This solar eclipse had the effect of reestablishing peace between Lydians and Medes.

Herodotus described this event like this. In the "middle of the battle" between Lydians and Medes, he says: "[D]ay suddenly became night. Thales of Miletus had predicted this loss of daylight to the Ionians by establishing in advance that it would happen within the limits of the year [585 BCE] in which it did in fact happen. When the Lydians and the Medes saw that night had replaced day, they did not just stop fighting; both sides also more actively wanted an end to the war."[483]

Thus, omens were not always bad, which explains their persistence among the Greeks and other people. Natural phenomena were signs of how the world works. They were also manifestations of possible divine interest in human affairs. The Greeks used omens to understand the gods, nature, and the heavens. They also included omens in their history.

Ploutarchos, commenting on Homer more than a millennium later, agreed that Homer was referring to a solar eclipse.[484] Modern astronomers have confirmed the accuracy of Theoklymenos and Ploutarchos's conjecture about it, suggesting that there was, in fact, a total solar eclipse over the Ionian Sea on April 16, 1178 BCE. This brings us near the date Herodotus assigns to the Trojan War. Herodotos was born about 485 BCE and he says the Trojan War took place around 800 years before his time,[485] that is, in approximately 1285 BCE.

The scientists who lined up astronomical clues with the narrative of the *Odyssey*, Marcelo O. Magnasco of the Rockefeller University in New York, and Constantino Baikouzis of the Observatorio Astronomico de La Plata in Argentina, reported the following: when Odysseus killed the suitors, there was a new Moon, a requirement for a solar eclipse; that day was also a festival day

[483] Herodotos, *The Histories* 1.74.

[484] Ploutarchos, *The Face on the Moon* 931d–932d.

[485] Hirodotos, *The Histories* 2.145.

of Apollo, a solar god; six days before the killing, Aphrodite (Venus) was high in the sky; twenty-nine days before the slaughter, Kalypso sent Odysseus off at sunset, telling him to watch the constellations Pleiades and Bootes and keep the Bear on his left. This meant Odysseus sailed east. By that time Hermes (Mercury) had delivered Zeus's message to Kalypso in the island of Ogygia, all the way to the far west. This meant that Hermes was in the western end of his travels standing high at dawn.

According to Baikouzis and Magnasco, the astronomical phenomena in the *Odyssey* "define the timeline of the epic, forming a layer akin to finding a perspective grid drawn in pencil behind a painting." Odysseus took nineteen years to return home, a time period that fits with the Metonic cycle, about which there is more later on. This Metonic cycle is "an eclipse repeat cycle." In other words, the stars dictated the story of Odysseus: when he built his raft, when Poseidon sunk it, and when he would reveal himself in his house in Ithaca. From the moment Odysseus left Kalypso to the time he met his wife was exactly one month. All in all, Odysseus and his story fit with the Sun's eclipse on April 16, 1178 BCE.[486]

Stavros Papamarinopoulos, a geologist at the University of Patras, and a team of Greek scientists, reject April 16, 1178 BCE, as the date of the Solar eclipse over the Ionian Islands. They argue, instead, the eclipse of the Sun over the Ionian Islands took place twenty-nine years earlier: at noon on October 30, 1207 BCE. This was a partial eclipse that coincided with Odysseus's killing of the suitors. The reading of Homer also favors their conclusion the eclipse was a fall phenomenon. Fruit like pears, figs, apples, and grapes grow during the waning year, not in the spring. The same is true with long nights.[487]

This pushes the astronomical knowledge of the Greeks back to the Bronze Age (late thirteenth century BCE). Baikouzis, Magnasco, and Papamarinopoulos recognized advanced astronomical knowledge in the *Odyssey*. But because scholars doubt the Greeks had interests in astronomy before the classical age, that is, the fifth century BCE, they largely wrapped Homer in myths. In fact, some scholars see nothing historical about the Trojan War and Homer. But those scholars who see history in the Homeric epics are also confused. They date Homer to the

[486]Constantino Baikouzis and Marcelo O. Magnasco, "Is an eclipse described in the Odyssey?" *Proceedings of the National Academy of Sciences of the United States of America*, July 1, 2008, Vol. 105, No 26, 8823–8838.

[487]Stavros Papamarinopoulos et al., "Astronomical Phenomena and the Dating of the Return of Odysseus," in *Ancient Greece and Contemporary World*, eds., Stephanos A. Paipetis and Chrestos Kechagias (Athens: Livanes Publishing for the University of Patras, 2017) 526–535 (in Greek).

eighth century BCE but the Trojan War four centuries before Homer. Inevitably, such scholars thought it implausible that Homer had that kind of astronomical knowledge. However, the Greeks of Homer's time learned practical astronomy from their ancestors or from their relatives in Minoan Crete of the second millennium BCE who did know how to read the heavens.[488] According to Minas Tsikritsis, expert in ancient Aegean archaeology, Bronze Age Crete, dated from about 2700 to 1500 BCE, had some kind of rudimentary analog computer for figuring out the eclipses of the Moon and the Sun. That device was a sundial and assisted in the calculation of geographical latitude as well.[489]

Like the Antikythera Mechanism, the surprise is not that the Greeks were so much more advanced in science and technology than we are willing to concede but that it has taken so long to reach this conclusion.

Turnings of the Sun

Even more than Homer, Hesiod, who wrote after Homer, probably in the late eighth century BCE, paid a lot of attention to the stars. He had to. He was a dirt-poor farmer who had to know how to tell time. He wrote *Works and Days* as a guide and agricultural calendar for the hard-working farmer. The message of Hesiod was the critical importance of paying attention to the weather, the seasons, and the signs of the heavens.

Hesiod also reports on solstices and equinoxes. The Greeks of Hesiod's time looked at the equinox as a season when the length of the day was approximately equal to the length of the night. As for the solstices, Hesiod describes them as *tropai helioio*, or turnings of the Sun. It is from this expression we have the word tropic. At the summer solstice the Sun is in the zenith at the Tropic of Cancer around June 21; at the winter solstice the Sun is over the Tropic of Capricorn around December 22.

Hesiod tells the farmers to figure out the time of the year by observing the risings and settings of stars just before the rise and setting of the Sun. He says: "When the Pleiades, daughters of Atlas are rising, start the harvest; when they set, begin plowing."[490] He warned the farmer not to delay sowing their crops until the winter solstice and the "turning of the Sun" because his harvest will

[488] Jack Dempsey, *Calendar House: Secrets of Time and Power in Ancient Crete's Great Year* (DVD, 2008).

[489] "Researcher cites ancient Minoan-era 'computer,'" *Athens News Agency*, April 7, 2011.

[490] Hesiod, *Works and Days* 383–384, tr. Apostolos N. Athanassakis (Baltimore: The Johns Hopkins University Press, 1993).

be poor: "If you plow the good Earth during the winter solstice, you will reap but tiny handfuls."[491]

Hesiod also highlights the days of the month, each having special significance for the life and work of the Greeks. Zeus the counselor sends the days; Hesiod tells the farmer, pay attention to them, particularly the thirtieth, which is the best for overseeing work that has to be done and for apportioning food. This is also the day that people sort out what is truth and what is not. The fifth day: good for opening wine jars; the seventh: threshing holy Demeter's grains as well as perfect for cutting wood for house and ships; the ninth: better towards evening, husband and wife ought to plant an offspring; the ninth is also an excellent day to be born; the twenty-seventh: yoke the necks of oxen, mules, and horses for field work or drag a ship to the wine-dark sea.[492]

Seasons

Some three centuries after Hesiod, the great Greek philosopher and physician Hippocrates stressed the vital role of seasons on human health and welfare. He urged the aspiring doctor to understand the different effects seasons have been having on life. The medical practitioner must grasp the significance of the risings of the Sun, the quality of water and land, and the universal and local hot winds and cold. With this knowledge of the seasons, including a thorough understanding of "the risings and settings of the stars," the physician or another person will be certain of the "nature of the year that is coming." In other words, Hippocrates was teaching preventive medical care, which, in this case, was intimately linked to both the natural world, the Earth, and the heavens.

If all this sounds too much like meteorology or environmental medicine, perhaps it is. This is because, Hippocrates said, astronomy is very important to medicine. Changing seasons make a difference to health, including diseases.[493] Hippocrates warned of potential violent and dangerous health effects of the following natural phenomena: both solstices, especially the summer solstice; both the equinoxes, especially the autumnal; the risings of the stars, especially of the Dog Star and the constellation Arcturus; the settings of the constellation Pleiades. During these natural phenomena diseases flourish.[494]

[491] Hesiod, *Works and Days* 479–480.
[492] Hesiod, *Works and Days* 765–828.
[493] Hippocrates, *Airs, Waters, Places* 1–2.
[494] Hippocrates, *Airs, Waters, Places* 11.

To ameliorate such crises, Hippocrates offered detailed advice on the best way of life one ought to follow during the different seasons. Food, exercise, sleep, work, the taking of baths, and even the appropriate time for sex were part of a scientific strategy in balancing astronomy, medicine, and health throughout the year. He explained he divided the year into four seasons: winter, spring, summer, and autumn. He said winter lasts from the setting of the Pleiades to the spring equinox; spring from the equinox to the rising of the Pleiades; summer from the Pleiades to the rising of Arcturus; and autumn from Arcturus to the setting of the Pleiades.[495]

Hippocrates, like Hesiod, touches on the agrarian and scientific nature of Greek civilization: respectful of the gods, and drawing on ancient oral and scientific traditions tied to the land, the memory of farmers, and to the decoding of the secrets and influence of the heavens.

Celebrating Nature and the Gods

Nature was the key to Greek civilization. Hippocrates urged the Greeks to fashion their lives to fit within the dictates of nature, even the cold of winter: It does not make sense, he says, for the body not to be exposed to the cold of winter. Trees, for example, that have not experienced the cold of winter can neither bear fruit nor themselves be vigorous. Besides, spring follows winter. Just forty-four days after the winter solstice, he says, the west wind blows and the season is milder. Arcturus then rises and the season for the swallow comes into being.[496]

Yes, the swallow remains firmly in my mind, bringing the season of renaissance in the natural world. The Greeks read nature and coordinated their lives and the working of the land with festivals and sacrifices honoring the gods whose benevolence was expected for food security and abundance—and much more. They celebrated the gods of *oikos* (home): Hestia, Hekate, Apollo Agyieus (guardian of the streets), and Zeus Ktesios (protector of property).

The Greeks also observed annual agricultural rituals, especially *Haloa*, the land festival honoring Demeter, goddess of agriculture, indeed, goddess Earth; Dionysos, god of wine and the theater; and, possibly, Poseidon, god of the sea. *Haloa* comes from *halos* or threshing floor and ploughed land. The purpose of the festival was for the gods' blessing and safeguarding of the germinating seed

[495] Hippocrates, *Περί διαίτης (Regimen)* 3.68. 10–17.

[496] Hippokrates, *Περί διαίτης (Regimen)* 3.63.61–64, 77–87. Ειτα δε αρκτουρου επιτολη, και χελιδονα ωρη ηδη φαινεσθαι (86–87).

in the soil. This was largely a women's celebration in the Athenian town of Eleusis during the twenty-sixth of the winter month Poseideon.

Another Panhellenic celebration of Demeter was the Eleusinian mysteries taking place in Eleusis in the month *Boedromion* (September–October). The other divinities of importance in the Eleusinian mysteries included the daughter of Demeter, Persephone, and Dionysos. The mysteries became a celebration of farming and life. Perhaps the mysteries promised life after death. We don't know. The Eleusinian mysteries, however, were so important they became the national religion of Greece. The mysteries also influenced *Thesmophoria*, another agricultural festival lasting for three days. In the Athenian calendar, this women-run celebration took place during the eleventh, twelfth, and thirteenth of the month *Pyanopsion* (October–November). Women reaped and sowed the crops and praised Demeter.

Goddess Demeter. Silver 4-drachmas coin from the Cycladic island of Syros. Mid-second century BCE. Courtesy Numismatic Collection, Alpha Bank, Athens.

Dionysos was the focus of the festival *Anthesteria*, which took place every year during the eleventh, twelfth, and thirteenth of the month *Anthesterion* (end of February). This was a celebration of spring flowers and wine. An additional spring festival, *Daphnephoria* or carrying of laurel, was all about nature and the seasons under the symbolism of Daphne and Apollo. Daphne was the nymph daughter of the river god Ladon in Arkadia or Peneios in Thessaly. Daphne was determined to stay a virgin just like Artemis, the goddess of nature.

Eros, however, struck her with a "lead" arrow against passion while striking Apollo with the golden arrow of love. "Marriage torches to Daphne were nothing less than anathema." But for Apollo the mere sight of Daphne was an explosion of passion. Rather than allowing Apollo to make love to her, she pleaded with her divine river father to intervene. Peneios promptly made her the laurel tree. Apollo was running after Daphne and almost touched her when she became a tree:

"Tree though she was, Apollo still loved her. Caressing the trunk with his hand, he could feel the heart still fluttering under the new bark. Seizing the branches, as though they were limbs, in his arms' embrace, he pressed his lips to the wood; but the wood still shrank from his kisses. Phoebus Apollo then said to her: Since you cannot be mine in wedlock, you must at least be Apollo's tree. It is you who will always be twined in my hair, on my tuneful lyre and my quiver of arrows ... As I, with my hair that is never cut, am eternally youthful, so you with your evergreen leaves are for glory and praise everlasting."[497]

The victors of the Pythian games at Delphi honoring Apollo received a laurel crown. Second, a maypole, decorated by 365 laurel garlands, was at the center of the *Daphnephoria*. The top of the maypole had metal spheres representing the stars. And the *Daphnephoria* took place either every nine years or every year. Pindar wrote that during the *Daphnephoria*, "The Seasons, daughters of Themis, have come to the horse-driving city of Thebes to bring a banquet for Apollo, lover of garlands."[498] Finally, the Greeks had a holiday of fun during the *Oschophoria*, a festival of wine and grapes. Farmers also honored Artemis and Hermes *Agathos* (the good).

Calendars

In fact, the Greeks invented calendars for bringing order to the celebration of their gods and for keeping record of the critical political activities of their governments. For instance, Athenians had to know when the Areopagos, Supreme Court, and the Ekklesia, the assembly and executive, met. The days of those

[497] Ovid, *Metamorphoses* 1.451–559, tr. David Raeburn (Penguin Books, 2004).
[498] Pindar, *The Odes* Paean 1, 6–10.

meetings affected their work and religious festivals. The calendar balanced the Greeks' sacred and political lives.

Hesiod was right. The calendar he described lasted for a very long time. Ploutarchos, writing some 800 years after Hesiod, considered the first day of every month (νουμηνία, *noumenia*, the new-Moon-day) the "holiest of days."[499] The second day of the month was the day of ἀγαθός δαῖμων, *agathos daimon*, good god. The third day was Athena's birthday. The fourth day was for the celebration of Herakles, Hermes, Aphrodite, and Eros, son of Aphrodite. The sixth day of the month all over the Greek world was dedicated to the birth of Artemis *Agrotera* (of the agrarian world). The eighth day was sacred to Poseidon and his son, Theseus. The sixteenth day honored Artemis. Athenians took cakes to the sanctuaries of the goddess and to the crossroads. Almost half of the calendar days of the Athenians were festival days.[500]

For Athenians as well as other Greeks, summer solstice coincided with the end of the grain harvest. This was also the time that most poleis began their new year. In addition, summer solstice brought to power the newly elected government of Athens. The Athenians celebrated the transition from one year to another with *Dipolieia* or *Bouphonia*, a festival in honor of Zeus *Polieus*, Zeus, defender of the polis. The altar of Zeus *Polieus* was on the Acropolis. The festival started with the celebration of the first field ploughed, which the Athenians called the Field of Hunger, behind the *Prytaneion*, the building housing the communal hearth of the polis and the *Prytany*, the Presidency of the governing Council. At the same time a trial for murder took place at the *Prytaneion* (main government building). The murder had to do with the killing of the ox responsible for the cultivation of that first piece of land. The excuse for the killing/sacrifice of such a sacred animal was its supposed consent: The sacrificial bull tried to eat the barley and wheat on the altar at which time the priest killed it. The priest then threw down the battle-ax and fled. Athenians then charged and tried the battle-ax with murder.

Athenian early history named *Bouzyges* (Ox-yoker) the first man who ever ploughed the land. A priest of Zeus *Teleios* (the perfect or fulfiller), also known as Zeus *Bouzyges*, cursed those who refused to share water and fire, undermining citizenship and social and political cooperation. This was the exact opposite

[499] Ploutarchos, *Moralia* 828a.

[500] Jon D. Mikalson, *The Sacred and Civil Calendar of the Athenian Year* (Princeton, NJ, 1975) 1–24, 201–203.

model of the monster Polyphemos, who would have nothing to do with laws, gods, and society. The murder trial of the battle-ax in the *Prytaneion* was symbolic of the importance of the bull in farming and, therefore, the organization of the polis.

Dipolieia prepared the ground for another celebration of what it meant to be Athenian. This was the *Panathenaia*, worshipping Athena as well as being the Athenian version of the Olympics and a fun holiday over the New Year. The *Panathenaia* was the greatest festival of Athens. It took place during the month of *Hekatombaion* (July/August), the birthday of Athena. Young women (*ergastinai*) wove images of Athena onto a new robe. A public procession that included the *ergastinai* as well as young girls serving Athena (*arrhephoroi*), and officials, priests, and priestesses carried the robe to the Acropolis and presented it to Athena. This great procession was immortalized in the freeze of the Parthenon.

Athenians and Greeks from all over the Greek world competed in athletic, poetic, musical, and nautical events and feasted in honor of Athena.

Despite the Athenians' piety and exuberance over Athena, they often made mistakes in their public expression of respect for the gods. Aristophanes makes fun of their cavalier attitude towards their calendar, often messing up their responsibilities to celebrate their gods with appropriate festivals.

Selene, the Moon-goddess and sister of Helios and Eos, Sun and Dawn, complains to the Athenians that her light saves them at least a drachma every month, which, otherwise, they would squander on buying torches. However, the Athenians insult the gods with their failure to sacrifice at the time gods expect the offerings; second, the Athenians compound their mischief with indulging in libations and food at the very moment when the gods are fasting in memory of [the sons of Zeus] Sarpedon and Memnon [killed by Achilles in Troy]. So, Aristophanes tells the Athenians, it pays to rely on the Moon-goddess to reckon their days and sacred calendar.[501]

Herodotus was also critical of Greek calendars. He found the "monthly system" of the Egyptians "cleverer" than that of the Greeks. He explained why: The progress of the seasons forces the Greeks to insert an intercalary month every other year. The Egyptians, however, have twelve months of thirty days

[501] Aristophanes, *Clouds* 615–626.

each. They add five extra days to every year.[502] The Egyptians also had the Nile, a divine river that inscribed the seasons on the land. According to Heliodoros, the Nile "announces the seasons of the year: the summer by his rising, the autumn by his ebbing, and the spring by the flowers growing by his side, and by the crocodiles laying their eggs. The Nile ... is exactly identical with the year ... for the letters of his name, if translated into numerals, will amount to three hundred and sixty-five units, equal to the days in the year."[503]

The Greeks had no Nile. The heavens and the land gave them the clues to fashion their calendars. Geminos is comprehensive on how the Greeks defined and understood the seasons. He says they reckoned the days and months by the Moon and the year by the Sun. One revolution of the Earth on her axis equals a day. Just like the Egyptians, the Greeks knew that the year had 365 days. However, the Earth's spinning around her axis 365 times is not exactly the time it takes the Earth to circle around the Sun, which is one year or 365.242199 days. The same reality governs the month, which turns out to be 29.53059 days. Nevertheless, time reckoning by the Greeks was not far off the mark. Seeing the crescent of the Moon was the beginning of a month. The time between one crescent and another was never more than thirty days and never less than twenty-nine days. The Sun year was vital for marking time, telling the Greeks when to sow and when to harvest; and the Moon cycles, averaging twenty-nine and a half days each, reminded them to join farming to civic life. According to Geminos, the astronomical calculation of the month and the year helped the Greeks to follow the tradition of their ancestors in sacrificing to the gods, which meant "the same sacrifices to the gods to be performed at the same times of the year."[504]

The month in Greece started with the new Moon. One month had twenty-nine days followed by a month of thirty days. The Greeks called the thirty-day months "full" and the twenty-nine-day months "hollow." The calendar year had twelve months and, from time to time, it had thirteen months. Adding a thirteenth month guaranteed that the calendar was in accord with the seasons.

Not all Greek poleis had the same calendar. In fact, some poleis had a variety of methods in reckoning time. Aristophanes was not kidding about Athens, which had more than one official calendar. More than a century before Aristophanes, when Solon was the archon of Athens in the early sixth century

[502] Herodotos, *The Histories* 2.4.

[503] Heliodoros, *Ethiopian Story* 9.22.

[504] Geminos, *Introduction to the Phenomena* VIII.15.

BCE, according to the third century Greek writer Diogenes Laertios, Solon convinced the Athenians to use a lunar month.[505] But by the fifth century BCE, the archon, the chief politician of Athens, had his own calendar, which was different than that of the *Prytany* calendar of ten months of thirty-six or thirty-five days each. Aristotle says this reckoning was the result of the *Prytany* having adopted a Moon year.[506]

Most poleis followed their own calendar—giving the months their own names. Athens, for example, started the year in July, its twelve months had the following names: 1. *Ekatobaion,* July–August; 2. *Metageitnion*, August–September; 3. *Boedromion*, September–October; 4. *Pyanopsion*, October–November; 5. *Maimakterion*, November–December; 6. *Poseideon*, December–January; 7. *Gamelion*, December–January; 8. *Anthesterion*, February–March; 9. *Elaphebolion*, March–April; 10. *Mounichion*, April–May; 11. *Thargelion*, May–June; and 12. *Skirophorion*, June–July.

Several poleis also started the New Year at different times, inserting the thirteenth month arbitrarily to suit their needs. The Greek calendar was simple and scientific but Greek chronology was difficult. Counting the Olympiads, and the four years within each Olympiad, was a Panhellenic custom that brought some order in chronology. There was also another calendar, one according to the goddess Selene, the *kata theon* calendar, which was more accurate than the political calendars of the poleis. This *kata theon* or *kata Selene* calendar served all Greek poleis as a common calendar.[507]

The most used Greek calendar took the name Metonic in honor of the fifth-century BCE Athenian astronomer Meton. He suggested inserting seven lunar months in nineteen years. Diodoros was a contemporary of the astronomer Geminos. He was enthusiastic about the innovation of Meton. He reported that Meton won fame for his knowledge of astronomy and his nineteen-year cycle. The Athenians recognized the value of Meton's proposal. They put it into effect on summer solstice, which was the beginning of the solar year. All this happened on the thirteenth day of the Athenian month Skirophorion in the year 432 BCE. Diodoros records that after nineteen years the stars return to the same place in the heavens, completing a Great Year, which some people named the Year of Meton. Even to our time, Diodoros says of the late first

[505] Diogenes Laertios, *Lives of Eminent Philosophers* 1.59.

[506] Aristotle, *Constitution of Athens* 43.2.

[507] Sacha Stern, *Calendars in Antiquity* (Oxford: Oxford University Press, 2012) 61.

century BCE, the greatest number of Greeks reckon time with the nineteen-year cycle and are not cheated of the truth.[508]

The background to the Great Year was this: Twelve Moon months need an additional 11 days to fit the length of a Sun year. However, 235 lunar months make 19 solar years of 6,940 days. Geminos said that each 19-year cycle included 12 years of 12 months each and 7 years of 13 months each; 125 full months (30 days) and 110 hollow months (29 days).[509]

Fragment 19: At the center we read "We know four"; in the next line below, we see the number 76 years [OC: omicron stigma] for the 76-year Kallippic calendar and 19 years [IΘ: iota theta] for the 19-year Metonic calendar. Then, on the line below, we read ΣΚΓ, sigma kappa gamma, 223 [months] of the Saros cycle. Courtesy Tom Malzbender and Hewlett Packard.

So, the Metonic cycle synchronized the Sun and the Moon, which meant, first of all, adjusting the calendar to the seasons, and second, harmonizing the agricultural and political life of the Greeks. We already mentioned that as part of that major function, the Antikythera Mechanism had an Olympiad dial that gave the names and identified the time for the Panhellenic games,

[508] Diodorus Siculus, *The Library of History* 12.36.2–3; 2.47.6.

[509] Geminos, *Introduction to the Phenomena* VIII. 50–57.

including the Olympics.[510] The Olympiad dial was within the Metonic calendar.

The Greeks made their astronomical computer more accurate, reconciling the Sun and the Moon cycles even more tightly, with the Kallippic cycle, which was four Metonic cycles minus one day. The cycle of the two- and four-year religious and athletic celebrations included not merely the Olympics in Olympia, but the Nemean in Nemea, the Pythian in Delphi, the Isthmian in Corinth, the NAA in Epiros, and the Halieia in Rhodes.

The Antikythera computer also predicted eclipses by means of the Saros cycle of 223 lunar months. This worked from long observations that, after an eclipse, one had to wait eighteen years, eleven days, and eight hours for another repeat eclipse. This means eclipses separated by 223 months have similar geometries: the Moon is nearly at the same distance from the Earth and the eclipse takes place in the same time of the year. There's a subsidiary dial within the Saros dial known as the Exeligmos (turn of the wheel) dial. This is a triple Saros dial of fifty-four years. The Saros cycle has 6,585 1/3 days. The Exeligmos, being three times longer than the Saros, yields a whole number of days. In addition, the accuracy of predicting the lunar eclipse with the Exeligmos dial increases to a specific time of the day.[511]

Meanwhile, the real but unknown designers of the Antikythera computer coded information in glyphs in selected months of the Saros dial about the prediction of eclipses of the Sun and the Moon: fifty-one glyphs showing thirty-eight lunar and twenty-seven solar eclipses. And, according to the 2008 scientific study of the Antikythera Mechanism, these eclipse predictions in the glyph sequence matched data on actual eclipses.[512]

To observe an eclipse of the Moon, for example, one has to know where to look in the sky. The Meton cycle of nineteen years guaranteed the reappearance of the same-phased Moon in the same spot in the sky. The Kallippic cycle of seventy-six years made the detection of the eclipses of the Sun and the Moon in the heavens more precise. Thus, the Metonic and Kallippic cycles complemented

---

[510] Freeth et al., "Calendars with Olympiad display and eclipse prediction on the Antikythera Mechanism," *Nature*, July 31, 2008, 454: 614–617. The Korinthian months included: (1) ΦΟΙΝΙΚΑΙΟΣ; (2) ΚΡΑΝΕΙΟΣ; (3) ΛΑΝΟΤΡΟΠΙΟΣ; (4) ΜΑΧΑΝΕΥΣ; (5) ΔΩΔΕΚΑΤΕΥΣ; (6) ΕΥΚΛΕΙΟΣ; (7) ΑΡΤΕΜΙΣΙΟΣ; (8) ΨΥΔΡΕΥΣ; (9) ΓΑΜΕΙΛΙΟΣ; (10) ΑΓΡΙΑΝΙΟΣ; (11) ΠΑΝΑΜΟΣ; (12) ΑΠΕΛΛΑΙΟΣ.

[511] Ptolemy, *Almagest* IV. 2.

[512] "Supplementary Information," 29 (*Nature*, July 31, 2008).

the Saros and the Exeligmos time prediction of the eclipses by giving the exact place in the heavens where the eclipses would take place. Not only that, but the 2016 decoding of the eclipses' inscriptions revealed that predicting the eclipses also came with the bonus of forecasting the color of the Moon and the local weather, especially the winds expected to follow the eclipses.[513]

If this sounds strange to us, it was normal to the Greeks. Eclipses are different phenomena affecting local weather and the color of the Moon. But, in reality, are they? The Greeks did not think so. As I have already discussed in this chapter, they observed the eclipses with immense curiosity, wrapped up in the fear of the gods—the Sun and the Moon were gods. This explains one of the reasons why the Greeks did so well in the design of the Antikythera computer. This was to be and, in fact, became a machine predicting the future, even the will of the gods. That's why the world's first computer captures so well all that knowledge about the heavens and society together.

The admirable working of the Antikythera machine tempted Xenophon Moussas to argue that he saw the signature of Archimedes and his students on the Antikythera computer.[514] Tony Freeth agrees. He, too, connects the Antikythera device to Archimedes, though he rightly admits, there's no historical evidence Archimedes, who was assassinated by the Romans in 212 BCE, had anything to do with the fabulous mechanical universe of the Antikythera computer. However, Freeth dates the Antikythera device to 205 BCE. Such an older astronomical calculating machine—he refused to call it computer until 2021—he would like to believe, had a higher likelihood of being the direct or indirect product of the mathematical genius of Archimedes and his younger contemporary Apollonios of Perga. Apollonios was, as I have already said, a truly great mathematician. His work and especially his "Conics" made a significant contribution to the advancement of geometry. The application of "Conics" was essential for the development and improvement of sundials, optics, and celestial mechanics.[515] The "Conics" today guide spacecraft to reach the planets. In addition, the epicycle theory of Apollonios explained the movement of the planets. Thus, Apollonios added a scientific perspective to geometric astronomy.

---

[513] M. Anastasiou et al., "The Back-Dial and the Back-Plate Inscriptions," *Almagest* 7–1, May 2016, pp. 138–215.

[514] Xenophon Moussas, *The Antikythera Mechanism,* 83–84.

[515] Swerdlow, "The Recovery of the Exact Sciences of Antiquity," in *Rome Reborn*, 136–137.

A Computer of Genius

Freeth can be forgiven for his preference for mathematicians. The Antikythera Mechanism, he says, "was essentially a mathematician's instrument." [516] He repeated his love for mathematics in a study he did in 2021,[517] in which he ironed out the ambiguities left from all previous researches of the Antikythera computer. He especially wanted to know more about the bronze gearing system in the front plate. With only a third left of the original machine, was there a way to bring back to light accuracy in the Cosmos embedded in the front plate? The back plate inscriptions described the motion of the Sun, the Moon, and the five planets, including their place in the front of the computer.

Like in 2005, Freeth put together an Antikythera Mechanism Research Team of multidisciplinary scientists at the mechanical engineering department of the University College London (UCL) to create a digital model of the device. These researchers relied on previous studies: the 1905–1906 investigation by the German philologist Albert Rehm; the pioneering 1974 study by Derek de Solla Price; and several reconstructions done in the twenty-first century. These studies, however, according to Freeth, failed to pay proper attention to the inscriptions and their descriptions of the Cosmos. "Our challenge," the UCL team said, "was to create a new model to match all the evidence." The scientists solved this "3D puzzle" and, in the process, they admitted the Antikythera computer was "a creation of genius—combining cycles from Babylonian astronomy, mathematics from Plato's academy and ancient Greek astronomical theories."

The Cosmos of the Antikythera computer had rings that carried the planets. The inscriptions revealed "little spheres," making the planets visible. The Sun god Helios was marked by a "little golden sphere." This is similar to the markings with gems of the planets and the Sun in the astronomical tablet of *The Alexander Romance* by Pseudo-Kallisthenes (chapter 6).

The UCL scientists decoded the twenty-three gears of fragment A. This brought the mechanisms of the gears in balance with the existing evidence brought to light by the inscriptions etched in the front bronze plate of the computer. Next, the effort to come up with a new model for the Cosmos of the front side of the Antikythera astronomic calculator received a boost by

[516] Tony Freeth, "Eclipse Prediction on the Ancient Greek Astronomical Calculating Machine Known as the Antikythera Mechanism," *PLOS ONE* 9(7), July 30, 2014.

[517] Tony Freeth et al., "A Model of the Cosmos in the ancient Greek Antikythera Mechanism," *Scientific Reports*, 11, 5821, March 12, 2021.

the 2016 discovery in the inscriptions of two numbers: 462 (ΥΞΒ) and 442 (ΥΜΒ). The first number represents a planetary cycle, the time of 462 years Venus takes to move back and forth among the stars and return to the same Earth spot in the night sky. Fragment D helped in tying 462 to the Venus cycle. Fragment D has sixty-three teeth.

The UCL researchers matched the evidence from both fragments, A and D, and succeeded in devising a mechanism that precisely worked with the 462-year cycle. The other number—442—mirrors the movement (cycle) of Saturn. These numbers are "factorizable," that is, they facilitate the mechanization of the movement of the planets through the gears. Thus, the mechanization, the gears, for tracking Venus and Saturn existed in the front side of the computer. No comparable evidence existed for the movement of the other planets: Mercury, Mars, and Jupiter. This probably happened because the section in the front plate that had those inscriptions was damaged or lost. The other heavenly body for which the Antikythera device had physical evidence for its movement was the Moon.

The inscription numbers 462 and 442 for the motion of Venus and Saturn among the stars came from Greek mathematics and the astronomical work of the early fifth century BCE philosopher Parmenides. I have already examined the work of this thinker in chapter 2. He believed the Earth was a sphere and the Moon received its light from the Sun. The UCL scientists used the math theory of Parmenides, his numbers for Venus and Saturn, to develop theoretical mechanisms, reverse-engineer gears with the appropriate number of teeth to replicate the heliocentric planetary motions described by the inscriptions.[518] In addition, they used modern computers in the extrapolation of the cycles of the other planets. They created the gears to realize the richness of the Cosmos embedded at the front plate of the Antikythera device.

The Wheel of the Sun god Helios (in fragment A) was the Cosmos, the focus of the theoretical reconstruction of the Antikythera computer by the UCL scientists. They chose the planetary cycles for accuracy, the fitting of most of the gears into tiny spaces measuring about 9.7 mm between the front and back plates, and the sharing of gears. This enabled them, they said, to build a model (gearing and display) that satisfied all existing physical evidence and, therefore, was as close to the original Greek computer as was humanly possible. Future research will judge that ambitious goal.

---

[518] Becky Ferreira, "Scientists Have Unlocked the Secrets of the Ancient 'Antikythera Mechanism,'" *Vice*, March 12, 2021.

They summarized their learning, including their admiration for the awe-inspiring capabilities of the Antikythera computer. Tony Freeth probably wrote these concluding remarks, which went beyond the computer's prediction of the eclipses of the Moon and the Sun and the phases of the Moon:

"Because of the loss of evidence, we cannot claim that our model is a replica of the original, but our solution to this convoluted 3D puzzle draws powerful support from the logic of our model and its exact match to the surviving evidence. The Antikythera Mechanism was a computational instrument for mathematical astronomy, incorporating cycles from Babylonian astronomy and the Greek flair for geometry .... It calculated ... the *heliacal risings and settings* of prominent stars and constellations, and the *Olympiad cycle*—an ancient Greek astronomical compendium of staggering ambition. It is the first known device that mechanized the predictions of scientific theories and it could have automated many of the calculations needed for its own design, the first steps to the mechanization of mathematics and science. Our work reveals the Antikythera Mechanism as a beautiful conception, translated by superb engineering into a device of genius. It challenges all our preconceptions about the technological capabilities of the ancient Greeks."

I could not put it better. Freeth is right saying that at the Cosmos of the front plate: "The Sun, Moon and planets are displayed in an impressive tour de force of ancient Greek brilliance."[519] Freeth and numerous other scientists admire the Antikythera computer. This admiration is a result of years of personal study—and after more than a century of investigations by their predecessors. The Freeth team admitted the device is an astronomical compendium of pure genius. It continues to astonish its observers. In fact, several years before Derek de Solla Price did his pioneering study of the Antikythera Mechanism and decades before Tony Freeth and his colleagues studied it in depth, it was a model for the brain of the April 12, 1961, *Vostok* Russian spacecraft that carried the first cosmonaut, Yuri Gagarin, into space. Kevin Knuth, physics professor at the University in Albany, said the *Vostok* was guided by a geared computer similar to the geared Antikythera computer.[520]

The next chapter compares the Greeks with their neighbors: Egyptians and Babylonians.

[519] University College London, Press Release: "Experts recreate a mechanical Cosmos for the world's first computer," March 12, 2021.

[520] Aristos Georgiou, "How Russia's Yuri Gagarin Became the First Man in Space Thanks to Steampunk Contraption," *Newsweek*, April 12, 2021.

Chapter 8
Seeing the World: Greeks and Non-Greeks

Greek philosophers and historians recognized the achievements of foreign science. The Greeks borrowed astronomical data and some applied mathematics from the Babylonians and the Egyptians. However, their approach to scientific thought was different. This chapter discusses the innovations that the Greeks contributed to the discipline of science, which superseded the accomplishments of other ancient cultures and influenced our modern world.

How did that happen? This chapter explains how the Greeks used reason to construct a Cosmos that included the Parthenon and advanced theoretical and applied science. The Parthenon is simply an example of Greeks at work. After the Parthenon, it was only a matter of time before the Greeks constructed a modern-like computer.

Did the Barbarians Invent Philosophy?

Diogenes Laertios, an Epicurean philosopher, complained that some non-Greeks and Christians in his time, the third century of our era, were slandering the Greeks with their suggestions that philosophy started with the barbarians. After all, the Persians gave birth to Magi; the Babylonians or Assyrians had their Chaldeans (astrologers); and the Indians had their Gymnosophists (naked wise men); while the Celts and Gauls had their Druids or Holy Men. The Egyptians even considered the Greek god Hephaistos to be Egyptian: being the son of Nile who came into being 48,863 years before Alexander the Great. It was Hephaistos, Egyptians say, who gave birth to philosophy.

Diogenes Laertios, however, dismisses these claims as baseless. He turns to the Greeks, rather than to the barbarians, for the discovery of philosophy. He says the pro-barbarian writers are in error to "attribute to the barbarians the achievements of the Greeks, with whom not merely philosophy but the human race came into being."[521] I am not qualified to judge this last claim. But one can excuse Laertios for possibly exaggerating the role of the Greeks. He was writing in the third century, clearly a time of stress and extreme danger for the Greeks. His writings were his defense. He was right about philosophy, however.

Greeks and Their Eastern Neighbors

Let's see what the Greeks thought about their Eastern neighbors. Thales of Miletos, one of the Seven Sages of Greece, visited Egypt and Babylonia and returned home with some knowledge of what the Egyptians and Babylonians had accomplished in geometry and astronomy. In fact, Eudemos of Rhodes, mid-fourth century–285 BCE, student of Aristotle and author of histories of astronomy and geometry, now lost, took this to mean that Thales brought geometry from Egypt to Greece.[522] Aristotle also thought the Egyptians invented geometry primarily because their priestly class had plenty of free time.[523] The historian Diodoros agrees with Aristotle. The Babylonian priests educated their sons at home, taking the time to drill knowledge and dogmas into their heads. In contrast, Greek pupils studied several subjects, though the need to earn a living ended their formal education after a few years.

According to Diodoros, "Very few [Greek students] really pursue advanced studies in order to prepare themselves for a profession. These students, however, are always attempting to offer new interpretations of the most important doctrines. They do not follow the ways of their predecessors. The barbarians [Babylonians, Egyptians] who always concentrate on the same subjects have sure knowledge of every detail of their specialty, while the Greeks, who hope to profit from their scientific expertise and who are always establishing new schools and quarreling with each other over the most important theorems and concepts, teach their pupils conflicting views and cause their minds to wander

[521] Diogenes Laertios, *Lives of Eminent Philosophers* 1.3.

[522] Proklos, *A Commentary on the First Book of Euclid's Elements* 65, tr. Glenn R. Morrow (Princeton, NJ: Princeton University Press, 1992).

[523] Aristotle, *Metaphysics* 981b23, tr. W. D. Ross in *The Basic Works of Aristotle*.

from one speculation to another. Their entire lives are spent in a state of intellectual confusion. They are unable to believe in anything at all with certainty. However, if a man were to investigate in earnest the most important schools of thought, he would discover that Greek scholars differ from one another considerably and that they hold opposing views regarding the most essential dogmas. The Chaldeans [Babylonians], however, say that the nature of the world is eternal; that the universe has neither a beginning nor an end: the universe will not suffer destruction at a later time. They also say that the order of all things and their arrangement has come into being by the gods. So, all things taking place in the heavens are not a result of chance but of some definite and firm judgment reached by the gods. And since they have made their observations of the stars over a long period of time and have known the movement and force of each of them more precisely than any other people, the Chaldeans have the ability to reveal to humanity many of the things that will occur in the future."[524]

The report of Diodoros is very interesting. We need to keep in mind, however, that he was writing at a time, the first century BCE, when the Greeks were living under Roman occupation. But, still, the first century BCE was not a bad time for Greek science. In fact, this was the time of Poseidonios and Geminos in Rhodes, one of the workshops for the Antikythera Mechanism. So, Diodoros' opinion was biased or misinformed. The first century BCE was part of the golden age of Greek science. Things also are rarely black or white. Like Herodotus, the great historian of the fifth century BCE, Diodoros gives too much credit to the Babylonians and other foreigners who possibly influenced the Greeks. Indeed, he makes the ludicrous claim that all that Greek thinkers, including Homer, did, and for which they achieved greatness in Greece, came from Egypt.[525]

Despite his bias, Diodoros is very useful. His universal history is valuable. Yes, the Babylonian priests had security and plenty of time to teach their children arithmetic for the study of the heavens. And, of course, the Greeks never lost a chance to raise questions, even opposing each other on scientific issues. But who says that priestly class societies and a settled dogma are fertile fields for science and innovation? Nevertheless, Babylonians, Egyptians, Chinese, and Indians accomplished a lot in the advancement of human culture. They prized mathematical knowledge primarily for solving practical problems related to

[524] Diodorus Siculus, *The Library of History* 2.29.4–30.2, tr. mine.

[525] Diodorus Siculus, *The Library of History* 1.96–98.

commerce, taxation, construction, government, and land measurement and ownership. Some of the intellectuals of these ancient civilizations extended mathematical knowledge to realms of thought approaching the speculative or theoretical.[526]

The Babylonians used arithmetic to sketch the revolution of the stars and the Greeks employed geometry for their models of the Cosmos. But sometimes the Greeks used the arithmetic calculations of the Babylonians in support of their celestial geometrical models. Greek and Babylonian methods of studying the heavens existed side by side down to the fourth century of our era.[527] That, however, does not make Greek and Babylonian astronomy of equal value in relation to science.

This was obvious a long time ago. Not all Greek intellectuals thought highly of the Egyptians and Babylonians. Plato, for example, had doubts about them. He knew the Egyptians well. He or one of his editors, Philip of Opus, said the Egyptians and Syrians observed and examined the stars for thousands of years, "an infinite time." Their astronomical knowledge spread everywhere, including Greece. There was reason for the antiquity of Egyptian astronomy: the Egyptians and Syrians had beautiful summers without clouds or rain; so, observing the heavens came naturally. In contrast, Greek summers are inferior to those of Egypt and Syria. Hence, Plato says, the Greeks observed the ordering of the star gods later than their neighbors. But this did not bother Plato because the Greeks improved whatever they received from foreigners. This was especially true in astronomy and even religion.[528]

Plato described the Egyptians and the Phoenicians more like businessmen, their love of money separating them from people like the Greeks who, for the most part, loved learning and, therefore, developed philosophy and science.[529]

The Egyptians and Babylonians were also very practical people. They studied the stars and developed a calendar to get clues on predicting the will of their gods. The Egyptians specifically wanted to know when the Nile would flood their land. And along with the Babylonians, the Egyptians developed geometrical knowledge for restoring property lines after flooding. The Egyptians also

[526] Richard G. Olson, *Technology and Science in Ancient Civilizations* (Santa Barbara, CA: ABC Clio, 2010) 52.

[527] Evans and Berggren in *Geminos's Introduction to the Phenomena*, 14–15. Evans and Berggren exaggerate the influence of the Babylonian "numerical parameters" on Greek astronomers.

[528] Plato, *Epinomis* 987a, 987d–988a.

[529] Plato, *Republic* 435e–436a.

used their science to build the tombs of their kings. According to American scholars writing in the late 1950s, Egyptian science "was devoid of speculation: their astronomy, geometry, and arithmetic were all directly related to practical matters of daily life." "The systematic study of the cosmos lay beyond their ken."[530]

In 1978, Arpad Szabo of the Mathematical Institute, Hungarian Academy of Sciences, suggested the Greeks might have borrowed some of their scientific knowledge from the Egyptians and Mesopotamians. But he saw a difference in the science of the Greeks and the science of Egyptians and Babylonians. He explained that difference as follows:

"The most substantial difference between Greek and Oriental sciences is that the former is an ingenious system of knowledge built up according to the method of logical deduction, whereas the latter is nothing more than a collection of instructions and rules of thumb, often accompanied by examples, having to do with how some particular mathematical tasks are to be carried out."[531]

In 1979, Charles Henry Edwards, an American mathematician, agreed with Szabo. He compared the Greeks and the Egyptians and Babylonians. He said the Greeks "assimilated" the Babylonian and Egyptian "lore of number and geometry." But the Greeks contributed to mathematics "the consciously logical and explicitly deductive approach that is now its distinguishing feature."[532]

In 1979, Geoffrey Lloyd, professor of ancient philosophy and science at Cambridge University, reached a similar conclusion about Greek science. However, like Szabo and Edwards, he felt obliged to set the record straight about Greek science in relation to other sciences in the Mediterranean. He was aware of the project of many academics to denigrate the originality of the Greeks, continuing the Christian slander that the Greeks plagiarized Egyptian and Mesopotamian achievements in science. He explained that Egypt and Babylonia also contributed to science. Yet there was a difference between the scientific contributions of the Egyptians and Babylonians, and that of the Greeks, especially in astronomy, mathematics, and medicine. Lloyd identified the distinctive Greek contributions to science.

---

[530] Wiener and Noland, eds., *Roots of Scientific Thought*, 5–6.

[531] Arpad Szabo, *The Beginnings of Greek Mathematics*, tr. from the German by A. M. Ungar (Boston: D. Reidel Publishing Company, 1978) 186.

[532] Charles Henry Edwards, Jr., *The Historical Development of the Calculus* (New York: Springer-Verlag, 1979) 5.

"The Greeks were certainly not the first to develop a complex mathematics —only the first to use, and then also to give a formal analysis of, a concept of rigorous mathematical demonstration. They were not the first to carry out careful observations in astronomy and medicine, only the first—eventually— to develop an explicit notion of empirical research and to debate its role in natural science. They were not the first to diagnose and treat some medical cases without reference to postulated divine or daemonic agencies, only the first to express a category of the 'magical' and to attempt to exclude it from medicine," Lloyd wrote.

In other words, the Greeks raised questions of fundamental importance to the nature of science. "Much as the Egyptians and Babylonians contributed to the content of ... [scientific] studies," Lloyd says, "the investigations only acquire self-conscious methodologies for the first time with the Greeks." With such contributions to science, it was natural Western science would be but a continuation of Greek science. Lloyd was certain of that. The Greeks invented science and started the debate on its aims and methods. This debate continues today, Lloyd wrote in 1979.[533]

The American mathematician George Simmons argues that what the Egyptians and Babylonians did in science was no more than "fragmentary rules of thumb." Thales, therefore, "did not find a science of geometry" in Egypt and Babylonia. According to Simmons: "The distinguishing feature of Greek geometry is precisely its abstract and theoretical character anchored in demonstrable truth, whereas what Thales found was an oriental mishmash of 'recipes'— partly right and partly wrong—for calculating various areas and volumes, and these areas and volumes were always conceived concretely, as the area of a piece of land or the volume of a grain storage bin. This older mathematics had been mindlessly transmitted from one generation of priests to the next from time immemorial; it was dead tradition, mostly computational procedures amounting to approximate formulas, without any logical connections or foundation, and with the underlying train of thoughts—if any—long forgotten."[534]

Another modern scholar, Benno Artmann, says the Greeks probably borrowed "some mathematical rules and procedures" from the Egyptians and

[533] Geoffrey Lloyd, *Magic, Reason and Experience: Studies in the Origin and Development of Greek Science* (London: Cambridge University Press, 1979) 232–233, 264, 267.

[534] George F. Simmons, *Calculus Gems: Brief Lives and Memorable Mathematics* (New York: McGraw-Hill, 1992) 3–8.

Mesopotamians. However, "nothing has been found in these [non-Greek] sources in the sense of mathematics as we know it today [in 1999] and that we encounter in Euclid's work. Mathematics was invented in [the] Greek classical period, beginning with Thales and Pythagoras about 530 B.C.E. and finding its final form with Euclid about 300 B.C.E. It stands on equal footing with the great Greek works in literature, sculpture, painting, architecture, historical and political writings, medicine, and philosophy."[535]

Simmons also had similar thoughts. He expressed enormous appreciation for the Greek achievement. "In all of history," he said, "nothing is so astonishing as the sudden appearance of intellectual civilization in Greece." And like others who praised the Greeks, Simmons brought the Egyptians, Babylonians, and Chinese in his reckoning, saying:

"No one has ever surpassed the Egyptians or Babylonians in monumental stone sculpture, or the Chinese in the production of incomparable works of art in bronze and ceramics... . But what the Greeks achieved in the realm of the intellect is greater still: they invented mathematics, science, and philosophy; they were the first to write genuine analytical history as opposed to mere descriptive annals; and they were the first to speculate freely about the nature of the world and the meaning of life, without being confined by the chains of any stultifying traditional orthodoxy."[536]

In chapter 3, I looked at how Plato advanced the cause of science. Indeed, he used mathematics for studying and understanding nature and the heavens. His model of the Cosmos had significant consequences in Greek thought. Plato warned that science was not just another body of tested knowledge for the benefit of all people. Science, and mathematics in particular, he said, would improve the intelligence and wellbeing of those learning for the sake of learning—people who are also lovers of freedom and free of the corruption of money. Failing that, according to Plato, one would expect rotten results: technically trained people scheming and favoring dishonesty, much like the Egyptians and Phoenicians and others who are caught in the selfishness of wealth.[537] Plato could have easily added Americans and Europeans. I raised that fact, of the scheming of scientists and their funders in modern times, in the Introduction.

[535] Artmann, *Euclid—The Creation of Mathematics*, 1–2.
[536] Simmons, *Calculus Gems*, 3.
[537] Plato, *The Laws* 747b–c.

So, when I speak of science, I am thinking of the great confidence and uneasiness of Plato; when I say the Greeks invented science, I am referring to what the Greeks considered their proudest achievement, all in the spirit of giving credit to the contributions of their neighbors.

Science among the Greeks

The Greeks developed science that was part of careful observation and experiment. Greek science had theory and testing of the truth. In other words, science was proved knowledge. And as I have already cited Aristotle, he would say science was all about the causes of things, including the pleasure of appreciating the awesome beauty of nature.

The distinguished British classical scholar, Hugh Lloyd-Jones, mirrors Plato and Aristotle when he says the Greeks invented science and the arts of civilization primarily because of their polytheistic religion, their gods being forces working in the real world and acting through nature.

Plato believed reason and the gods were running the heavens, the world being alive and in the image of the gods. In fact, in the vision of Plato, heaven was a god the Greeks called *Ouranos*. This god decorates himself with stars, which are also gods. In addition, *Ouranos* makes it possible for the stars to revolve in his space around their orbits. *Ouranos* also brings the seasons and nourishment to the Earth and all living beings, especially humans to whom *Ouranos* made the "gift of numbers," intelligence, and everything that is good. With the divine gift of numbers, men are perfectly capable to examine the heavens and all celestial and terrestrial phenomena. Men without numbers, Plato says, would not be intelligent in anything: they would be "like animals incapable of rational thought." Even courage and moderation would not help the person without knowledge of numbers because that person would not know how to put his perception and memory to good use. He would be a person without wisdom, which is the greatest part of all virtue. Such a person would also never be good or happy. Yet, according to Plato, *Ouranos* sowed humans with the capacity to count things and understand what they see:

"Of the things *Ouranos* shows us, taken one by one, what can we behold more beautiful than the day? Later, when we come to see the night, everything appears different to our vision. Since heaven never stops making these bodies ply their course night after night and day after day, he never stops teaching humans one and two, until even the slowest person learns well enough to count," Plato wrote.

Plato also assigned *Ouranos* the role of constructing the Moon and establishing the months in relation to the year. We should be grateful to the heavens, he said; we have crops, the Earth bears food for all living things, and the winds that blow and the rains that fall are beneficial to humans and the natural world.[538]

Plato had a special affection for the Earth, which he saw as our foster-mother, spinning around the axis of the Cosmos while being the "guardian and maker of day and night." In addition, Plato described the Earth as the "first and oldest of the gods."[539] For Plato, the star gods in the sky, the heavens, were something to behold, seeds of knowledge and understanding. He extolled the supreme value of seeing the heavens with our own eyes, the heavens becoming our first textbook to enlightenment and reasoning, making sense of numbers and the Cosmos. He was stunned by the beauty of the Earth.

"Had we never seen the stars," Plato said, "the Sun god Helios, and the heavens, none of what we say about the Cosmos would make much sense. In fact, we would have nothing to say about the Cosmos. However, the sight of day and night, and our understanding of months and returning years, of equinoxes and solstices, have created numbers, and have given us the conception of time, which opens the path to inquiry into the nature of the Cosmos. These pursuits have led us to philosophy, by far the greatest gift the gods have ever made or will give to mortals."[540]

Plato and the Greeks included science whenever they spoke about philosophy. That's why they used the gods and, especially, reason, in explaining the natural phenomena on Earth and the Cosmos. This was an ordered and just system, governed by causal laws. "Reason," Plato said, "is our king in both heaven and Earth."[541]

"Such a notion," says Lloyd-Jones, "was a prerequisite of rational speculation about cosmology, science and philosophy; this explains why it was the Greeks rather than their neighbours who began such speculation."[542]

---

[538] Plato, *Epinomis* 978c–979b.

[539] Plato, *The Laws* 897; and *Timaios* 27–31, 39–41.

[540] Plato, *Timaios* 47a–b.

[541] Plato, *Philebos* 28c.

[542] Hugh Lloyd-Jones, *Greek in a Cold Climate* (Savage, Maryland: Barnes and Noble Books, 1991) 223, 226–227.

Another modern scholar, Arnold Hermann, complements Lloyd-Jones when he says Greek thinkers advanced science because of the reformation they brought about on how they thought of the gods. This happened primarily in the sixth and fifth centuries BCE when a few early thinkers made the gods less anthropomorphic, giving "a more sublime characterization of the divine." This, according to Hermann, turned out to be "a most revolutionary idea." Divine qualities, when abstracted, became the first concepts or ideas, which, in turn, became Love, Strife, Being, Change, Mind, Justice, Good, and Necessity. The Greeks used these principles to explain the workings of the Cosmos without necessarily putting gods behind them. These ideas of inventive speculation became the "building blocks" of theoretical science or, as the Greeks called it, philosophia: love of wisdom.[543]

It's possible there's a grain of truth in this modern speculation, which springs from the early Greek thinkers. However, the evidence is almost non-existent that the Greeks followed on the path of a few philosophers about their gods, though most of those thinkers did not abandon the gods while making a difference in the evolution of science and philosophy in Greece.

The Greeks never separated Love from Aphrodite or other such assets from their gods. At the same time, they were attracted to the natural world by their own impulses. Plato and Aristotle reported that thinkers embarked on endless inquiries concerning the natural phenomena because of the wonder and curiosity they had about nature and the Cosmos.

It is also possible, looking at the gods more like gods rather than superior beings in almost human-like form, that the Greeks discovered an additional ladder to the secrets of the heavens and Earth. The divine attribute of Love of Aphrodite or Justice of Themis or Justice of Zeus became paradigms for a good society. In that process of methodical speculation, Greek thinkers explained the workings of the Cosmos, which, as I have said more than once, means an ordered and beautiful and just world.

It was this effort of getting closer to the truly divine character of their gods in nature and the universe that inspired them to get closer to science. Plato and Aristotle are right that the interest of the Greeks in the workings of the natural world and the heavens was an additional attraction to asking questions and building a body of knowledge significant to science. Plato reasoned that the

[543] Hermann, *The Illustrated To Think Like God*, 4.

gods loved especially those persons who love wisdom.[544] Aristotle put it more accurately. He said whoever exercises and cultivates his reason is in the best state of mind and beloved by the gods. If the gods take any interest in humans, as we assume they do, they are bound to be delighted when they come across humans who are virtuous and rely on reason. These philosophers are upright and noble. "They are beloved by the gods."[545]

This theological, natural, and philosophical reality enriches our understanding of science among the Greeks. The gods also helped the Greeks to be in charge of their lives. Greek religion, says Mary Lefkowitz, a classicist at Wellesley College, "is a religion for adults, and it offers responsibilities rather than rewards. Yet despite its realism, and its clear differences from any of the religions we are now used to, these same stories [of the gods] can still offer a reliable guide to life in our own time."[546]

Building a Cosmos within a Cosmos erected more than the Parthenon, the temple of the goddess Athena. Similar thinking and practice accelerated technological advancement in a variety of frontiers. The Greeks used chemical technology in constructing fire cauldrons, fire ships, and incendiary arrows since 430 BCE. By the eighth century, this chemical weapon, Greek fire, was used effectively in war.[547] Archimedes destroyed the Roman fleet with machines and solar weapons (chapter 4). The fragments of the Antikythera Mechanism, however, are by far the best proof the Greeks had advanced technology. The geared astronomical machine was a Cosmos within a Cosmos.

Another example of sophisticated Greek science is the Tower of the Winds in the ancient agora of Athens. The Tower of the Winds is about the birth of science and mathematics while showing the form of technology the Greeks developed and used in the second and first centuries BCE. Andronikos Kyrrhestes, an astronomer from Macedonia, built the Tower of the Winds sometime after the manufacture of the Antikythera Mechanism, that is, in the late second century BCE. The Greeks called the Tower of the Winds *Horologion* (Clock). Vitruvius described it as "an octagonal marble tower."[548] It is indeed a thirteen-meter-high octagonal planetarium with sundials and a wind vane with

[544] Plato, *Phaedo* 82b–c.

[545] Aristotle, *Nicomachean Ethics* 10.1179a24–30.

[546] Mary Lefkowitz, *Greek Gods, Human Lives: What We Can Learn from Myths* (New Haven: Yale University Press, 2003) 239.

[547] Partington, *A History of Greek Fire and Gunpowder*, 28.

[548] Vitruvius, *On Architecture* 1.6.4.

a clepsydra or water clock, which was powered by a stream of water pouring into a semi-circular reservoir attached to the south face of the tower. There is a frieze of the eight wind gods at the top of the tower: Boreas (north); Kaikias (northeast); Euros (east); Apeliotes (southeast); Notos (south); Livas (southwest); Zephyros (west); and Skiron (northwest). There were nine sundials under the frieze. The Tower of the Winds was "a tour de force of timekeeping instruments."[549]

Price's 1974 report, *Gears from the Greeks*, made clear that Western scholars must reeducate themselves on the Greeks: bring their view on the Greek past up to date, harmonizing it in accord with the findings from the study of the Antikythera Mechanism. The blunt language of Price and his insistence the Greeks had developed advanced technology did not sit well with the scientific community. Fortunately, there are some exceptions.

In the early twentieth century, the British scholar Richard Livingstone explained Greek science as a product of truthfulness among the Greeks. He said this was not because the Greeks told fewer lies than other people. No, the reason was different. According to Livingstone:

"[The Greeks] had the desire and the power to see the world as it is. By this essential quality they gave Europe the conception of philosophy and science. These we inherit from them alone. Palestine and our German ancestors neither created them, nor show any signs of the temper that creates them, and Rome received her share from Greece... . The Greek Truthfulness is spontaneous, natural, and effortless—the native quality of the artist, who sees, and forgets himself in the vision... . Little thinking is needed to see that the direct, detached, objective temper is the generative principle of the Greek achievement, for it is the parent of science and philosophy, which are the children of a desire to see things in themselves as they are, and not as the seer might wish them to be."[550]

Writing at about the same time as Livingstone, D'Arcy Thompson, one of the best translators of the zoological works of Aristotle, gave us his view of Greek science as follows:

"When we consider the science of the Greeks, and come quickly to love it and slowly to see how great it was, we likewise see that it was restricted as

---

549 Robert Hannah, "Timekeeping," in *The Oxford Handbook of Engineering and Technology in the Classical World*, 753.

550 Richard Livingstone, "Literature," in *The Legacy of Greece*, 267–274.

compared with our own, curiously partial or particular in its limitations. The practical and 'useful' sciences of chemistry, mechanics, and engineering, which in our modern world crowd the others to the wall, are absent altogether, or so concealed that we forget and pass them by. Mathematics is enthroned high over all, as ... she should be; and of uncontested right she occupies her throne century after century from Pythagoras to Proclus, from the scattered schools of early Hellenistic civilization to the rise and fall of the great Alexandrian University. Near beside her sits, from of old, the daughter-science of Astronomy; and these twain [two] were worshipped by the greatest scientific intellects of the Greeks. But though we do not hear of them nor read of them, we must not suppose for a moment that the practical or technical sciences were lacking in so rich and complex a civilization."[551]

Certainly, Thompson was right, the Antikythera Mechanism being the most spectacular evidence of the reign of the practical or technical sciences in the Greek world. But for the Greeks the practical and theoretical did not occupy separate departments. Remember the Greeks called actors as well as metallurgists *technitai* or craftsmen. The important thing was the employment of reason in deciphering the heavens above, the natural world on Earth and, just as complicated, the nature of man. The Greeks chose science to gain understanding of both the heavens and Earth.

At about the time Thompson was translating Aristotle, another British classical scholar, John Burnet, outraged many of his academic class with his praise of the Greeks, especially their propensity to bring science into being. In the "Preface" of the third edition of his *Early Greek Philosophy*, he explained the purpose of his book was to spread the good news of the Greek beginnings of Western science. He put it this way: "My aim has been to show that a new thing came into the world with the early Ionian [Greek] teachers—the thing we call science—and that they first pointed the way which Europe has followed ever since, so that ... it is an adequate description of science to say that it is 'thinking about the world in the Greek way.' That is why science has never existed except among peoples who have come under the influence of Greece."[552]

About thirty years later, in the early 1950s, the German scholar Bruno Snell gave credit to the Greeks for inventing scientific thought, indeed tracing science out of the Greek language. He explained how that happened:

[551] Thompson, "Natural Science," in *The Legacy of Greece*, 140–141.

[552] John Burnet, *Early Greek Philosophy*, fourth edition (London: Adam and Charles Black, 1930) v.

"Greek is the only language which allows us to trace the true relation between speech and the rise of science; for in no other tongue did the concepts of science grow straight from the body of the language. In Greece, and only in Greece, did theoretic thought emerge without outside influence, and nowhere else was there an autochthonous formation of scientific terms. All other languages are derivative; they have borrowed or translated or got their terms by some other devious route from the Greeks. And it was only with the help of the unique achievement of the Greeks that the other societies were able to progress beyond their own pace of conceptual development."[553]

The American scholars Philip Wiener and Aaron Noland, also writing in the 1950s, were convinced the Greek achievement was "without precedent in human history." They admired the Greeks for being respectful but unafraid of their gods, seeking to know the why and how of the working of the laws of nature. "Thus, it was that the 'fluid' mythology of the proud and self-reliant Greeks," they wrote, "encouraged the sort of controlled and rigorous speculation and incisive inquiry that heralded the great classical heritage not only in the arts and philosophy and political theory but in the theoretical sciences."[554]

I already cited Dijksterhuis about the Greek beginnings of Western thought. He was also the author of a scholarly book on Archimedes.[555] He explained the connections of modern science to Greek science this way: "Every inquiry about the sources of present-day knowledge is bound to lead ultimately to Hellas; it does so immediately where the foundation of mathematics and natural science are concerned, and in the latter case it will undoubtedly come up against Aristotle, who like no other Hellene, perhaps like no other scholar of any age, dominated the evolution of scientific thought."[556]

In the late twentieth century, George Simmons reached a similar conclusion: "The Greeks were a self-confident people with scarcely a trace of false modesty; and if they themselves could have surveyed the history of the world over the past 2500 years, up to our own time [1990s], they would certainly

---

[553] Bruno Snell, *The Discovery of the Mind: The Greek Origins of European Thought*, tr. T. G. Rosenmeyer (Oxford: Blackwell, 1953) 227.

[554] Wiener and Noland, ed., *Roots of Scientific Thought*, 7.

[555] *Archimedes*, tr. C. Dikshoorn, first published in 1956 (Princeton, NJ: Princeton University Press, 1987).

[556] E. J. Dijksterhuis, "The Origins of Classical Mechanics from Aristotle to Newton" in *Critical Problems in the History of Science*, ed. Marshall Clagett (Madison, WI: University of Wisconsin Press, 1969) 164.

have had no hesitation in saying that other cultures emerged from barbarism in direct proportion to the degree to which they absorbed this [Greek] civilization."[557]

The French scholar, P. H. Michel, and the Jewish physicist, Samuel Sambursky, also agreed with Burnet. In mid-1950s, Sambursky defended an "undeniable historical fact": the roots of modern science are Greek because "the basic principles of the scientific approach, which are still as valid as ever to-day, were discovered in Ancient Greece."[558] Sambursky connected the first natural philosophers of ancient Greece to the science of his day. Those philosophers also inspired and complemented the craftsmen, architects, and engineers who built the technological infrastructure and the machines of the Greeks, including the Antikythera Mechanism. But Sambursky, like so many modern scientists, dismissed the technical dimensions and nature of Greek culture. He somehow thought the Greeks did not embark on any "technical undertaking and the bold practical application of science." He also had the illusion—common among Western scientists—that the Greeks did not experiment and made no "technical invention."

Despite these philosophical and scientific blind spots, which distort the achievement of the Greeks, Sambursky was right in his overall scientific understanding of Greek science. His insight on the link between ancient Greek and modern science borders on the lyrical. He put it this way:

"That those ... [Greek] philosophers were among the spiritual ancestors of our own era will not be doubted by anyone who compares the heritage of Greek science—its methodical approach, the vigour of its imagination and inspiration, its associative strength and power of inference—with the science of our own time .... The [Greeks'] intuitive grasp of the atomic theory is such as to arouse our astonishment no less than their mathematical devices for explaining the movements of the heavenly bodies. The same inspiration is the source of their clearsighted vision in the framing of mechanistic cosmogonies at a time when the machine was still unknown, as also their penetrating analysis of various epistemological problems, such as the relation between the human senses and the human mind. Whoever makes a close study of the scientific

---

[557] Simmons, *Calculus Gems*, 11.

[558] S. Sambursky, *The Physical World of the Greeks*, tr. from the Hebrew by Merton Dagut (New York: Macmillan, 1956) v, 2–3.

world of Ancient Greece cannot but be filled with veneration and his veneration will but increase, the more he realizes that beyond all differences and changes, the cosmos of the Greeks is still the rock from which our own cosmos has been hewn."[559]

Michel also wrote in the late 1950s that science, as we know it "is undoubtedly a creation of the Greek genius."[560] The Dutch mathematician and scholar, Dijksterhuis, the Italian mathematician and historian of science, Russo, and the British classical scholar, Lloyd-Jones, all of whom I cited already, also considered science originating with the Greeks. Lewis Wolpert, professor of biology and applied medicine at the University College, London, also concluded that science has had "its origins in Greece."[561]

In the late 1950s, Bertrand Russell, British philosopher and Nobel Prize winner for literature, judged the Greeks this way: "Philosophy and science, as we now know them," he said, "are Greek inventions. The rise of Greek civilization, which produced this outburst of intellectual activity, is one of the most spectacular events in history. Nothing like it has ever occurred before or since. Within the short space of two centuries, the Greeks poured forth in art, literature, science and philosophy, an astonishing stream of masterpieces which have set the general standards for Western civilization."[562]

If you doubt Bertrand Russell, visit the National Archaeological Museum in Athens. For me, entering this magnificent Temple of the Muses is like entering a time machine. Suddenly, I find myself surveying the surviving masterpieces of Greek art. I am not an art expert, but that has little to do with my appreciation of the good and the beautiful and my feelings, now at their peak of excitement and passion. Anywhere I look, I see statues, amphorae, and jewelry whose history I want to know, and through history, know them more intimately. Who made them, and when, and for what purpose? Why don't we make things as beautiful as these Greek artifacts? Take a look, for instance, at the bronze head of an athlete from Delos dated early first century BCE—not far behind the manufacture of the Antikythera computer. What does it tell

[559] Sambursky, *The Physical World of the Greeks*, 229, 242, 244.

[560] P. H. Michel, "Greek Science" in *History of Science: Ancient and Medieval Science from the Beginnings to 1450*, ed. Rene Taton, tr. A. J. Pomerans (New York: Basic Books, 1963) 181.

[561] Lewis Wolpert in *Science and Mathematics in Ancient Greek Culture*, ed. C. J. Tuplin and T. E. Rihll (New York: Oxford University Press, 2002) xii.

[562] Bertrand Russell, *Wisdom of the West* (London: Crescent Books, 1959) 10.

you? I see the struggle and agony of this boxer in his face. This face also reveals "the brilliant casting techniques of the ancients."[563] These casting techniques were so perfect that, to this day, one sees a drop of sweat exactly below his nose shining on his upper lip. The sculptor transformed human anguish, athletic competition, and beauty into bronze that still talks to us. This is superb craftsmanship and art.

Or what would you say about the golden head of Artemis? I am usually stunned in front of such proof of craftsmanship and beauty, nay, perfection. Artemis is in a reflective mood, ready to talk to you and me, make us think about us, our societies, and the Cosmos. I go away from the museum full of pleasure and thoughts, determined to come back. Or take a close look at early fifth-century BCE Kore statues in the Acropolis Museum.

Young Athenian woman (Kore with almond eyes). *ACR 674* About 500 BCE. Courtesy Acropolis Museum.

[563] John Boardman, *Greek Art*, fifth edition (London: Thames and Hudson, 2016) 18.

These Korai are young Athenian women. One of these Korai has thick interwoven hair, which are like the rays of Helios surrounding a gorgeous Aphrodite-like face shining with beauty and goodness. The beauty of craftsmanship is not limited to female figures. Look at Moschophoros, a young shepherd carrying a calf dedicated to Athena on the Acropolis of Athens.

Young shepherd carrying a calf (Moschophoros). ACR 624. About 570 BCE.
Photo: Nikos Daniilidis. Courtesy Acropolis Museum.

Then three centuries later in Alexandria, Egypt, there's Berenice II, sister-wife of Ptolemy III of Egypt. I see outstanding coin craftsmanship; the solemnity and natural beauty of Berenice join in a lasting icon of the beautiful and good.

Queen Berenice. Courtesy NM Collection Demetriou, Numismatic Museum, Athens.

The emphasis on beauty is not accidental. To the Greeks, it was almost everything. And in the case of marble and bronze statues, or in the case of humble clay for the creation of myriads of very practical vases, stamping beauty was a mark of more than affection for the beautiful. The surviving pots, including fragments of vases, write their own history of Hellas.

Potters created vases for the holding and pouring of water, wine, and olive oil. The most useful vases included amphora for wine; skyphos for drinking water and wine; hydria for carrying water; krater for mixing water and wine, the drinking wine of the Greeks; kylix was another drinking cup, more delicate than skyphos; kantharos was still another drinking cup with two large upright handles for lifting a bowl; lekythos was a tall cylindrical vase for olive oil; pithoi were large storage jars. These vases were more than practical: they were paradigmatic of outstanding craftsmanship and iconic beauty. They were exported all over the Mediterranean, especially France, Spain, Germany, Italy, Corsica, South Russia, Asia Minor, Mesopotamia, and Sudan. The pottery of Greece, says Joseph Veach Noble, a great American expert on Greek ceramic art, "stands as a witness to an era of perfection in the union of technique and beauty."[564] In addition, these vases document the history of science and technology.

George Simmons, writing in the early 1990s, focused on Greek science. He says Thales "originated abstract geometry as the study of the properties of idealized geometric figures" as well as "he created the concepts of theorem and proof." This kind of inventiveness helped Thales appreciate criticism. He and the other Ionian philosophers insisted that truth demanded criticism. Simmons wisely extrapolates this critical turn of mind of Thales as an illustration of skepticism and proof.

"[A]t the heart of the Greek way of doing philosophy," Simmons wrote, "which is unique in the history of thought; it led to the concept of mathematical proof—which arose nowhere else—because skepticism and demonstrative proof are opposite sides of the same coin; and this notion of proof is the essence of genuine mathematics, which gave rise in turn to its companion, mathematical science, both of which also appeared nowhere else. This is the way intellectual civilization was born in ancient Greece."[565]

True, the Greeks applied mathematics to make sense of their observations of nature, inventing the scientific method. They then created the scientific disciplines we have today. For example, mathematical physics is forever connected with Archimedes; the heliocentric theory with Aristarchos of Samos; zoology with Aristotle; the atomic theory with Democritus; philosophy with Plato and Aristotle; medicine with Hippocrates and Galen; history with Herodotus and Thucydides; and astronomy with Eudoxos, Aristarchos of Samos, Hipparchos, and Ptolemaios.

In the late 1990s, a group of French, Italian, British, and American scholars wrote a guide to classical knowledge, which they called "Greek Thought." In 2000, this book was translated from the French and published in English.

[564] Joseph Veach Noble, *The Techniques of Painted Attic Pottery* (New York: Metropolitan Museum of Art, 1965) xi.

[565] Simmons, *Calculus Gems*, 9–11.

In their introduction, the two editors, Jacques Brunschwig and Geoffrey Lloyd, one French and the other British, wrote about the "unparalleled originality" of the Greeks. The Greek past, they say, is also the past of the Western people; our past. The Greeks nourished us with their countless texts. With the Greeks, they say, "We are on home ground in a distant land; we are traveling without leaving our own room. All our thinking, in one way or another, passes through reflection on the Greeks."[566]

Even the history of science came out of Greek science. According to the Russian classical scholar Leonid Zhmud: "The history of science was born as the history of ancient [Greek] science. It is the theories and discoveries of Greek scientists that provide the material on which the methods of the history of science were worked out over the centuries."[567]

The American historian of science, Frank Egerton, would agree. In 2012, he summed up the Greek contribution to civilization. The Greeks, he said, "invented natural philosophy during the 500s BC and theoretical and practical science during the 300s BC. Greek zoology and botany included substantial natural history information about animals and plants in relation to their surroundings. That information was highly prized by later civilizations." In a sense, Egerton wrote, the Greeks gifted us with "the critical mind—perhaps the first giant intellectual step for mankind."[568]

Western Resistance to Greek Scientific Achievements

Yet, in actual practice, these noble sentiments and correct evaluation of Greek contributions to science and Western civilization find not much fertile ground in Western science. Western scientists spent more than the twentieth century studying the Antikythera Mechanism. There was resistance to, and ridicule of, the idea that the Greeks had developed advanced technology some 2,200 years ago.

However, the gradual realization—that the Greeks were almost modern in their technologies—caught scholars off guard because they did not believe

[566] Jacques Brunschwig and Geoffrey E. R. Lloyd, eds., *Greek Thought*, ix.

[567] Leonid Zhmud, *The Origin of the History of Science in Classical Antiquity*, tr. from the Russian by Alexander Chernoglazov (Berlin: Walter de Gruyter, 2006) 1.

[568] Frank N. Egerton, *Roots of Ecology: Antiquity to Haeckel* (Berkeley: University of California Press, 2012) xii, 1.

that was possible. Remember the discussion in the Introduction: Westerners are brought up to distrust the Greeks. The monotheistic religion and education of Westerners are obstacles to critical thinking. Since the establishment of Christianity as the state religion of the Roman Empire in the fourth century, Westerners have felt unease with "pagan" or "Christian" Greeks.

Friedrich Nietzsche, 1844–1900, the greatest philosopher of modern times, captured the feelings of modern people towards the Greeks. He wrote:

"Almost every age and stage of culture has attempted at some point to free itself, with deep feelings of anger, from the Greeks, because, in comparison with them, all one's own achievements, although apparently completely original and quite sincerely admired, suddenly seemed to lose colour and life and to shrivel into an unsuccessful copy or even a caricature. Thus people feel shame and fear in the face of the Greeks—unless there be one individual who reveres truth above all else and is therefore able to admit even this truth to himself: that the Greeks are chariot-drivers who hold the reins of our culture, and every other culture in their hands, yet the chariot and the horses are almost always made of too-puny stuff and unequal to the glory of their drivers, who then regard it as a joke to drive a vehicle into the abyss—and then jump across it with the leap of Achilles."[569] There's a sculpture of a bronze horse and a boy rider/ charioteer at the National Archaeological Museum in Athens. The boy and the horse speak the language of Friedrich Nietzsche. These masterpieces of craftsmanship are dated around 140 BCE, the probable time when the Antikythera computer came into being. They were found in the waters near Euboea.

Nietzsche spoke truth about the Greeks in 1872. He loved them. His honesty speaks to us as well. Westerners, especially scientists, continue to resent the Greeks. When it comes to the "seminal reestablishment of the experimental method," for example, modern scientists, according to Russo, deny "any indebtedness to ancient [Greek] science."[570] Did the ancient Greeks have wooden gears, but mechanical gears in gear trains carrying motion from one drive shaft to another? Yes, the Antikythera Mechanism worked with bronze gear trains, which did not go down well with scholars steeped in deep prejudices against the Greeks. The majority of the technical men and women of the West, including most scholars on Greek culture, ignored the wisdom of a few

[569] Friedrich Nietzsche, *The Birth of Tragedy and Other Writings*, ed. Raymond Geuss and Ronald Speirs, tr. Ronald Speirs (Cambridge: Cambridge University Press, 1999) 72.

[570] Russo, *The Forgotten Revolution*, 351.

scholars who highlighted the Greek beginnings of Western science. They kept the ancient Greek scientists in a theoretical purgatory. This was a question of jealousy and identity. They wanted to believe that fire and gears belonged to them alone. They were under the illusion that modern scientists since the sixteenth century invented science and technology. They resent anyone reminding them that the Greeks invented "modern" science and "advanced" technology more than two millennia ago.

For example, Archimedes is still an ambiguous historical personality. Yes, several scholars in 2010 called him a genius (chapter 4). But this was not the case only a few decades ago. Despite the fundamental role of Archimedes in the making of Greek and modern science, he is mostly remembered as a kind of mythical scientist running naked in the streets of Syracuse shouting "Eureka! Eureka!" ("I found it! I found it!") Westerners also remember him as a scientist who was so absentminded that while the war was raging around him, he was playing in a sandbox, drawing geometrical forms. Along with a few Greek scientists of the golden age of Greek science, Archimedes exists in the clouds. Russo argues that the West simply erased the Alexandrian or "Hellenistic" civilization from its "collective historical conscience, not unlike the phenomenon of repressed memories. Our culture, though built on the twin foundations of history and science, resorts to various expedients to hide from itself the historical importance of the birth of science." As for Archimedes, says Russo, certainly, he is remembered "but as a legendary character, outside of history."[571]

This is the metaphysical icon, the reasons, why Price—in the 1970s—had such a hard time carrying out a conversation with his colleagues about the Antikythera Mechanism. He was a messenger pointing to the shortcomings of how the West treats its intellectual ancestors, the Greeks, at a time the West did not want to hear about them. Or, when the West heard about the Greeks, the messages were not straightforward, entirely objective, or sincere. In 1974, Price published his "Gears from the Greeks." In 1974, Americans and British blessed the Turkish invasion of the Greek island of Cyprus. Certainly, in 1974, Western scientists did not wish to talk about the Greeks.

But even before 1974, the atmosphere was cloudy. In 1966, the German scholar H. Blummer reported that, in reading Greek literature, one gets the impression the Greeks "depreciated" hand labor and handicrafts. This was an

[571] Russo, *The Forgotten Revolution*, 6.

obvious reference to the smear perpetuated by Ploutarchos about Archimedes: the man was a great mathematician but had contempt for manual laborers and their crafts. Nevertheless, Blummer was honest, admitting that, no matter who made Greek handicrafts, they were beautiful. He praised "the perfection of the technical arts of Greece." He explained: "[Greek] trades attained a high degree of perfection in antiquity ... there are numerous productions of ancient handicraft which even our modern trades cannot rival ... with the exception of such trades as bakers, butchers, or fullers, Greek handicraft in almost every branch developed into art."[572]

However, not many people read Blummer. In the real world, fantasies and propaganda have been obscuring and denigrating Greek achievements.

In 1968, the Swiss writer Erich von Daniken published *Chariots of the Gods*, in which he spread the falsehood that aliens had sowed the Earth with advanced technology. These aliens, he says, were "gods," "extraterrestrial intelligences" or "astronauts" who "visited the primeval Earth in manned spaceships." Daniken's book sold millions of copies. But even more damaging than Daniken's science fiction has been the "scholarship" of several writers on ancient technology. These "scholars" willfully denigrate the technological marvels of the ancient world. Like Daniken, they are suggesting that "ancient astronauts" discovered those technologies rather than Greeks, Chinese, Romans, or other ancient people.[573] The implication of that nonsense being, Price noted with disgust, that the Greek device from the waters of Antikythera was taken to be the work of "alien astronauts coming from outer space and visiting our civilization."[574]

Such dangerous attitudes mirror hubris. It's like the propaganda one hears almost daily from US Government officials that, despite the grave political, economic, militaristic, and ecological ills of the United States, America has the best economic system, the best food, the best environmental protection, and the best of everything. In the same blatant fashion, scholars and scientists preach that our time, the twentieth and twenty-first centuries—an epoch of mass murder, nuclear bombs, massive contamination of food and drinking water, global ecocide, extinction of species, and global warming—is the "best" and most "advanced" the world has ever seen. As for science, the mere suggestion that it was born and developed substantially in ancient Greece leaves most modern scientists cold.

---

[572] H. Blummer, *The Home Life of the Ancient Greeks*, tr. from the German by Alice Zimmern (New York: Cooper Square Publishers, 1966) 500.

[573] Peter James and Nick Thorpe, *Ancient Inventions* (New York: Ballantine Books, 1994) XVI.

[574] Price, "Gears from the Greeks," 12.

For example, in 1980 or 1981, the American physicist Richard Feynman saw the fragments of the Antikythera Mechanism in the National Archaeological Museum in Athens. The modern-like device astonished Feynman to the point of rejecting its authenticity. In a letter he recounts his experience in the museum: he thought his daughter would love the bronze statues of horses but, he said, bronze horses bored him because he had seen "so much of that stuff before." Yet coming across the Antikythera computer was a real surprise, which he described like this:

"[A]mong all those art objects [in the Greek National Archaeological Museum in Athens] there was one thing so entirely different and strange that it is nearly impossible. It was recovered from the sea in 1900 and is some kind of machine with gear trains, very much like the inside of a modern wind-up alarm clock. The teeth are very regular and many wheels are fitted closely together. There are graduated circles and Greek inscriptions. I wonder if it is some kind of fake."

This means that, in the mind of Feynman, the Greeks could not have invented science; much less create and possess advanced technology. He also showed his ignorance by suggesting that Greek philosophy was shallow and that the Renaissance, made possible by the recovery of Greek knowledge, "was the freeing of man from the intimidation of the ancients."[575]

Other scientists and engineers had an incomplete education. They simply connect gears to the Renaissance. In 2012, Mathias Buttet, research and development and production director at the Hublot Manufacture, a premier Swiss watchmaker, admitted that, when he was in school, he was a "model student and a qualified engineer." Despite that, he said, he had accepted that "gear train theories had been devised during the Italian Renaissance." The Antikythera computer, he said, made him "question all that." He was impressed by the "level of knowledge shown by our predecessors—the engineers and astronomers of [Greek] Antiquity" who manufactured the Antikythera Mechanism. Buttet then did everything possible to "recreate this Mechanism in miniature without modifying any of its functions."[576] Indeed, the watch Hublot built on the basis of the design and engineering of the Antikythera computer looks beautiful and, to me, seems to be a faithful copy of the accepted modern representation of the Greek original Antikythera Mechanism. This work was much

[575] Richard P. Feynman, *What Do You Care What Other People Think?* (New York: W. W. Norton, 1988) 94–95.

[576] Mathias Buttet, "Hublot" in Nikolaos Kaltsas et al., eds., *The Antikythera Shipwreck: the ship, the treasures, the mechanism* (Athens: National Archaeological Museum and Kapon Editions, 2012) 3.

more than earning money in producing a fancy watch. It was a result of passion and respect for Greek antiquity. Buttet proudly announced, "The Antikythera computer was part of our roots and part of the history of watchmaking. It was also a masterpiece of our Greek ancestors. It came to light once again in a fantastic way to remind us we are their descendants."[577]

Despite this example of genuine good will and self-interest, there are those attacking the Greeks because the Greeks had slaves. Slavery legitimizes putting down the Greeks. Benjamin Farrington, a British scholar who published a book in 1944 on "Greek Science," argued that the supremacy of slavery in the ancient world did science in. He conveniently forgot the supremacy of slavery in Europe and America. He verbalized his confused thoughts like this: "[A]ncient society had set in a mould which precluded the possibility of an effective search for power other than the muscles of slaves.... Graeco-Roman science was good seed, but could not grow on the stony ground of ancient slave society."[578]

Imagine calling "stony ground" the land that gave birth to Eudoxos, Aristotle, Euclid, Archimedes, Apollonios of Perga, Ktesibios, Philon of Byzantium, Hipparchos, Poseidonios, Geminos, Heron of Alexandria, Galen, and Ptolemaios? This deeply flawed argument, indeed slander of the Greeks that gave birth to science, has been causing damage to classical studies and, more important than that, to our appreciation of Greek culture.

Geymonat praises the originality of Archimedes but subscribes to this biased opinion that ancient Greeks and Romans did not industrialize because they had the "natural machine" of slavery.[579] It's the same thing with the Canadian classical scholar Stephen Bertman. In 2010, he argued that slavery ("a ready supply of manpower") prevented the Greeks ("ancient scientists") from developing an industrial revolution. Yet, Bertman credits the Greeks with inventing science. He repeats Leonardo da Vinci's acknowledgment that he drunk from the well of Greek learning. Indeed, he says that scientists in the twenty-first century ought to do the same. They ought to continue doing science like the Greeks: following their uncompromising model, leaving almost nothing off-limits. He rightly praises their "scientific attitude" as "the most

[577] Mathias Buttet, "An Innovation 2,100 Years Old," in Stekas, ed., *The Antikythera Mechanism* 176.

[578] Benjamin Farrington, *Greek Science* (Nottingham: Spokesman, 1980) 304–308.

[579] Mario Geymonat, *The Great Archimedes*, tr. R. Alden Smith (Waco, TX: Baylor University Press, 2010) 8–9.

visible contribution of ancient Greek civilization to the modern world, a contribution that stands as its most influential living legacy."[580]

Like Bertman, and equally enthusiastic about the Greeks' affection for the new, Armand D'Angour, a British classical scholar, fell into the trap of Greek slavery. He is also under the delusion that only the modern era is a prosperous ground for useful gadgets. "The development of gadgets and devices that might be considered useful," he wrote, "belongs to a post-industrial mindset, and seems to have little purchase on the Greek imagination." D'Angour focused his study on classical Greece so he ignored the scientific revolution of the Alexandrian Era, though he mentioned in passing the Antikythera computer. Nevertheless, slavery saved the phenomena for D'Angour. "The widespread and largely unquestioned use of slaves for physical, mental and mechanical tasks [among the Greeks]," he said in 2011, "will in any case have removed much of the incentive to practical inventiveness in commercial and technological arenas."[581]

In contrast to Geymonat, Bertman and D'Angour, who erred on slavery but, otherwise, they, too, drunk from the well of Greek learning, there are scholars who do not take the Greeks seriously, or fail to understand them. One prominent example of such behavior was Gregory Vlastos, who taught Greek philosophy at Princeton University. In 1975, he criticized the early Greek thinkers, as well as Plato and Aristotle, for failing "to grasp the essential genius of the scientific method."

I was astonished by such a harsh judgment by a fellow Greek who, as a philosopher, ought to have been better informed. Yet he gives the impression he knows the Greeks. His books mirror a man who spent his life studying Plato. He faced Plato, however, with his head full of Christianity, never friendly to Plato or any other Greek philosopher or scientist. In addition, Vlastos embraced the Greeks with the cold hands of an early Christian convert. He abandoned Orthodox Christianity for Protestantism or he was converted to Protestantism, his ticket for America and education. He came to formal academic philosophy through the road of a preacher, which he learned at the theology school of the University of Chicago. Only this Christian background explains why Vlastos

---

[580] Stephen Bertman, *The Genesis of Science: The Story of Greek Imagination* (Amherst, NY: Prometheus Books, 2010) 240.

[581] Armand D'Angour, *The Greeks and the New: Novelty in Ancient Greek Imagination and Experience* (Cambridge: Cambridge University Press, 2011) 181.

appeared to be contradictory about the Greeks, being an extremely sophisticated Christian apologist. Second, what gave him the expertise in science so he could talk authoritatively about the scientific method, much less the "genius" of the scientific method? All he became in America was a Protestant preacher and a "philosopher," which in the America of the 1930s meant a pedant dissecting Greek philosophical texts outside of their Hellenic philosophical and historical context.

But having insulted the best of the Greeks with his meaningless statement, that, supposedly, they had missed the boat of the scientific method, he immediately rushed to diminish the damage he had already done by crediting the Greeks with the discovery of something else "which may still be reckoned one of the triumphs of the rational imagination: the conception of the Cosmos that is *presupposed* by the idea of natural science and by its practice."[582]

Following the intentional ambiguity of Vlastos, an American historian, J. Donald Hughes, made a fool of himself in 1994 by denying that "innovation" went anywhere in Greece or Rome.[583]

It is this "scholarly" ignorance and denigration of the Greeks that nearly dismissed Price's work. The Antikythera Mechanism sunk into oblivion for the second time.[584] But, as we have seen, not for long. Price made it irresistible.

Translating Greek Texts, Adopting the Grammar of Heavens

However, in the extremely long and murky history of the survival of Greek thought, the decipherment of the Antikythera Mechanism—discussed in chapters 5 to 7—was a victory against a ceaseless obscurantism trying to keep the Greeks buried in palimpsests, remnants of their scientific technology hiding under the sea.

For that reason, we ought to be thankful to the medieval Greeks for copying the works of their ancient ancestors like Archimedes. We should also be grateful to the Nestorian Christians for translating Greek philosophical and scientific texts into Arabic during the ninth and tenth centuries. During that

[582] Gregory Vlastos, *Plato's Universe* (Parmenides Publishing, 2005) xxiv.

[583] J. Donald Hughes, *Pan's Travail* (Baltimore: The Johns Hopkins University Press, 1994) 137.

[584] Rob S. Rice, "The Antikythera Mechanism: Physical and Intellectual Salvage from the 1st Century B.C." (Unpublished paper, 1995, 4).

era, the Arabs used Greek thought to develop their culture. Finally, in modern times, we should be grateful to Johan Ludvig Heiberg and Thomas Little Heath. The "Herculean contributions" of these two great classical scholars of the twentieth century brought Archimedes into the light of day,[585] making it easier, in fact, possible to decipher the palimpsest hiding the thought of Archimedes. Netz is particularly grateful to Heiberg. "We cannot express forcefully enough our debt to Heiberg," he wrote in 2011. "We would certainly not have been able to read as much as we have without his transcription as our guide ... He did, after all, edit the bulk of the contents of Codex C in less than a decade, working alone, and with no technological aid beyond an (excellent) set of photographs."[586] The $2 million the American philhellene Mr. B spent for the palimpsest was the best investment he ever made.

According to its 2008 podcast series, the American Mathematical Society said this about Archimedes: "Archimedes was one of the most brilliant people ever, on par with Einstein and Newton." Yes, but Archimedes was more than on par with Newton and Einstein. He was their teacher. He was fundamental, creating the mathematics and physics underpinning science, including the technology of X-ray fluorescence used in revealing his writing under the Christian prayers. The ancient ink of the Archimedes text had iron in it, which the X-rays caught, thus deciphering the buried words of Archimedes.

There was Archimedean science in the contributions Hewlett-Packard and X-Tek Systems made in the reading of the complicated gearwork and inscriptions of the fragmentary first computer. These speak volumes of the genuine affection Hellenic culture has for some people and institutions in the West.

Now that we have a chance to read the excavated Archimedes, we appreciate his reason, science, and humanism even more. Archimedes speaks with a voice that sheds even more light on the advanced technology of the Antikythera computer, which was the logical outcome of his thought and much more—Hellenic civilization.

[585] Shirley B. Gray, "A Centennial Celebration of Two Great Scholars: Heiberg's Translation of the Lost Palimpsest of Archimedes—1907 Heath's Publication on Euclid's Elements—1908," *Notices of the American Mathematical Society*, Vol. 55, No. 7, August 2008, 776–783. See also: Erik Petersen, "Itinera Archimedea: On Heiberg in Constantinople and Archimedes in Copenhagen" in *The Archimedes Palimpsest*, ed. Reviel Netz et al., 2 vols (New York: Published for the Walters Art Museum by Cambridge University Press, 2011) I, 119–127.

[586] Reviel Netz, "The Place of Codex C in Archimedes Scholarship" in *The Archimedes Palimpsest*, I, 272.

Greek thinkers studied the natural world and the Cosmos above all for knowledge and models of science and behavior. Early thinkers like Anaximander and Empedocles saw equality and justice in the actual workings of the heavens. It was this cosmic democracy that made a touchdown in Athens and other Greek poleis. The early thinkers brought their celestial concepts of cosmic equality and justice down to Earth, supporting truth and reason and equality and justice to be common for all Greeks.

These aspirations found the best and most effective expression in law. Herodotus recounts a story that illustrates the willful commitment of the Greeks to the dictates of a common law. This story is about Demaratos, a traitor Spartan king advising the Persian king Xerxes during Persia's war against Greece in the early fifth century BCE. Xerxes ridiculed Demaratos, saying that the few free Greeks would dare oppose his vast armies. Demaratos then tells Xerxes that his own people, the Spartans or Lakedaimonians, are free, but not entirely free: "Their master is the law," he says, "and they're far more afraid of this than your men are of you."[587]

The other application of cosmic principles in Greek society was in agriculture (chapter 7). To repeat: The Greek farmer paid a lot of attention to astronomy, botany, meteorology, and the weather. The gods were in nature, which was divine, and the Greeks had to be certain when to offer their sacrifices to those gods. From Homer in the eighth century BCE to Ptolemaios in the second century of our era, nature and the universe were the common threads of Greek religion, philosophy, and science.

[587] Herodotos, *The Histories* 7.104.

Chapter 9

Why Did Greek Science and Civilization Almost Disappear?

While many of us take the advancements of ancient Greece as key elements of the history of Western civilization, it was not always the case. Once the Roman Empire adopted Christianity as its official religion in the fourth century and sought to spread it throughout the empire and beyond, the Greek people, particularly its intellectuals, became suspect for their non-Christian polytheistic roots. The burgeoning church in the East and West began a systematic destruction of the rich trove of knowledge the ancient Greeks had accumulated. Monks torched libraries and schools. Christian authorities and owners of precious works of art and science melted them down. Thinkers and teachers ran for their lives. Christianity attempted to snuff out the bright light of knowledge that was Hellas. This chapter gives the historical context that explains the collapse of Greek science. It highlights why so many of the scientific and technological discoveries of the Greeks, such as the Antikythera Mechanism, were, in all likelihood, lost for centuries because Christianity made their existence impossible.

Murdering Homer

Western intellectuals are right in crediting the Greeks for the origins of their civilization. I did the same thing in this book. Yet, despite the Renaissance, which formally integrated Greek thought into Western culture, Christianity is still having a stifling effect on Greek thought. The ancient Greeks become an ambivalent value in the Western tradition. In fact, during the late twentieth century in America, an industry of academic classicists lost all shame and came out in the open against the Greeks. Their attack on Greek civilization was so ludicrous and fake that a couple of American classicists, Victor Davis Hanson and John Heath, were compelled to come to the defense of ancient Greeks. They wrote a book about the "murder" of Homer at the hands of elite philologists and theorists who prefer "therapeutics, moral relativism, blind allegiance to progress, and the glorification of material culture" to Greek ideas and values.[588] This is true, but underneath the superficial rhetoric of these anti-Greek scholars, I suspect there's monotheistic metaphysics clashing with Greek ideas. That one-god metaphysics proclaims an omnipotent god creating the world out of nothing in a few days; giving humans powers of life and death over the natural world: all antithetical to the polytheistic metaphysics of the Greeks, who thought the world alive, beautiful, eternal, and divine. Thus, I suspect, monotheism (Christianity, Judaism, and Islam), not trivial philological pursuits, has been fueling the anti-Greek impulse in the West.

The other historical reason for the surviving antagonism between all things Greek and monotheistic comes from the secrecy surrounding the early conflict between Christianity and Hellenism. What Christianity did to the Hellenes and their culture in the fourth century and after remains, by and large, secret. This is because Christians and other monotheists wrote that history.[589] And despite the Renaissance, Christians refuse to acknowledge their atrocity against the Greeks—and non-Greek polytheists and their culture all over Europe.

There's no party today representing ancient Greeks. With the exception of a few thousand Greek polytheists, most modern Greeks are Christians. So, there has never been reconciliation between Christian Greeks and polytheistic Greeks. The secret history remains secret.[590]

This secret history throws light on the origins, nature, and political and global purposes of Christianity. It also explains why modern Greece is facing an identity crisis of major proportions. Christian Greeks don't know who they are. They may study what the ancient Greeks did, but their cultural environment is filled with alien images of a Jewish god. Some of the saints, however,

[588] Victor Davis Hanson and John Heath, *Who Killed Homer?* (New York: The Free Press, 1998) 158–159.

[589] For a clue of what happened in the Western Roman empire, see Charles Freeman, *The Closing of the Western Mind: The Rise of Faith and the Fall of Reason* (New York: Alfred A. Knopf, 2004) and Catherine Nixey, *The Darkening Age: The Christian Destruction of the Classical World* (Boston: Houghton Mifflin Harcourt, 2018). Freeman's study is more scholarly and comprehensive of the Christian impact.

[590] Evaggelos Vallianatos, *The Passion of the Greeks: Christianity and the Rape of the Hellenes* (Harwich Port, MA: Clock and Rose Press, 2006).

are Greek and have taken over the virtues of Hellenic gods. But the general consequence of the Christianization of Hellas has been bad for Hellas and the world. Christianity made Greece a palimpsest, that is, Christianity erased most of Hellenic culture from Greece and on the bloody skin of the country wrote its own Scriptures. In other words, Christianity remade Greece into something antithetical to its very being. Christianity forced Greece to become a country where Christian prayers supplanted Homer, Plato, Aristotle, and Archimedes; Christian writers wrote their stories on top of scraped Hellenic literature.

Burning Books and Libraries

This is no place for a full treatment of the Christianization of Greece, save for those events that connect Christianity to the destruction of those institutions responsible for the invention and creation of the Antikythera computer. The pressure became unbearable in the late fourth century.

During the reign of Emperor Theodosius, the reach of Christianity embraced Alexandria. The deadliest blow of the Christian administration of Theodosius hit the Mouseion and the Library of Alexandria, the jewels of science and knowledge of the Greek world. These institutions of advanced studies survived and thrived as long as the Greeks were in charge of their fortunes, especially in Egypt. But with the rise of Rome and the loss of Greek freedom, decline undermined Greek culture in Egypt as well. The Romans stepped into the world of Egypt and Alexandria and the flames of war approached the library as well.

The library escaped the destructive effects of the contest between the Roman forces of Julius Caesar and the Greeks of Alexandria in 48 BCE. The prolific Greek scholar, Didymos, credited with something like 3,500 to 4,000 books, did his writing after 48 BCE in Alexandria. Didymos astonished his contemporaries so much that they called him *Chalkenteros* or Bronze-Guts. This bronze-gutsed scholar could not possibly write so many books without the unparalleled resources of the Library of Alexandria. We also have the report of Strabo, who was in Alexandria some twenty years after the Alexandrian war of Caesar. Strabo had good things to say about the Mouseion and the library in the royal palaces. The Mouseion had a public walk, a theater for lecturing, and a large house with a dining hall for its resident scholars.[591]

[591] Strabo, *Geography* 17.1.8.

Luciano Canfora, Italian classical scholar and author of *The Vanished Library*, an insightful and valuable study of the fate of the Library of Alexandria, says that the damage to the library by Caesar was minimal; the scrolls burned "were no more than a tiny part of the vast collection at Alexandria. We must agree, then, that the history of the classical textual tradition never suffered the grave blow that would have been inflicted by the loss of such a library, had that loss actually taken place."[592]

The great Library of Alexandria faced, for the second time, fires and destruction in the late third century when Emperor Aurelian, 270–275, carried his war against the Syrian queen Zenobia into Alexandria. Zenobia, who claimed to be related to the last Greek queen of Egypt, Kleopatra, tried to restore the Greek kingdom of Egypt.

In 363, the Greek historian Ammianus Marcellinus reports that the *Broucheion* (Bruchion) was in ruins. This was the quarter of Alexandria that used to house the Alexandrian Library and the "distinguished men" who worked at the Mouseion. This destruction bypassed the smaller library located in the temple of Zeus Serapis (*Serapeion* or Serapeum), which impressed Marcellinus. He says that the Serapeum was "conspicuous" in the midst of the temples of Alexandria: "no description can do it justice," he says, "yet is so adorned with extensive columned halls, with almost breathing statues, and a great number of other works of art, that next to the Capitolium, which elevates revered Rome to eternity, the whole world beholds nothing more magnificent." [593]

In 391, the Christians finally did away with the magnificent Serapeum. Christian soldiers raided the shrines of the gods and demolished the temples of Zeus Serapis. The Christians finished off the library of Serapeum when they attacked the temples in the fashion of fighting a war against statues, stones, and votive offerings. The Greek philosopher Eunapios, who was a contemporary witness to this destruction, reports the Christian soldiers plundered the world of learning of Alexandria and stole whatever they pleased. They boasted they had defeated the gods, gloating over their impiety and sacrilege.

After the government troops had their fill of vandalism of the sacred and learned world of Alexandria, the authorities imported monks to finish the atrocity they started. These men, says Eunapios, were men in appearance, but

---

[592] Luciano Canfora, *The Vanished Library: A Wonder of the Ancient World* (Berkeley: University of California Press, 1990) 144.

[593] Ammianus Marcellinus, *History* 22.16.15, 12, tr. John C. Rolfe (Loeb, 1937).

in reality, they were "swine, living the lives of pigs," committing countless unspeakable barbarisms and crimes. They felt proud and pious whenever they could do something that expressed their utter contempt for the gods and things divine. These monks, or men wearing black robes and behaving in crude manners, had the power of tyrants. The government settled them at Canobus, a community at the mouth of the Nile. The monks then forced the residents of Canobus to abandon the worship of the gods for the worship of slaves, and those not even honest slaves but criminals. The monks would collect the bones and skulls of criminals, says Eunapios, and would force the people to venerate those bones and skulls. The monks called the dead criminals martyrs and ambassadors to the new god.[594]

According to Edward Gibbon, the British historian of the eighteenth century who spent his life studying the Roman Empire, the decision of Emperor Theodosius in 391 to destroy the "idols" of Alexandria "sent up a shout of joy and exultation" among the Christians. The Greeks, or as Gibbon calls them, "the unfortunate Pagans," "[R]etired with hasty and silent steps, and eluded, by their flight or obscurity, the resentment of their enemies. [Archbishop] Theophilus proceeded to demolish the temple of Serapis, without any other difficulties than those which he found in the weight and solidity of the materials; but these obstacles proved so insuperable that he was obliged to leave the foundations, and to content himself with reducing the edifice to a heap of rubbish, a part of which was soon afterwards cleared away, to make room for a church erected in honour of the Christian martyrs. The valuable Library of Alexandria was pillaged or destroyed; and near twenty years afterwards, the appearance of the empty shelves excited the regret and indignation of every spectator whose mind was not totally darkened by religious prejudice."

Gibbon rightly puts much of the blame for this crime on the Christians' leader, Archbishop Theophilos: "The perpetual enemy of peace and virtue; a bold, bad man, whose hands were alternately polluted with gold and with blood."[595] In the twentieth century, the classical scholar James Westfall Thompson accuses the Christians, not the Arabs, of destroying the library and literature in Alexandria.[596] The American historian, Ramsay MacMullen of Yale

[594] Eunapios, *Lives of the Philosophers and Sophists* 465–473.

[595] Edward Gibbon, *The Decline and Fall of the Roman Empire*, ed. Hugh Trevor-Roper, 6 vols. (New York: Alfred A. Knopf, 1993) III, 151–152.

[596] James Westfall Thompson, *Ancient Libraries* (Berkeley, CA: University of California Press, 1940) 24.

University, says that the Christians silenced, burned, and destroyed Greek civilization as a form of "theological demonstration."[597] Luciano Canfora argues that the "burning of books was part of the advent and imposition of Christianity."[598]

Fires of the Mint, Blazes of the Smelters

With the burning of books, there was little hope for science and technology. By means of terror and propaganda, Christianity devalued Greek culture so much that the arts of civilization, including technology and science, fell into a precipitous decline. This forced the owners of the Antikythera Mechanism and other such devices to do away with them, melting them down for their metal. Tassios explains the rarity of these devices differently. He gives two reasons why we have the fragments of only one Antikythera machine: "All non-artistic metal objects in [Christian] antiquity were recasted—a destiny that even statues had not always escaped," he wrote. And then he added something about "possible insufficiency of technical awareness of some archaeologists in the past."[599] Christianity is nowhere to be found in Tassios's speculation. Like other researchers on the Antikythera Mechanism, he does not venture a theory on why a popular machine nearly disappeared from the face of the Greek Earth or why even statues also were melted down. He may be right about archaeologists, however. Archaeologists, no less than other scientists, have been loaded with biases and misinformation. Many of them, in the past and now, have not been well trained in detecting ancient technology. These archaeologists ought to know that, as a matter of course, the Christians for several centuries buried ancient Greek civilization and melted down or burned the bronze devices and statues of the Greeks, converting them into pots and frying pans.

For example, in the mid-fourth century, Firmicus Maternus, a Christian apologist, appealed to the brother emperors, Constantius and Constans, to "take away ... the adornments of the temples. Let the fire of the mint or the

[597] Ramsay MacMullen, *Christianizing the Roman Empire A.D. 100–400* (New Haven: Yale University Press, 1984) 119.

[598] Canfora, *The Vanished Library*, 192.

[599] Tassios, "Prerequisites for the Antikythera Mechanism to be produced in the 2nd Century BC," in *The Antikythera Shipwreck* 251.

blaze of the smelters melt them down."[600] Palladas, an Alexandrian Greek schoolteacher and poet of the fifth century, witnessed a smith transforming a statue of Eros into "a frying-pan." In the early fifth century, the Christian historian Socrates documented the Christians' recasting of the Greek bronze statues of the library (Serapeion) of Alexandria into "pots and other convenient utensils for the use of the Alexandrian church."[601]

So, it is very possible the fires of the mint and the blazes of the smelters ate Antikythera Mechanism-like machines, which in the Christian society of the fourth century and after, lost all utility and meaning. In the next chapter, I document similar destruction by fire of ancient Greek treasures by Western Christian crusaders when they captured Constantinople in 1204.

Christianity's greatest orator, the "golden-mouth" Ioannes Chrysostom, was straightforward about his dislike for the Greeks. He was the patriarch of the Eastern Roman Empire from 398 to 404. He preached that fishermen, tent-makers, and illiterate men were by far superior to Hellenic philosophers, including Plato. He had nothing but contempt and hatred for Hellenes. Their philosophy, he said, was just talk and fables without any trace of wisdom; Hellenic philosophers were no better, he said. They were full of ashes and dust. Everything about them reeked with foulness and ulcer. Their teachings, according to Chrysostom, were worm-eaten.[602]

The Murder of Hypatia

In such a climate of terror and theological frenzy, the black-robed monks murdered Hypatia. She was born around 370 and died a horrible death in 415. She was a brilliant Platonic philosopher and mathematician.

Socrates, c. 380–post 439, a lawyer and an ecclesiastical historian, was a contemporary of Hypatia. He says that the monks: "waylaid her returning home. They dragged her from her carriage, took her to the church called Kaisareion (from the palace Kleopatra had built for Caesar), where they

[600] Firmicus Maternus, *The Error of the Pagan Religions* 28.6, tr. Clarence A. Forbes (New York: Newman Press, 1970).

[601] Palladas in *The Greek Anthology* 9.773. Socrates, *Ecclesiastical History* 5.16, tr. A. C. Zenos, Nicene and Post-Nicene Fathers, Second Series, vol. 2 (Peabody, MA: Hendrickson, 1995).

[602] John Chrysostom, *Commentary on Saint John the Apostle and Evangelist, Homily* 66, tr. Sister Thomas Aquinas Goggin (New York: Fathers of the Church, 1960) 224–225.

completely stripped her naked, and then murdered her with tiles. After tearing her body to pieces, they took her mangled limbs to a place called Kinaron, where they burnt them."[603]

Another Greek who probably knew Hypatia was Palladas. Like Hypatia, Palladas was a Hellene. His epigram on Hypatia speaks of his passionate love for her only because her wisdom had made her almost divine. He says: "Revered Hypatia, ornament of learning, undefiled star of wise teaching, when I see you and your wisdom, I worship you, seeing the starry house of the virgin. Your enterprise is in heaven."[604]

Theon was the father of Hypatia. He was a philosopher, mathematician, and astronomer. He edited the popular book *Handy Tables* of Ptolemaios. This book was the source for astronomical tables for Greeks, Arabs, and Latins for centuries. Theon taught Hypatia mathematics and philosophy. She left scholarly commentary on the "Conics" geometry of Apollonius of Perga and algebra of Diophantos of Alexandria. Diophantos made contributions to mathematics and, especially, algebra. He flourished in the mid-third century. Hypatia taught the philosophy of Plato and Aristotle at Athens and Alexandria. Archbishop Cyril of Alexandria disliked Hypatia. He could not stand seeing her teaching Greek philosophy and mathematics. Besides, he convinced himself she was an obstacle to his efforts to get the attention of the governor of Egypt, Orestes, who was a student of Hypatia. Cyril orchestrated the murder of Hypatia. Gibbon summarizes how the killing took place:

"On a fatal day, in the holy season of Lent, Hypatia was torn from her chariot, stripped naked, dragged to the church, and inhumanly butchered by the hands of Peter the reader and a troop of savage and merciless fanatics: her flesh was scraped from her bones with sharp oyster-shells, and her quivering limbs were delivered to the flames. The just progress of inquiry and punishment was stopped by seasonable gifts; but the murder of Hypatia has imprinted an indelible stain on the character and religion of Cyril of Alexandria."[605]

The murder of Hypatia was one of countless other crimes the Christians committed against the Greeks. The Christians dug their Greek enemies into obscurity and nothingness, nearly destroying their spirit and certainly wiping

[603] Socrates, *Ecclesiastical History* 7.14.12–15.7, ed. Gunther Christian Hansen (Berlin: Akademie Verlag, 1995) 360–361; also, in *Creeds, Councils and Controversies*, ed. J. Stevenson, rev edition W.H. C. Frend (London: SPCK, 2000) 284.

[604] *The Greek Anthology* 11.400, tr. mine.

[605] Gibbon, *The Decline and Fall of the Roman Empire*, V, 17–18.

out the outward signs and material culture of their civilization. The Roman state and the Christian Church punished by death whoever admitted to being a Hellene. Murderous anti-Hellenic laws dominated the legal system of medieval Greece at least until the mid-fourteenth century.[606] The British classical scholar, C. M. Bowra, equated the triumph of the Christian Church in Greece to a cataclysm.[607]

Christianity and Science

In the Introduction, I drew a picture of potential medieval Christian contributions to the origins and development of modern science and technology. However, that picture becomes murky when one looks at early Christianity, which all but cursed humans interested in the natural world and science. Tertullian (160–225), the North African apologist who hated the Greeks, spoke with contempt about "the useless affectation of scrupulous curiosity." He made fun of Thales of Miletus:

"[W]hen, star-gazing as he walked with all the eyes he had, he had the mortification of falling into a well.... His fall ... is a figurative picture of the philosophers; of those, I mean, who persist in applying their studies to a vain purpose, since they indulge a stupid curiosity on natural objects, which they ought rather (intelligently to direct) to their Creator and Governor."[608]

Tertullian also asked: Is there any likeness between the Christian and the philosopher? Or where is the likeness between the disciple of Greece and the disciple of heaven?[609] Then he lashed out more against the Greeks, particularly Aristotle: "Unhappy Aristotle!" he wrote. "Who invented ... dialectics, the art of building up and pulling down; an art so evasive in its propositions, so far-fetched in its conjectures, so harsh, in its arguments, so productive of contentions—embarrassing even to itself, retracting everything, and really treating of nothing! ..."

[606] Constantine Armenopoulos, *Procheiron Nomon* 6.11.4–6, ed. Konstantinos G. Pitsakes (Athens: Dodone, 1971).

[607] Bowra, *The Greek Experience*, 2.

[608] Tertullian, *Ad Nationes* 2.4, tr. Peter Holmes in *The Ante-Nicene Fathers*, ed. Alexander Roberts and James Donaldson, 10 vols. (New York: Scribners, 1896–1903) III, 133 (reprint, Peabody, MA: Hendrickson, 1995).

[609] Tertullian, *Apology* 46.

What indeed has Athens to do with Jerusalem? What concord is there between the [Platonic] Academy and the Church? What between heretics and Christians? Our instruction comes from 'the porch of Solomon,' who had himself taught that 'the Lord should be sought in simplicity of heart.' Away with all attempts to produce a mottled Christianity of Stoic, Platonic and dialectic composition!"[610]

Augustine, 354–430, lived about two centuries after Tertullian. His attitude towards the Greeks did not differ from that of Tertullian. He showed contempt for the Greeks and urged violence against them. The irony is that he was versed in Greek learning; in fact, he had some likening of Plato and Platonists if only because, he wanted to believe, Plato and the Platonists were monotheists and did not waste their time studying nature. But Plato was neither a monotheist nor one who hated the natural world. Indeed, Plato loved and studied nature and the Cosmos for knowledge and truth. He was both a moral philosopher and a mathematician. He worshipped the Greek gods, which angered Augustine, who believed the "gods were thoroughly evil."[611] In time, he forgot Plato and decided he did not like the Greeks after all. His commitment was to Jesus and the Scriptures alone.

Augustine stressed that nothing in human ingenuity could compare to the power of the "Scripture whose divine authority puts it above the literature of all other people and brings under its sway every type of human genius—and that, not by some casual intellectual reaction, but by a disposition of Divine Providence."[612] His discovery of the Christian god led him to an incredible self-flagellation, a fake apology for the sins of youth and Greek and Latin studies. He disliked educated people as much as he disliked learning. Nevertheless, he enjoyed showing off the smattering of science he learned during his Greek studies. For example, he spoke about natural science as follows: "When it is asked what we ought to believe in matters of religion, the answer is not to be sought in the exploration of the nature of things [*rerum natura*], after the manner of those whom the Greeks called 'physicists.' Nor should we be dismayed if Christians are ignorant about the properties and the number of the basic elements of nature, or about the motion, order, and deviations of the

[610] Tertullian, *On Prescription Against Heretics* 7, tr. Peter Holmes in *The Ante-Nicene Fathers*, III, 246.

[611] Augustine, *City of God* 8.10–13, tr. Gerald G. Walsh et al., ed. Vernon J. Bourke (Garden City, NY: Image Books, 1958) 159–163.

[612] Augustine, *City of God* 11.1.

stars, the map of the heavens, the kinds and nature of animals, plants, stones, springs, rivers, and mountains; about the divisions of space and time, about the signs of impending storms, and the myriad other things which these 'physicists' have come to understand, or think they have. For even these men, gifted with such superior insight, with their ardor in study and their abundant leisure, exploring some of these matters by human conjecture and others through historical inquiry, have not yet learned everything there is to know. For that matter, many of the things they are so proud to have discovered are more often matters of opinion than of verified knowledge. For the Christian, it is enough to believe that the cause of all created things, whether in heaven or on Earth, whether visible or invisible, is nothing other than the goodness of the Creator, who is the one and true God."[613]

John William Draper, a nineteenth-century scholar, knew that the church incorporated Augustine's theology into its doctrine. He was furious. He accused the church of becoming "a stumbling block in the intellectual advancement of Europe for more than a thousand years."[614] Like Draper, the twentieth-century scholar Andrew Dickson White concluded that the church almost crushed science. He pointed out that Augustine had a pernicious effect on science for the following reasons: "No treatise was safe thereafter which did not breathe the spirit and conform to the letter of this maxim [of Augustine]. Unfortunately, what was generally understood by the 'authority of Scripture' was the tyranny of sacred books imperfectly transcribed, viewed through distorting superstitions, and frequently interpreted by party spirit. Following this precept of St. Augustine there were developed, in every field, theological views of science which have never led to a single truth—which, without exception, have forced mankind away from the truth, and have caused Christendom to stumble for centuries into abysses of error and sorrow."[615]

Lucio Russo reported that after the fourth century "obscurantism and stasis" smothered Europe, blocking "most avenues of intellectual development for a thousand years."[616] And Geoffrey Lloyd reached a similar conclusion: it was

---

[613] Augustine, *Enchiridion* 3.9, tr. Albert C. Outler (The Library of Christian Classics, 7) (Philadelphia: Westminster Press, 1955) 341–342.

[614] John William Draper, *The Conflict between Religion and Science*, 15th ed. (London: C. Kegan Paul, 1881) 51–52.

[615] Andrew D. White, *A History of the Warfare of Science with Theology in Christendom*, first published in 1896, 2 vols. (New York: George Braziller, 1955) I, 325.

[616] Russo, *The Forgotten Revolution*, 1.

Christianity that yielded the force responsible for obscurantism and stasis. All this started with the Christianization of Greece and Rome. This is how Lloyd put it: "What finally sealed the fate of many key institutions of pagan culture was the conversion of the Roman empire to Christianity, bringing more overwhelming changes both in values and social structures than the Mediterranean world had experienced previously."[617]

This "conversion," which would never have happened without the overt and systematic war Christianity fought against the Greeks and the classics, ushered in an era of darkness in Western Europe while it was responsible for the loss of most of the scientific, literary, artistic, philosophical, and technological works of ancient Greek thinkers, scholars, and engineers. As a result, several of the works of Archimedes did not make it to our day.

In the early twenty-first century, two more scholars, Charles Freeman and Richard Billows, confirmed the incompatibility of Christianity with Greek thought. Freeman explained that the combination of church and state authority in forcing theological consensus, "backed by Christian suspicions of scientific argument, crushed all forms of reasoned thinking." The result was the Dark Ages—a thousand years during which "serious mathematical and scientific thinking" became extinct in Europe.[618] Billows, professor of classical studies at Columbia University, sees the triumph of Christianity spelling "the beginning of the end" of the civilization of the Alexandrian Age.[619]

To this condition of intellectual impoverishment, Edward Gibbon added the habit of Christians to worship relics, making the bones of saints and countless nameless skeletons "more valuable than gold or precious stones." Such "religion fiction" and "superstitious practice," says Gibbon, "tended to increase the temptations of fraud and credulity, insensibly extinguished the light of history and of reason in the Christian world."[620]

The Roman state adopted Christianity, including Augustine's severe version, making slaves of the bishops and they, using political power, not reason, settled the theological arguments in arriving at certain "convenient truths." But more insidious than that, church authorities forced reasonable people out of reason, boosting the decline of science, a near disappearance for several centuries.

[617] Geoffrey Lloyd and Nathan Sivin, *The Way and the Word* (New Haven: Yale University Press, 2002) 89.

[618] Freeman, *The Closing of the Western Mind*, 337, 340.

[619] Richard Billows, *Before and After Alexander: The Legend and Legacy of Alexander the Great* (New York: Overlook Duckworth, 2018) 286.

[620] Gibbon, *The Decline and Fall of the Roman Empire*, III, 163.

So, the suppression of Greek thought by Christianity shut down the Western mind until the emergence and flowering of the European Renaissance in the fifteenth and sixteenth centuries. The Renaissance restored reason to European civilization, giving birth to our world. Two events made that possible. Greeks in Greece kept studying and copying the ancient texts. And the Arabs fell in love with Hellenic scientific and philosophical culture in the eighth to tenth centuries. I examine these two events in the next chapter.

Chapter 10
The Rebirth of Greek Science and Thought

After the Christianization of the Roman Empire, the great works of Greek science and thought were almost destroyed. However, thanks to the efforts of the great potentates of the Islamic world, these seminal intellectual achievements survived through translation into Arabic. Even more important than the Arabic translation of Greek texts was the role of Greek scholars in the survival and transmission of Greek culture to the rest of the world. Greek scholars brought the surviving texts of their ancestors to the Italian city-states in the fifteenth century. Greek scholars then translated those texts into Latin, triggering the Renaissance in Europe.

The Survival of Greek Learning in Medieval Greece

Fortunately, the Greeks of New Rome—Christians of medieval Greece or Byzantium—did not have an Inquisition. The campaigns of the Orthodox Church against the classics were severe but sporadic. The clerics could cite the apostles directing all Christians to "Stay clear of all Greek books."[621] The clerics could be brutal. But, in general, they were selective in their opposition or support of Greek studies. The church watched carefully who did what in the study of ancient Greek culture. It always worried that the unfettered pursuit of Hellenic studies was bound "to fill men's heads with pagan ideas."

[621] *Didascalia et Constitutiones Apostolorum* 1.6, ed. F. X. Funk (Originally published in 1905, Torino, 1964) 13–15.

In addition, the ceaseless oversight of the Hellenic tradition by Christian authorities worsened the internal security of medieval Greece, which had to fight perpetual wars against countless enemies in the East and West. This anti-Hellenic state policy explains the paradox of why a Renaissance in medieval Greece failed to take place. Wars and religious authoritarianism rotted the Hellenic impulse of medieval Greece.[622] After all, medieval Greece had the world's only depository of ancient Greek texts, which, once leaked to the Muslims and Westerners, triggered bursts of scientific and humanistic rebirths.

The educated people of medieval Greece, however, considered themselves Greek. Through clerical intolerance and wars, they protected enough of the classical texts for the continuation of their culture. Indeed, the Greek texts that survived to our times survived primarily because they were school textbooks. Speros Vryonis, an eminent Greek scholar on medieval Hellenism, put it like this: "The heritage of Greek learning and letters [in medieval Greece] was perpetuated through a system of education that emerged from the Greek world in late antiquity and which was crystallized in Alexandria. This educational system was spread throughout the [medieval Greek] empire, and the representatives of local urban, provincial, and central government were largely trained in this system. The school curriculum thus guaranteed the survival of the kernel of the Greek authors and texts: Homer, many of the lyric poets, the dramatists, historians, philosophers, mathematicians, physicians, and scientists. Only that part of the classical heritage survives which had been incorporated into the school textbooks."[623]

In addition, some medieval Graeco-Roman emperors and senior ecclesiastics were less fanatical than others. During those truces in the war against the *ethnikoi*, "Greeks who worshipped the gods," wealthy lovers of antiquity or philhellenic emperors and patriarchs funded the copying of texts. But, overall, Christian Greeks did not exactly feel at home in the presence of their ancient Greek legacy. According to another modern scholar, Colin Wells, Christian Greeks kept ancient Hellenic literature at arm's length but under control. This was a literature "at once too barbed with secular reason and other dangers for the Christian Byzantines to embrace too closely, yet too imposing, too downright gorgeous to dismiss altogether."[624]

---

[622] R. R. Bolgar, *The Classical Heritage and its Beneficiaries* (London: Cambridge University Press, 1973) 88–90.

[623] Speros Vryonis, "Byzantine Civilization, a World Civilization," in *Byzantium: A World Civilization*, ed. Angeliki E. Laiou and Henry Maguire (Washington, DC: Dumbarton Oaks, 1992) 20–21.

[624] Colin Wells, *Sailing from Byzantium: How a Lost Empire Shaped the World* (New York: Bantam Dell, 2007) 36.

In the sixth century, Emperor Justinian hated the Greeks but borrowed from Greek culture. He built St. Sophia, Christendom's greatest and most beautiful church modeled after the best traditions of Greek temples. Anthemios of Tralles and Isidoros of Miletus, Greek architects who followed the science of Archimedes, designed and supervised the construction of St. Sophia. Anthemios's brother, Alexander, was a physician who enriched the Greek medical tradition with his own experience.[625] One of the friends of Anthemios and Isidoros, a mathematician named Eutokios, edited the first four books of the "Conics" of Apollonios of Perga. Apollonios was a mathematical genius and a junior colleague of Archimedes. Eutokios also wrote commentaries on some of the works of Archimedes, including *On the Sphere and Cylinder*, *On the Equilibrium of Planes*, and *Measurement of a Circle*. Isidoros used Eutokios's commentaries and prepared new editions of Archimedes's works, a vital step in the survival and transmission of the ideas of Archimedes to our time.

The other sixth-century scientist who continued with the Greek traditions of reason and inquiry was Ioannes Philoponos. He wrote on mathematics and astronomy. But his original contributions were in physics and mechanics, where he challenged Aristotle's ideas on motion that it is air which moves the projectile. On the contrary, Philoponos said, air resists the moving projectile. When we throw a ball, he said, we impart to the ball energy embedded in the movement itself, which continues to propel it after it has left our hand.[626] Philoponos anticipated the concept of inertia and demonstrated that falling bodies of different weights fall pretty much with the same speed, anticipating Galileo's similar conclusion.

The seventh and the eighth centuries were times of trouble for medieval Greece. The rise of Islam in the seventh century challenged the monotheistic monopoly of Christianity and put medieval Greece on notice. Aggressive policies by Muslim Arabs captured the Greek provinces of Syria and Egypt. This sudden appearance of a superpower next to Greece threatened the very survival of Christianity and medieval Greece. The Christians of Greece faced Islam as an existential and theological danger. After all, what was the meaning and promise of Christianity if Muslims were victorious? The response to that mortal challenge was civil war over icons, which, in fact, included much more

[625] Petros Bouras-Vallianatos, "Clinical Experience in Late Antiquity: Alexander of Tralles and the Therapy of Epilepsy," *Medical History* (2014), vol. 58(3), 337–353.

[626] John Philoponos, *In Aristotelis Physicorum libros quinque posteriores commentaria*, ed. H. Vitelli (Berlin, 1888) IV, 641, 7–30; 642, 1–26.

than icons and lasted till the middle ninth century. The iconoclasts were simply accommodating to the enormous influence of the Islamic superpower next door. In such a hostile and uncertain political reality, learning diminished until the iconoclastic wars were over. The late ninth century witnessed a cultural revival that set the foundations for a new beginning with what was left of the rich Greek cultural traditions of the empire.

In the late ninth century, libraries in Constantinople, the capital of Greece, began to replenish their ancient Greek collections. The copying of some of the works of Archimedes was part of that cultural revival.

Greek Thought, Arabic Culture

During the same period, Arabs in Baghdad started something unprecedented in world history—building their culture on the thought of the Greeks. They did that by massive borrowing. These were the Arabs of the early Abbasid society in Baghdad. Their caliphs purchased most of the Hellenic philosophic and scientific works circulating in Greece.

The caliphs hired Nestorian Christians who translated into Arabic those works during the eighth to the tenth centuries.

The Nestorian Christians were followers of Nestorius, c. 381–451, the bishop of Constantinople from 428 to 431. This bishop, a fanatical Christian who hated the Hellenes or pagans, refused to accept that the mother of Jesus, Mary, was also "Theotokos" or mother of god. Eventually, he withdrew his objection to "Theotokos," only warning that such a belief would confuse people in recognizing a difference between the human and divine in Christ. However, Nestorius was also a ruthless theologian and politician. He got into trouble with Empress Pulcheria, and he offended Pope Celestine and bishop Cyril of Alexandria, the man who murdered Hypatia in 415. With the consent of Emperor Theodosius II, he was anathematized by the Council of Ephesos in 431 and spent twenty years in exile. But his followers, the Nestorians, had to get out of Greece. They found hospitality in Persia and Iraq. The memories of Alexander the Great and Greek civilization were still alive in the Middle East and Mesopotamia. Indeed, when the Christian emperor Justinian shut down the Platonic Academy of Athens in 529, Greek philosophers fled to Persia where King Khursaw I offered them hospitality and support. The Persian king even built a school modeled on the Mouseion and Library of Alexandria. Nestorians carried with them Greek books.

The Nestorians used the logical works of Aristotle to explain or legitimize Christian texts. In addition to Aristotelian logic, they were interested in medical knowledge, astronomy, and mathematics. Thus, they translated from the Greek into Syriac and, eventually, into Arabic, the *Timaeus* of Plato, the *Elements* of Euclid, the *Conics* of Apollonios of Perga, the *Almagest* of Ptolemaios, works of Hippocrates, Aristotle, and Galen, and the *Isagoge* of Porphyry. This was an introduction to the *Categories* of Aristotle. Porphyry was a philosopher who flourished in the third century.

With the founding of Baghdad in 762, Nestorians found themselves in a privileged position. The rulers of Islam wanted what they had. The Abbasid caliphs, especially Abu Jafar al-Mansur, 754–755, al-Mahdi, 775–785, Harun al-Rashid, 786–809, and al-Mamun, 813–833, did their best to introduce Greek science to Arabs and Persians. To do so they had to have the Greeks' scientific and philosophical texts translated into Arabic. Excellent salaries brought Nestorian translators to their courts. In 765, caliph al-Mansur appointed Gurgis ibn-Gibril ibn-Buhtisu, a Persian Nestorian physician, to be his personal doctor. The Buhtisu family served the caliphs for several generations as court physicians and advisors who were also instrumental in the translation of Greek scientific texts into Arabic. Hunyan ibn Ishaq, 809–873, a Nestorian and one of the best translators of Greek texts into Arabic, was appointed the director of the House of Wisdom, a research institute and library in support of the Greek project of the Abbasid Caliphate.[627]

Dimitri Gutas, professor of Arabic Language and Literature at Yale University, describes such a translation movement as "an astounding achievement." It was supported by the entire Abbasid society for well more than two centuries. It was a social movement related to the founding of Baghdad and to the Abbasids as the managers of a world empire. In addition, Gutas says, the translation of the Hellenic scientific and philosophical works into Arabic between the eighth and the tenth centuries was a major intellectual movement in world history comparable to the golden age of Greek history, the Renaissance, and the scientific revolution of the sixteenth and seventeenth centuries.[628]

These Arabic translations of Hellenic works of science and philosophy—after four hundred years of Christian warfare against them—was like a flower

[627] De Lacy O'Leary, *How Greek Science Passed to the Arabs* (London: Rutledge and Kegan Paul, 1951) 61–72; Freely, *Aladdin's Lamp*, 72–82.

[628] Dimitri Gutas, *Greek Thought, Arabic Culture: The Graeco-Arabic Translation Movement in Baghdad and Early Abbasid Society (2nd–4th/8th–10th centuries)* (New York: Routledge, 1998) 1–8.

blooming in the desert. It was the first international effort to protect the Hellenes' legacy for humankind. The Arabs put the Christian Greeks (Rum) on notice. Their feverish quest for Greek texts got the Christian Greeks out of their usual indifference toward their treasured Hellenic inheritance.

In the seventh century, medieval Greece suffered humiliating defeats at the hands of the Arabs. And, as we mentioned, from 730 to 843 the Christian Greeks had sunk to an almost dark age of constant warfare and preoccupation over smashing and restoring Christian icons. But with the beginning of the ninth century they started seeing the Abbasids take a serious interest in Hellenic texts of science and philosophy. They knew, however, that the pay for copying manuscripts was good. So, they started copying the Greek works for the Abbasids.

Aside from buying the medieval Greeks' cultural legacy, the Arabs were perplexed by the Christian Greeks' connection to the ancient Greeks. Could they be related? After all, the ancient Greeks were definitely not Christian. Al-Gahiz (d. 868), a chief propagandist of the Abbasids, was convinced the Christians of Greece were just artisans, clearly not philosophers. He explained his perplexity as follows: "[W]orks like the Organon, On Coming to Be and Passing Away, and Meteorology were written by Aristotle, and he is neither Byzantine nor Christian; the Almagest was written by Ptolemy, and he is neither Byzantine nor Christian; the Elements was written by Euclid, and he is neither Byzantine nor Christian; medical books were written by Galen, who was neither Byzantine nor Christian; and similarly with the books by Democritus, Hippocrates, Plato, and on and on. All these are individuals of one nation; they have perished but the traces of their minds live on: they are the Greeks. Their religion was different from the religion of the Byzantines, and their culture was different from the culture of the Byzantines. They were scientists, while these people [the Byzantines] are artisans who appropriated the books of the Greeks on account of the geographical proximity.... [The Byzantines] used their religious beliefs to boast superiority over the Jews, to display arrogance toward the Arabs, and to wax haughty over the Indians to the point that they actually claimed that our sages are followers of theirs, and that our philosophers have followed their example."[629]

Another Arab intellectual, Al-Mas'udi, a cultural historian of the tenth century, said Christianity was primarily responsible for the differences

[629] Quoted in Gutas, *Greek Thought, Arabic Culture*, 87.

between the Byzantines and the Greeks, the reason why the Byzantines are inferior to the Greeks. Al-Mas'udi explained the difference as follows: "During the time of the ancient Greeks and for a little while during the Byzantine [i.e., in this case, Roman] empire, the philosophical sciences kept on growing and developing, and scholars and philosophers were respected and honored. They developed their theories on natural science—on the body, the intellect, and the soul—and on the quadrivium, i.e., on **arithmetike**, the science of numbers, on **geometrike**, the science of surfaces and geometry, on **astronomia**, the science of the stars, and on **musike**, the science of harmonious composition of melodies. The sciences continued to be in great demand and intensely cultivated until the religion of Christianity appeared among the Byzantines; they then effaced the signs of philosophy, eliminated its traces, destroyed its paths, and they changed and corrupted what the ancient Greeks had set forth in clear exposition."[630]

This is not that far from the truth: the Christians in fact wrecked Greek philosophy. The Abbasids' philhellenism was a measure of their differences with the Christian Byzantines—the decisive factor of civilization. The Abbasids' Islam saw no trouble with the civilization of ancient Greeks. But the Christians established themselves on the conversion of the ancient Greeks. They built their churches on the ashes of Greek civilization. The Abbasids probably knew they were superior to the Byzantines for that very reason—their affection for Hellenic science.

The Hellenic studies of the Abbasids erected the first stone, outside of Christian Greece, in rebuilding the smashed and burned Hellenic culture. The Abbasids' love for Aristotle; the mathematician Euclid; Archimedes; Pappos; the mathematician and cosmographer Ptolemaios; the physician and botanist Dioskourides (first century); and medicine's masters, Hippocrates and Galenos (Galen), was a boon for the survival of Hellenic culture. It gave to the few surviving lovers of wisdom a taste of the pleasures of Hellenic science and philosophy. The Arabs made Aristotle the supreme philosopher (chapter 3). Medieval Christians tried to do the same thing. But as I will explain later on, the scientific views of Aristotle caused enormous grief to the Christians, who alternately hated and respected him. Yet, Aristotle educated the select few in Christian Western Europe for several hundred years. He was the authority in the sciences and philosophy.

---

[630] Quoted in Gutas, *Greek Thought, Arabic Culture*, 89.

Greek Palimpsests

In addition, for several centuries, Europeans would visit Christian Greece and either steal or buy Hellenic manuscripts and books from illiterate monks. These manuscripts and books found safekeeping in the Vatican Library, the Laurentian Library of Florence, the Library of San Marco in Venice, the Ambrosian Library of Milan, the Bibliothèque Nationale in Paris, the British Museum, and the libraries of Oxford, Cambridge, Paris, Padua, Rome, Berlin, Turin, Naples, Ravenna, Leiden, Budapest, St. Petersburg, Vienna, and other great European universities where, slowly, scholars studied them intensely, meticulously, trying to unlock their Hellenic wisdom. What these European scholars often discovered, particularly after the microscope came into being in the late sixteenth century, was that the Christian parchment books had a secret story to tell. Under their hymns and prayers, there often lay hidden from view gems of Hellenic thought, sometimes entire texts of classical works.

Hellenic books were written on expensive parchment. Christian monks, who could have cared less about Homer or any other ancient writer, would scrape the Hellenic text of a parchment, for example, and would create a *Palimpsestos*, a palimpsest, a book made by writing over an erased original text. The monks would write hymns over the scraped work of Hellenic books.

In 1998, such a palimpsest surfaced in New York and was sold to an anonymous collector. This 800-year-old Christian palimpsest was a tenth-century manuscript preserving a more complete text of the "Mechanical Theorems" of Archimedes's mathematical physics.[631] It is in this work that Archimedes brings together concepts of straight, curved, physical, and geometrical. He compared two infinitely large sets, stating they have an equal number of members. He combined finite and infinite, anticipating and searching for rigorous means of establishing calculus.[632]

Archimedes said, "Give me a place to stand, and I shall move the Earth." His thought has, as I have been showing, moved the world (chapter 4). And yet, by the 1200s, Christian Greece, which had been dismembered and looted by Western Christian armies searching for booty, had sunk into such

---

[631] Ned Potter, "The Mystery of Archimedes' Palimpsest," *ABC News*, Oct. 20, 2000; Reviel Netz, "The Origins of Mathematical Physics: New Light on an Old Question," *Physics Today*, May 2000; William J. Cannon, "Archimedes Unbound," *American Scientist*, July–August 1999.

[632] Reviel Netz, "Proof, Amazement, and the Unexpected," *Science*, November 1, 2002, pp. 967–68.

impoverishment that monks would have no second thoughts about scraping the works of Archimedes to preserve Christian prayers.

The Europeans knew of the tragedy of the Christian Greeks firsthand. They had caused a lot of it. When they conquered Constantinople in 1204, only eleven years before the Fourth Lateran Council issued its dark-age orders against freedom of thought, they burned down the libraries of that huge metropolis and behaved like savage barbarians. So, those who had any culture knew that Greece still had plenty of books hidden in the most unexpected places. No wonder they kept scouring Christian Greece. They knew what they were after. And the massive Arab borrowing from the Greeks, not a secret in the West, kept the plundering of Greek culture in high gear.

Science in Medieval Greece: The Work of Archimedes

However, like all states, medieval Greece did not willingly give out its secrets or science. For example, in the ninth century, the caliph of Baghdad, Mamoun, learned of the wisdom of Leo the Mathematician (c. 790–c. 869) living in Constantinople. Mamoun wrote to Leo, offering him gifts of gold and honors to join his staff in Baghdad. Leo showed Mamoun's letter to Emperor Theophilos, 829–842. Theophilos refused to allow Leo to work for Mamoun. But Leo answered another letter from Mamoun, solving all the mathematical problems in the letter. Mamoun was so pleased he wrote straight to Theophilos, urging him to send Leo to him. He offered the Greek emperor 2,000 pounds of gold and "peace and perpetual and endless treaties." Theophilos turned Mamoun down.[633] Instead, he made Leo archbishop of Thessalonike and appointed him to a teaching position and funded his research.[634]

Leo served Emperors Theophilos, and Michael III, 842–867. Leo's greatest contribution to the revival of classical studies was his supervision of the copying of the works of Archimedes, thus assuring that some of those works would reach us.[635] In fact, Leo transcribed six of the works of Archimedes: *On the Sphere and Cylinder*, *Measurement of a Circle*, *On Spirals*, *On the Equilibrium of Planes*, *The Sand-Reckoner*, and *Quadrature of the Parabola*, including the commentaries of Eutokios on three of these works. These studies

[633] *Theophanes Continuatus*, ed. I. Bekker (Bonn, 1838), 188–190.
[634] N. G. Wilson, *Scholars of Byzantium* (London: Duckworth, 1983) 79–84.
[635] Marshall Clagett, *Greek Science in Antiquity* (New York: Collier Books, 1963) 207.

of Archimedes passed to the West as Codex A. They became the archetype of all of the Archimedes manuscripts circulating in the West up to 1906.

In 1269, William of Moerbeke, a Flemish Dominican monk who learned Greek in Greece, translated Codex A into Latin. In addition, Moerbeke translated *On Floating Bodies*, which came from another collection of texts on mechanics and optics. Scholars of Greece prepared the texts of Codex A and the texts on mechanics and optics.

Close to two centuries after Moerbeke translated Codex A, the seven works of Archimedes made it to Italy. In 1450, the philhellene and humanist Pope Nicholas V funded the translation of the works of Archimedes into Latin by Jacob of Cremona, a priest who knew Greek. In addition, Greek scribes working in Italy made several copies of Codex A. One of the copies of the works of Archimedes and Eutokios's commentaries was done exclusively for Cardinal Bessarion sometime before 1468. Bessarion, 1403–1473, a Greek bishop who had played an important role in the failed effort to unite the Catholic and Orthodox Churches, moved to Rome with his huge Greek library. In 1439, Pope Nicholas V made Bessarion a cardinal, thus assuring Bessarion's influence in spreading Hellenism in the West, becoming a major spark for the Renaissance. Finally, Ioannes Hervagius published the works of Archimedes at Basel in 1544. However, for close to two millennia, the transmission of the works of Archimedes had been a matter of accident and the occasional intervention of one or more devoted lovers of Hellenic thought. N. G. Wilson, professor of classics at Oxford, said the books of Archimedes reached us hanging by "a slender thread." The problems surfaced when Diokles, a Greek mathematician working in the early second century BCE, a generation after the death of Archimedes, could not find a copy of Archimedes's *On the Sphere and Cylinder*.[636]

Those of the works of Archimedes that no longer exist were probably destroyed when the Christians set fire to the Library of Alexandria and its daughter library, Serapeion in 391. Second, no one of those receiving the treatises of Archimedes from Archimedes put them together so that they could circulate as a single tome. Also, ancient mathematicians studied works like "On the Sphere and Cylinder," "Measurement of a Circle," and "On the Equilibrium of Planes" and ignored the more complex books with the result that those "profounder investigations fell more or less into oblivion." Even Eutokios, who was a serious student of Archimedes, studied the above works and knew

[636] N. G. Wilson, "Archimedes: The Palimpsest and the Tradition," *Byzantinische Zeitschrift* 92 (1999) 95. See also G. J. Toomer, ed., *Diocles on Burning Mirrors* (New York: Springer, 1976).

very little about Archimedes's famous books: *Quadrature of the Parabola* and *On Spirals*.[637] However, enough of the works of Archimedes survived to our time—70 percent—that we have no trouble in judging him for the scientific genius he was (chapter 4). Besides, we have the testimony of great scientists and scholars who read his writings. Among those of his writings that survived in partial or fragmentary form included *Mechanical Theorems*, *On Floating Bodies*, and *Stomachion*. Scribes copied them in 975 in Constantinople, producing a book made of parchment, promising a long life.

Greek Technology

From late tenth to most of the eleventh century, the Greek world, or, more precisely, the imperial government that funded the copying of ancient Greek books, including the works of Archimedes, was at the height of its power. The Greek emperor was the leader of the civilized world. The imperial palace was graced by Greek technological gadgets designed primarily to impress or frighten foreign visitors. Power—military and intellectual—was written all over the palace and Constantinople.

We have the testimony of a tenth-century foreign visitor to Constantinople, Liudprand of Cremona, a bishop who was the envoy of the king of Italy, Berengar, to the Greek court. On entering the emperor's throne room, Liudprand said, one would notice the emperor dressed in luxurious vestments sitting on a huge throne, which moved from being on the floor to great height. Second, right in front of the throne, there was a bronze tree covered with gold leaf, its branches filled with golden birds singing like real birds. Close to the tree there were two enormous gilded lions guarding the emperor, opening their large mouths and roaring, their tails sweeping the floor.[638]

These automata were related to machines designed during the golden age of Greek science; medieval Greek engineers probably built the imperial robots from manuals from Heron of Alexandria.

We have already discussed the use of "Greek fire," a secret chemical weapon made from naphtha or liquid petroleum, burning pitch, and sulphur. Anna Comnena, 1083–1153, daughter of Emperor Alexios Comnenos, 1081–1118, said the

[637] E. J. Dijksterhuis, *Archimedes*, tr. from the Dutch by C. Dikshoorn (Princeton, NJ: Princeton University Press, 1987) 33–41.

[638] Liudprand, "Antapodosis," *Migne Patrologie Latina* 136: col. 895.

prow of Greek warships had bronze heads of lions with their mouths open. A thin layer of gold covered these lion heads in order to make them terrifying. During battle, the tubes carrying the liquid Greek fire moved through the mouths of the lions "in such way that they appeared to be belching out the fire."[639]

Medieval Greek scientists were also making tenuous forays into the scientific achievements of their ancestors, especially the exploration of the heavens.

The Orthodox Church Attacks the "Ungodly" Hellenes

Despite the cultural and political superiority of the Greek medieval empire, the perpetual conflict between Hellenism and Christianity sapped the Hellenic identity and vision of the Greeks. The results were unpleasant and destructive. Christian Greeks failed to know themselves. They did the bare minimum to the copying, editing, and studying of the works of their ancestors. But, unlike the Arabs, they failed to follow up and advance the sciences. They made themselves vulnerable to the barbarians waiting at the gates of the empire.

Michael Psellos mirrors this dramatic crisis. His life covered the eleventh century, 1018–1096. He was an exceptional intellectual caught in the maws of chaotic state policies concerning the Greek heritage of Christian Greece. He was a man of letters serving power. He was an advisor to several emperors. He was a prolific writer, a lover of Hellenic culture, especially Plato. But he was also a politically correct Christian; he had to be. This art was by no means insignificant, helping him to survive and even thrive in an environment rarely known for its philhellenism.

Psellos made a substantial contribution to reviving the University of Constantinople and, therefore, the copying of Greek texts. He was a Renaissance man centuries before the Renaissance. He fought the good fight all his life. He supported classical studies and Greek learning. In his *Chronographia* (Chronography, history), probably his best work about his times, Psellos bemoans the dramatic cultural decline of ancient cities like Athens, Alexandria, and the old and new Rome (Constantinople). He talks about the complete Christian damming of ancient Greek civilization, the golden streams of the Hellenic past.[640]

---

[639] Anna Comnena, *The Alexiad* 11.ix–x, tr. E. R. A Sewter (New York: Penguin Books, 1982).

[640] Michael Psellos, *Chronographia* 6.43, ed. Vrasidas Karales, 2 vols (Athens: Kanakes, 2004). For an excellent English translation of the Chronographia, see: *Fourteen Byzantine Rulers*, tr. E. R. A. Sewter (New York: Penguin Books, 1966).

The authorities did not threaten him, though his fear of one of the emperors drove him to "retire" in the silences and anonymity of a monastery.

A modern scholar of medieval Greece, Michael Angold, says that the secular education and philosophy of Psellos made him think of himself as the "guide of emperors and the moral arbiter of society."[641] That's probably true. Another modern scholar, Antony Kaldellis, gives Psellos more influence than he ever had. He says that Psellos pushed the Christian god "out of the world," creating "space for autonomous scientific inquiry based on Greek notions."[642] I would not go that far. Psellos did not push the Christian god out of his thinking, much less the world. But he defended the Hellenic tradition at the price of having to humiliate himself with confessions of faith.

The defenders of orthodoxy, including Emperor Alexios Comnenos, put on trial Psellos's star pupil, John Italos, because Italos had the audacity of taking Plato seriously. Italos was professor of Hellenic philosophy at the University of Constantinople. Anna Comnena, who was an outstanding historian, reports that Italos, a Greek refugee from southern Italy, lectured on Plato and Aristotle, as well as later Hellenic philosophers like Porphyry, Iamblichos, and Proklos, whose lives spanned the third to the fifth centuries. Anna accused Italos as having a "barbaric, stupid temperament." The man could not control himself, raising his voice, even resorting to fist fights with his critics. In addition, he spoke Greek with "defective pronunciation," often "mutilating syllables." He was a bad influence on the young, Anna said, even subverting the dogmas of the Christian faith. Italos kept causing "trouble," leading many to sedition. Anna's father, Emperor Alexios Comnenos, questioned Italos and, unsatisfied, referred him to an ecclesiastical court. In 1082, that court found Italos guilty of heresy. In fact, even patriarch Garidas and several nobles sided with Italos, the patriarch becoming the "dedicated disciple" of Italos. This caused a crisis in Constantinople and the church anathematized Italos and his ideas.[643]

The curse of the church against Italos was also a curse, an anathema, against Hellenism. The church attacked the "impious" dogmas of the Greeks and their civilization. The church also anathematized those who study the Greeks, believing in Greek ideas and values, spreading them to others openly

[641] Michael Angold, *Church and Society in Byzantium under the Comneni 1081–1261* (Cambridge: Cambridge University Press, 2000) 51.

[642] Antony Kaldellis, *Hellenism in Byzantium* (Cambridge University Press, 2007) 203.

[643] Comnena, *The Alexiad* 5.viii–ix.

or in secret. The anathema is especially vengeful of Plato and those studying Plato's ideas. The church inserted its anathema of Hellenism into the "Synodikon of Orthodoxy," a liturgical text meant to celebrate the triumph of orthodoxy over iconoclasm. That victory took place in 843. The curse of the church against the Greeks became dogma in 1082. It included the following:

"Anathema on those who claim to be pious but shamelessly or rather impiously introduce the ungodly teachings of the Hellenes into the orthodox Catholic Church concerning human souls and heaven and Earth and other created objects. Anathema on those who go through a course of Hellenic studies and are taught not simply for the sake of education but follow these empty notions and believe in them as the truth, upholding them as a firm foundation to such an extent that they lead others to them, sometimes secretly, sometimes openly, and teach them without hesitation. Anathema on those who of their own accord invent an account of our creation along with other myths, who accept the Platonic forms as true, who say that matter possesses independent substance and is shaped by the forms, who openly question the power of the creator to bring all things from non-existence to existence, and as their creator to impose a beginning and end on all things in the manner of their lord and master."[644]

This anathema against Hellenism has been read aloud in Orthodox churches since 1082. Such abominable practice takes place during the Feast of Orthodoxy in the first Sunday of Lent. We don't know if all Orthodox priests read this shameful text to the Orthodox faithful—in the twenty-first century. But we do know some priests keep alive the hatred for the ancient Greeks.[645]

Psellos grasped the enormity of the anathema. He defended his student, saying that he had little doubt the Hellenes of his time had become barbarians and the barbarian Muslims had become Hellenes. He explained the meaning of this as follows:

"Suppose a Hellene goes to [the Persian cities of] Susa or Ecbatana, the ancient palace of Darius, or talks to the inhabitants of Babylon: he will hear of things which despite his Hellenic culture he had never heard of, he will admire everyone he meets, and he will perhaps then learn for the first time that wisdom

---

[644] Wilson, *Scholars of Byzantium*, 154.

[645] See, for example, the 2010 translation of the anathema against Italos on the website of a Greek American Orthodox priest: The "Synodikon of Orthodoxy" in *Mystagogy Resource Center: An International Orthodox Christian Ministry* by John Sanidopoulos.

has directed their affairs. But if a pretentious barbarian visited us and talked to people in Hellas or any part of our continent, he would treat the majority of men not as asses but as of mulish stupidity. The greater part of the population knows nothing of the world of nature or of what lies beyond it, the remainder think that they know everything but do not in fact even know the route towards knowledge. Some claim to be philosophers and a great many more are anxious to learn. But the teachers sit with smug faces and long beards, looking pale and grim, with a frown, shabbily dressed. They dig up Aristotle from the underworld, from the depths of Hades, and give the impression of passing judgment on everything that he covered in a cloud of obscurity ... they give ... many brief explanations of the broad range of his researches. Our barbarian visitor is convinced that this is a childish game, gloats over our incompetence, and departs, with no addition to his knowledge but reduced to a state worse than ignorance."[646]

Psellos made this extraordinary statement in the late eleventh century. He must have been livid over the incoherent and outrageous anathema the church hurled against Hellenic culture, the primary passion of his life. But his angry comment also shows that despite the university he revived and his classical studies, and despite the wealth and power of medieval Greece, most of the Christian Greeks of his age were illiterate, living in a society flirting with barbarism, one of pretense and delusion.

The Outside World Raged against the Greeks, 1204

In addition to this internal rage over Hellenic culture and the identity crisis the clash of civilizations caused in medieval Greece, Anna Comnena admitted, "the whole outside world raged against us."[647] In fact, Christian and Muslim enemies surrounded Greece. Not only was the empire largely Greek in culture, but it was also wealthy and, in contrast to Rome and Western Europe that had been taken over by barbarians, Greece had kept the barbarians at bay—until 1204.

During that year, 1204, Christian armies from Western Europe, especially France, Germany, and Venice, sacked Constantinople, burning and looting

[646] Wilson, *Scholars of Byzantium*, 155–156.
[647] Comnena, *The Alexiad* 12.v.

the city. They used "debt" as an excuse for their plunder. A Greek noble with ambitions for the imperial throne promised Venice to pay the costs of the crusade, if only the crusaders sided with him against the legitimate emperor. But the pretender, put in the imperial palace with the weapons of Germans, French, and Venetians, could not fulfill his promise. The Greek treasury was empty. This gave the crusaders the excuse for storming and capturing Constantinople.

Niketas Choniates, 1140–1213, a Greek scholar present at the conquest of Constantinople, recorded the horrendous atrocities the crusaders of Italy, Germany, and France inflicted on the Greeks. He was certain the Europeans wanted what the Greeks had, their vicious plunder making reconciliation next to impossible. They even dug up the graves of the emperors searching for gold, pearls, and gems. Choniates denounced such barbarian practices, saying the "Western nations spared neither the living nor the dead."

The Latin crusaders also melted the Greeks' surviving classical sculpture for making silver and bronze coins. That way, Venice got its money back. The huge bronze statue of Hera and the statues of Paris Alexander and Aphrodite ended up in the furnace. The crusaders also melted a bronze "mechanical device," which Choniates described as a large and beautiful work of art, tall as a column. He is silent about the functions of this device, but from what he says about the "intricacy of its ornamentation," it's possible this device or machine had something to do with agriculture or rural life or the decipherment of the heavens. The sculpture of this device must have been ancient, modeled on the Shield of Achilles crafted by Hephaistos. Choniates reports the art of this device as follows:

"Every melodious bird, warbling its springtime tunes, was carved upon it; the tasks of husbandmen, the pipes and milk pails, and the bleating sheep and bounding lambs were depicted. The widespread sea and schools of fish were to be seen, some caught and others shown breaking out of the nets to swim free again in the deep. There were Erotes [Loves, sons of the goddess Aphrodite], shown in pairs and groups of three; innocent of clothing but armed with apples, they shook with sweet laughter as they threw these or were pelted by them. This four-sided monument terminated in a point like a pyramid, and above was suspended a female figure which turned with the first blowings of the wind, whence it was called *Anemodoulion* [Wind Servant]."[648]

[648] *O City of Byzantium, Annals of Niketas Choniates* 648, tr. Harry J. Magoulias (Detroit: Wayne State University Press, 1984).

Choniates called the crusaders "unlettered" barbarians who hated the beautiful. Among the statues they melted was that of Herakles, which, according to Choniates, was an "incomparable masterpiece." Another great work of art the crusaders destroyed was a bronze eagle dedicated to the first-century philosopher Apollonios of Tyana. This sculpture graced the Hippodrome of Constantinople. The eagle, depicted as if in flight, was holding a serpent in its sharp claws. The eagle, Niketas Choniates tells us, was so magnificent that it had a spellbinding impact on humans. But the Venetians, Germans, and Franks threw the eagle of Apollonios, along with other masterpieces of ancient Greek sculpture, into the flames.[649]

Greeks, Arabs, and Mongols After the Crusades

The surviving ancient Greek culture in Constantinople took a devastating blow from the European crusaders. It never recovered. The Greeks recaptured Constantinople in 1261. However, the reestablishment of Greek rule was now over a diminished state. The crusaders dismembered Greece, making the Greeks vulnerable to the Mongol Turks who, in 1071, had taken over most of Asia Minor, the heart of Greek Ionia, and the medieval Greek Empire. Steven Runciman, one of the best European historians of the twentieth century, was right that the fourth crusade had bad global consequences, including being a mortal blow to Greece and Christian unity. "There was never a greater crime against humanity than the Fourth Crusade," Runciman wrote. "Not only did it cause the destruction or dispersal of all the treasures of the past that Byzantium had devotedly stored, and the mortal wounding of a civilization that was still active and great; but it was also an act of gigantic political folly ... it upset the whole defense of Christendom.... In the wide sweep of world history, the effects were wholly disastrous."[650]

The Greek emperors in the thirteenth and fourteenth centuries tried to contain the crusading disaster by stemming the Mongol tide. They kept bribing the Mongol kings of Persia with marriages to their illegitimate daughters. They also funded the translation of Arabic texts into Greek, a desperate measure of studying their enemies under the stress of war.

[649] Niketas Choniates, *Historia* in *Patrologiae cursus completes: Series graeco-latina*, ed. J. P. Migne (Paris, 1857–1866), 139: cols. 1049–1052; *O City of Byzantium, Annals of Niketas Choniates* 650–651, 653.

[650] Steven Runciman, *A History of the Crusades*, 3 vols (New York: Harper Torchbooks, 1954) III, 130.

And the tragedy of what appeared to be a scholarly exchange between the Greeks and the Mongol-dominated Middle East was for the Greeks rediscovering Greek scientific texts, already "lost" among them, in their Arabic translations or adaptations to the needs of the Mongol and Islamic Middle East. Such changes, which often emasculated the ancient Greek works from their original polytheistic culture, did not bother the Christian Greeks who, like the Arabs, believed in one god. Christian Greeks also got in touch with Arabic texts through their contacts with Western Christendom that had already been translating Arabic texts into Latin. Those Latin translations of Arabic texts, which had been translations of ancient Greek texts, were now (in the thirteenth and fourteenth centuries) being translated into Greek.

In the early fourteenth century, Emperor Andronikos II urged his prime minister, Theodore Metochites, to study astronomy. Metochites followed up and studied under a Greek scientist, Manuel Vryenios, who had learned his astronomy from another Greek educated in Persia. This, and a few Greek contacts with two Mongol observatories (one in Maragha in east Azerbaijan and another in Tabriz, Persia) convinced a modern medievalist to describe Persia as "the forefront of Islamic astronomical research."[651]

Clearly, the Arabs benefited enormously from the wisdom and science of the Greeks. But the glory days of their translation into Arabic of many of the Greeks' philosophical and scientific works ended in the tenth century, four centuries before the perceived "astronomical research" in Mongol-occupied Persia.

Nikephoros Gregoras, 1290–1360, a distinguished Greek scholar, did not think much about Persia or the Mongols. He was a historian, rhetorician, and astronomer who learned his science from Metochites. He predicted solar eclipses.[652] He wrote in the mid-fourteenth century, rejecting the notion that the Mongols could have brought with them anything but destruction. Gregoras accuses the Mongols of genocide because they wiped out the ancient culture of Persia, Egypt, Mesopotamia, and the entire Middle Eastern region. In the twenty-fourth and twenty-fifth books of his "Roman History,"[653] he says there was no way that science could flourish in the land of the Mongols who

[651] Maria Mavroudi, "Exchanges with Arabic Writers during the Late Byzantine Period" in *Byzantium: Faith and Power (1261–1557): Perspectives on Late Byzantine Art and Culture*, ed. Sarah T. Brooks (New York: The Metropolitan Museum of Art and New Haven: Yale University Press, 2007) 66.

[652] Nicéphore Grégoras, *Correspondance*, ed. R. Guilland (Paris, 1927), 137, 153.

[653] Nikephoros Gregoras, *Roman History* in *Patrologiae cursus completus: Series Graeca posterior*, ed. J. P. Migne (Paris, 1857–1866), vol. 148: book 24; vol. 149: book 25.

killed the ancient wisdom of the Middle East. He likened the Mongols to the violent waters of a great arctic sea drowning Persia and all nations in Asia and the Middle East.[654] Gregoras also knew the Arabs well. He was fully aware that the Arabs were indebted to the Greeks for their science. But the Mongols took care of the Arabs, too.

Gregoras was a proud Greek who resented the presence of Muslims in countries steeped in Hellenic culture. At that moment, late in the decline of political medieval Greece, Gregoras mirrored a Renaissance-like movement in Greece where scholars like him were digging in the ruins of Hellas, identifying Homeric names of cities and trying to preserve the historical memory of the Greek past. But this hopeful dream struggled with the legacy of 1204. The catastrophe of the fourth crusade was too deep in Greece. The plunder and occupation of Constantinople for fifty years and the occupation of divided Greece for much longer by Christian Europeans hating Christian Greeks had dramatic consequences, like the bombardment and destruction of the Parthenon by Venetians in the seventeenth century.

Second, the Greek-European religious and political conflict, an anathema the churches hurled at each other in 1051, prepared the onslaught of the Mongol Turks surrounding Greece. The perpetual cold war between Christian Greeks and the Christian crusaders of the West and the Turkish Mongols of the East led to rapid cultural and political decline of the Greeks. They began exporting Greek books to other "safe" havens.

Archimedes in a Tomb

One such ancient Greek book contained the "Mechanical Theorems," "On Floating Bodies," and the "Stomachion" of Archimedes. This book was in the library of a Greek monastery in Jerusalem. Yet, the fate of the book of Archimedes, copied in 975 in Constantinople, would be almost worse than death. In 1229, monks ripped the Archimedes book and other ancient Greek books apart, folding their large vellum folios to create an Euchologion, or prayer book. The monks used orange juice to delete the original writing on the

[654] Gregoras, *Roman History*, xxv, 1–20. A modern historian, J. M. Roberts, captured the vision of Gregoras about the Mongols. The Mongols, he says, "blew up like a hurricane to terrify half a dozen civilizations, slaughtered and destroyed on a scale the twentieth century alone has emulated." *History of the World* (New York: Oxford University Press, 1993) 299.

vellum or scraped off the ancient Greek text with a knife. Then the monks used the "clean" folios for writing their hymns and prayers.

We already mentioned that the Christian book that resulted out of this operation was a cultural nightmare that, in Greek, is known as a palimpsest: creating a book by the violence of rubbing or scraping off the original writing. In this case, a prayer book palimpsest came into being in 1229 after its clerical publishers trashed the wisdom of Archimedes and other Greeks.

The next thing that happened, three hundred years later, in the sixteenth century, is that the Archimedes palimpsest ended up in the St. Sabas monastery in the Judean desert only fifteen miles from Jerusalem. Another 250 years passed in silence and, suddenly, someone took the Euchologion and brought it to the Metochion monastery in Constantinople. In mid-1840s, a German scholar, Constantin Tischendorff, examined the prayer book in the Metochion library in Constantinople and noticed the mathematics under the prayers. Tischendorff left the Metochion library with a folio from the palimpsest. When Tischendorff died, the executors of his estate sold to the Cambridge University Library, in 1876, this Archimedes fragment and some forty-three additional folios the "great" Biblical scholar had ripped off from Greek manuscripts.

About fifty more years passed and the palimpsest remained untouched in the manuscript and book collection of the Metochion library. Finally, in the late 1890s, Athanasios Papadopoulos-Kerameus, a Greek scholar, catalogued 890 manuscripts in the collection of the patriarchal library.[655] One of those catalogued manuscripts was the Archimedes palimpsest. Papadopoulos-Kerameus immediately detected the science under the prayers and said that much, thus revealing to the world the secret of the palimpsest. Perhaps, Papadopoulos-Kerameus, who was a secretary of the Patriarchate of Jerusalem and then professor at St. Petersburg University in Russia, had something to do with bringing the Archimedes palimpsest to the church of the Holy Sepulcher in Constantinople; yet, incredibly enough, this Greek scholar never thought of taking this unique treasure of Archimedes to Greece. The works of Archimedes left this man cold.

But the news of Archimedes in a prayer book touched profoundly another man from Denmark. This was a famous philologist and historian of Greek

[655] Athanasios Papadopoulos-Kerameus, ed., Jerusalem Library: A Catalogue of Greek Codices Found in the Libraries of the Holy and Apostolic Orthodox Patriarchate of Jerusalem and All Palestine, 5 vols. in 6, in Greek (St. Petersburg: V. Kirspaoum, 1891–1915).

science by the name of Johan Ludvig Heiberg, 1854–1928. In 1906, tipped off by a German philosopher, Hermann Schone, about the study of Papadopoulos-Kerameus, Heiberg visited Constantinople in 1906 and 1908, where he examined the Archimedes palimpsest very carefully. For the first time since 1229, an expert in classical studies looked at the folios of the prayer book with a magnifying glass, confirming the findings of Papadopoulos-Kerameus that underneath the prayers lay the thought of Archimedes. Heiberg transcribed as much of the Archimedes text as he could. Then he took pictures of two-thirds of the folios. Heiberg edited the works of Archimedes.[656] This was of inestimable value, especially for any future decipherment of the palimpsest text.

The Plunder of the Buried Archimedes

We don't know what happened to the Archimedes palimpsest after Heiberg studied it briefly in the summers of 1906 and 1908. Heiberg incorporated his palimpsest findings in the final edition of the works of Archimedes in the 1910s. World War I, 1914–1918, and the war between Greece and Turkey in the early 1920s provided the perfect excuse for the continued indifference of the scholarly community towards Archimedes and the horrendous treatment of the palimpsest at the hands of its "owners." According to Reviel Netz and William Noel, the worst damage to the Archimedes palimpsest occurred during the twentieth century. They describe a story they stitched together from court documents relating to the legal contest on the "ownership" of the Archimedes book, and from interviews with art dealers, professors, antiquity thieves, and prestigious houses cashing in on the shady trade in stolen treasures.

In the 1920s, Salomon Guerson, a French Jewish merchant in rare carpets and antique tapestries working in Paris, joined Dikran Kelekian, an Armenian dealer of antiquities, in stealing or buying the palimpsest from the monks of the Metochion monastery in Constantinople. Guerson brought the manuscript to Paris where, sometime after 1938, he seriously damaged four of its folios by painting on them pictures of the Evangelists looking like pictures he had seen in Greek manuscripts kept at the National Library of France in Paris. Guerson needed money badly, so he painted the forgeries on the manuscript to

[656] Johan Ludvig Heiberg, ed., *Archimedis opera omnia cum commentariis Eutocii*, 3 vols. (Leipzig: Teubner, 1910–1915/1972).

make it more attractive to Christian buyers. He did sell it to Marie Louis Sirieix, a Frenchman who, after hiding the book in his basement for decades, passed it on to his daughter, Anne Guerson, who finally sold it.

Professor Reviel Netz, a scholar from Israel, studied Greek science at Cambridge University, becoming an expert on Archimedes. He and William Noel quite by accident collaborated on a project of taking care of the Archimedes palimpsest that, suddenly, came as a loan for their artistic and scholarly study. An anonymous American bought the Archimedes palimpsest for $2 million when it surfaced in an October 29, 1998, auction at Christie's in New York. William Noel had the admirable insight of convincing this very wealthy man to let the Walters Art Museum exhibit the palimpsest and, above all, use the latest imaging technologies in reading the Archimedes text under the Christian prayers.

The goodwill, almost Renaissance-like philhellenism, and generosity of the rich Archimedes patron and, especially, the working together of Netz and Noel, resulted in the 2007 book, *The Archimedes Codex*, a very interesting and extremely important study[657] about the Archimedes palimpsest and the technological trials, and they were extensive and difficult, in deciphering a manuscript in an appalling state of disintegration and disappearance. Its 174 parchment folios and 8 paper pages were showing the effects of mildew and possible contact with fire. For example, Abigail Quandt, a colleague of William Noel at the Walters Art Museum, took four years in disassembling the folios of the manuscript. Noel documented how imaging experts and classical scholars, following the leads of the Danish philologist Heiberg, managed to decode the hidden Archimedes text.

The Archimedes Palimpsest

The Christians "erased" Archimedes because they despised the Greeks. It is also possible that the monks erased the Archimedes text because they had to have parchment for their prayer book. In Christian Greece, there was a ceaseless need for parchment for book production, which the Christians met by their destructive palimpsest method of reusing the parchment of old Greek books, including books several centuries old. The Archimedes palimpsest was probably

[657] Reviel Netz and William Noel, *The Archimedes Codex: How a Medieval Prayer Book Is Revealing the True Genius of Antiquity's Greatest Scientist* (New York: Da Capo Press, 2007).

just another product of the Christians' excommunication of Greek culture. Monks created their prayer books, even if that meant the prayers would be written over the scrubbed text of famous ancient Greek authors. They could have cared less about Greek books.

So, this palimpsest adds another piece of evidence of how complex, elaborate, and violent the Christian ideological and political system against non-Christian views was, especially Greek culture. And considering that Christianity borrowed so much from Greek polytheism and used Greek to spread its message, its hatred for the Greeks becomes even more distasteful.

The Archimedes palimpsest shatters my dreams of a lost world, a cold reminder of what is, rather than what could have been. That palimpsest, a model of barbarism, kept some of the ideas of Archimedes and the ideas of other Greek writers buried for centuries.

Some thirty folios of the Archimedes palimpsest have nothing to do with Archimedes. The monks, who ripped apart the Archimedes book, tore apart other books as well. Ten folios come from the work of Hyperides, a famous Athenian orator born in 389 BCE. Natalie Tchernetska, a Latvian graduate student at Cambridge University, made the discovery: that a fragment in the palimpsest came out of the speeches of Hyperides. We know that Hyperides delivered seventy-seven speeches in the courts of Athens. He defended Phryne, a beautiful Hetaira accused of offending the Eleusinian mysteries. To make his point of how beautiful Phryne was, Hyperides tore her robe in the court, revealing her breasts to the jury, which acquitted Phryne. The Hyperides text in the Archimedes palimpsest adds light to the history of the Persian Wars—suggesting that Athens manned 220 ships in the Salamis battle against Xerxes in 480 BCE. This text will add 20 percent to the surviving writings of Hyperides. Like Demosthenes, Hyperides opposed the Macedonian King Philip's hegemony over Greece, including Philip's son, Alexander the Great. When Alexander died in 323 BCE, Hyperides participated in a failed rebellion against the Macedonian Greeks. He was executed. The Archimedes palimpsest also includes seven folios containing "ancient views" on Aristotle.

Netz and Noel speculate that the palimpsest was completed months after Frederick II, Holy Roman Emperor of Sicily, Cyprus, Jerusalem, and Germany, liberated Jerusalem on February 18, 1229, from Muslim rule.

"The scribe of the [Archimedes palimpsest] book [Ioannes Myronas]," they write, "had joy in his heart as he wrote his prayers. Now that we knew that we could recover the texts in the Palimpsest we had nothing but understanding for the scribe and thanks for the fact that he had used such treasures to write his book."[658]

Reviel Netz and William Noel added Ioannes Myronas to those to whom they dedicated their book. It is possible, of course, that "Mr. B," the American "owner" of the palimpsest, had something to say about Myronas. I don't know.

Aristotle among Western Christians and Muslims

Just like Archimedes, Aristotle deserves lots of praise. His work was so powerful and attractive that, it is worth repeating, both the Arabs and medieval Christian churchmen referred to him as *The Philosopher*. However, the Christians approached Aristotle for purely selfish reasons. They read their doctrines into Aristotle, thinking that by borrowing the vocabulary of Aristotle they would boost their irrational beliefs. The Arabs did the same thing.

Here's how it happened: Starting in the twelfth century, some of the works of Aristotle that arrived in Western Europe through the Arabs became the object of anxiety among the leaders of Christianity—and with a "good" theological reason.

The ideas of Aristotle about the eternity of the Earth and the universe threatened to inflict a mortal blow against the Christian god who supposedly created the world from nothing; this Christian god was also expected to call his followers to account in a final day of judgment. So, in 1210, 1215, and 1231, the Christians repudiated the natural and cosmological works of Aristotle. In fact, in 1231, they removed the works of Aristotle from publication. But Aristotle's ideas about the natural world and the universe were so rich that, by 1255, his natural philosophical works were textbooks in medieval universities. However, theologians, once again, denounced Aristotle in 1277.

In 1210, the bishop of Paris ordered that, "Neither the books of Aristotle on natural philosophy nor their commentaries are to be read at Paris in public or secret, and this we forbid under penalty of excommunication." In 1270, the bishop of Paris condemned the views of Aristotle on the eternity of the world. In 1277, Pope John XXI ordered the bishop of Paris to investigate the influence of the works of Aristotle at the University of Paris. The bishop did not like

[658] Netz and Noel, *The Archimedes Codex*, 232.

what he discovered. He issued a broad frontal attack against the metaphysical and cosmological works of Aristotle.[659]

The Greek text of Aristotle arrived in Western Europe from the Greek East in the twelfth century. Europeans studied Aristotle from better Latin translations of the Greek text, especially the translations done by William of Moerbeke between 1260 and 1286. Greek scholars living in the West translated Aristotle into Latin. One of the Greek successful translators was Theodore Gaza. He translated Aristotle's biological works for Pope Sixtus IV in the 1470s.

The Western Europeans also had access to the Arabic commentators of Aristotle, especially the comprehensive commentaries of the Persian Muslim Avicenna (Abu Ali al-Husayn ibn Abdullah ibn Sina) and the Spanish Muslim Averroes (Ibn Rushd). Avicenna flourished in the eleventh century, 980–1037, and Averroes in the twelfth century, 1126–1198. Avicenna looked at Aristotle through the eyes of Plato and Averroes with less Platonic influence. But between them they presented Aristotle as the greatest philosopher who ever lived. Their commentaries on Aristotle were translated into Latin and found a wide audience in Western Europe.

All of this improved understanding, respect, and support for Aristotle, including that of students and professors at medieval universities. The result was a "juggernaut of Aristotelian philosophy" that was "unstoppable."[660]

Nevertheless, the church kept a wary and excommunicating eye on Aristotle and any other "pagan" threatening its obsolete dogmas. Creating the universe out of nothing has been fine for Christians.

Theoretical Purgatory for Aristotle in Modern Times

However, Western intellectuals gradually put aside the ambivalence of Christianity towards the Greeks. They started reaping the fruits of Greek science in the sixteenth and seventeenth centuries and thereafter. The Latin translations of the mathematical works of Euclid, Archimedes, Apollonios of Perga, Ptolemaios, and Pappos sparked the scientific revolution of

[659] "The Reaction of the Universities and Theological Authorities to Aristotelian Science and Natural Philosophy" in *A Source Book in Medieval Science*, ed. Edward Grant (Cambridge, MA: Harvard University Press, 1974) 42–50.

[660] David C. Lindberg, *The Beginnings of Western Science*, second edition (Chicago: University of Chicago Press, 2007) 227.

the seventeenth century. This was the century of Galileo, Kepler, Fermat, Descartes, Huygens, and Newton—the "heirs and beneficiaries" of the recovery of Greek science and mathematics.[661] The same thing happened with the rebirth of medicine. The translation into Latin of the works of almost the complete Hippocratic corpus and Galen set the foundations for modern medicine in Renaissance Europe.[662]

Yet, like all newly empowered people, they tried to raise their self-esteem by taking all the credit for science. For example, European but especially French astronomers in the eighteenth century attacked the Greek astronomer Ptolemaios, whose books educated Europeans and Arabs on the heavens for 1,500 years. French astronomers who knew little history robbed Ptolemaios of his original contributions to astronomy. Their attacks essentially degraded Ptolemaios. Yet, according to the twentieth century Danish scholar Olaf Pedersen, Ptolemaios "remains forever one of the most prominent figures in the history of astronomy."[663]

Giorgio de Santillana, the Italian American historian of science, said the epigones of Aristotle in modern times "degraded" him, making Aristotle "a master of quibbles."[664] Pierre Pellegrin, a French scholar who made his career by attacking Aristotle, in a later study complemented the argument of Santillana that the birth of physics in the seventeenth century "provoked" a "theoretical frenzy" against Aristotle, putting him in a "theoretic purgatory" for three centuries.[665] Yet, Galileo Galilei, 1564–1642, Isaac Newton, 1642–1727, and a few other great scientists remained faithful to the original Greek spark that kept them going.

[661] Swerdlow, "The Recovery of the Exact Sciences of Antiquity," in *Rome Reborn*, 165–167.
[662] Nancy G. Siraisi, "Life Sciences and Medicine in the Renaissance World," in *Rome Reborn*, 178.
[663] Olaf Pedersen, *A Survey of the Almagest* (Denmark: Odense University Press, 1974) 22–25.
[664] Giorgio de Santillana, *The Crime of Galileo* (New York: *Time*, 1962) 21.
[665] Pierre Pellegrin, "Aristotle" in *Greek Thought*, 574.

Conclusion

The Legacy of Greek Science and Civilization

This fascinating and complex story of science, suffering, and resilience explains how and why the Greeks created this marvel of heavens and Earth we call the Antikythera Mechanism. It's a complex but immaculate geared device of genius. This story also explains the near miraculous survival of Greek civilization out of which, in 1900, Greek sponge divers unearthed a stone-like artifact that turned out to be the fabulous Tablet. This celestial machine came out of Greek polytheism, democratic politics, the empire of Alexander the Great, the Mouseion and Library of Alexandria, the science and philosophy of Parmenides, Aristotle, Euclid, Eudoxos, Archimedes, Apollonios of Perga, Ktesibios, Aristarchos of Samos, Hipparchos, Poseidonios, and Geminos. In other words, philosophy, cosmology, astronomy, mathematics, and scientific technology brought the Antikythera computer to life. It's like a miniature galaxy engulfing the technological legacy and philosophy of the golden age of Greek science, fourth century BCE to the second century of our era.

The Tablet—the laptop Cosmos of the geared computer—came into being from the widespread tradition of *techne,* the application of the sciences and crafts coming out of geometry purposefully applied in the design of machines for the benefit of all Greeks. It was a reliable religious, athletic, and agricultural calendar. It connected celestial phenomena to the seasons, sowing and harvest, sacrifices to the gods, and the two- and four-year cycles of religious and athletic celebrations in the Greek world. Because of its predictive function, it served astronomers, farmers, priests, and athletes. It revealed the secrets of the stars by exhibiting the order of the whole heavens: it predicted the will of the gods.

Or, in the eyes of an Indian journalist writing in 2015, the Antikythera Mechanism "was indeed one-of-its-kind and truly an immaculately designed one that perfectly entailed complex wheels and mechanics to predict nature's doing. By bringing in the whole of the cosmos molded into the bronze structure the Greeks went on to pave the way for what entails modern scientific predictions and astronomical purviews."[666]

Indeed, the Tablet, itself a mechanical universe, worked through the Cosmos. It stood on the shoulders of great philosophers and scientists. Its complex interlocking gears made it the everyday workhorse of Greek astronomical and calendrical knowledge. It probably became the portable laptop computer of late antiquity and inspired similar practical machines. Moreover, it was used as a scientific instrument to check the applicability or improvements of theories.[667] It may also have been used for instructing students about the heavens. In addition, the Antikythera computer reported the heliacal risings and settings of great stars and constellations. It measured the angular distances of the Moon, the Sun, and the planets from the Earth. It mechanized the predictions of the eclipses of the Moon and the Sun, the faces of the Moon, and those of scientific theories. It gave us clues for the mechanization of mathematics and science. It was a computer of genius. It was the culmination of science and civilization among the Greeks.

The Library of Alexandria and its sibling institution, the Mouseion, captured that intensive quest for knowledge and its application for the well-being of the Greek world. The library and the Mouseion in Alexandria made the Aristotelian model of science Panhellenic and global. For the first time in world history, in the Alexandrian Era, we have a concerted effort to study and understand the world and the Cosmos. The library, with its fantastic collections of Greek books and books from other cultures, opened the human mind in ways reminiscent of our times. The Alexandrian Library sowed the seeds for our culture and civilization.

The Tablet was the supreme product of the Alexandrian Library and the philosophy and civilization that library represented. That was an age based on knowledge. And the Greeks of that age sponsored and invented modern-like science and technology, sowing the roots of our world.

---

[666] Preetam Kaushik, "Theories on Antikythera Mechanism," *Business Insider*, November 13, 2015.
[667] Tassios, "Prerequisites for the Antikythera Mechanism."

The Tablet was an instrument of the heavens and Earth. Its sophistication, indeed, modern technology, could not have been developed for the indulgence of a few rich people. It was the high-tech version of a technology that took generations to mature. It was a creation born out of the ancient Greeks' love and teaching of natural philosophy, astronomy, mathematics, physics, and engineering. It was dedicated to the practical and beneficial application of science.

Most of the wonders of ancient Greek thought and culture were lost through the suppression of Christianity. With the banning of the Olympics, the oracle of Apollo at Delphi, the Eleusinian Mysteries, and the burning of the great Library of Alexandria, the adherents of this new religion drove this rich world of philosophy and science into the shadows. Existing copies of ancient works became scarce or lost, or palimpsests left to be discovered hundreds of years later by scholars. According to Andrew White: "[T]he greatest thinkers in the [Christian] Church generally poured contempt upon all investigations into a science of Nature, and insisted that everything except the saving of souls was folly. ... For twelve centuries ... the physical sciences were ... discouraged or perverted by the dominant orthodoxy."[668]

Now, at the beginning of the third decade of the twenty-first century, the Antikythera Mechanism/Tablet, a fragment of the treasures of the past, once hidden or obscured, is available to us as an inspiration to continue to advance science and culture. To Kostes Stekas, a Modern Greek writer, editor, and photographer with deep interests in watchmaking, the Antikythera computer is an introduction to where we come from. It elucidates the origins of our Greek ancestors and illustrates what they could do and did.[669] This is essential, and not just for Modern Greeks. Just like Mathias Buttet, in one way or another, we in the Western world are children of the ancient Greeks, or, at least, some of us like to think that's the case.

The distinguished American academic, Allan Bloom, urged America educators to return to the Greeks because the Greeks teach wisdom. They are also "the models for modern achievement."[670]

In 1993, Bernard Knox, director emeritus of the Harvard Center for Hellenic Studies and author of several insightful writings on Greek literature and theater, reminded us not to forget or ignore "the astonishing originality

[668] White, *A History of the Warfare of Science with Theology in Christendom*, I, 375–376.

[669] Stekas, *The Antikythera Mechanism* 9.

[670] Allan Bloom, *The Closing of the American Mind* (Penguin Books, 1987) 369.

that sets ... [the Greeks] apart, that makes them unique. They invented democracy more than two thousand years before any modern Western nation took the first steps towards it; they invented not only philosophy and the theater but also the model of a national literature, with its canon of great writers, its critics and communicators, its libraries; they invented organized competitive athletics ... they invented political theory, rhetoric, biology, zoology, the atomic theory ... Above all, in startling contrast to the magnificent but static civilizations of the great Eastern river valleys—Tigris, Euphrates, and Nile—they created in the restless turbulence of their tiny city-states that impatient rhythm of competition characteristic of Western civilization ever since... we must always acknowledge how greatly, how deeply, how irrevocably, we are in their debt."[671]

Another American, historian and Librarian of Congress, Daniel Boorstin, reminds us about the need to be in constant touch with the Greeks. The Greeks, he said, represented the best and the greatest the world had to offer. "We are disciples of all of them," he said. "They all have become our contemporaries."[672]

Bruce Thornton, professor of classics at California State University-Fresno, is certain that the Greeks "started the human race down the long, difficult road of its liberation. ... That is why today we must once more listen to that brilliant, fractious discussion started by the Greeks, learn again to speak with their phrases and to see with their eyes. We must listen to the Greeks not because they will give us the answers, but because they first formulated the questions and problems, and they first identified whence we too believe the answers are likely to come: not from the revelations of the priest but from the minds of free rational human beings."[673]

In 2016, Sir John Boardman, emeritus professor of classical archaeology and art at Oxford University, confirmed the confidence of Bloom, Knox, Boorstin, and Thornton in the genius of Greek civilization. He said Greek culture and art were "the expression of a people unsurpassed for their purity of thought, behaviour and design, democrats, philosophers, poets and historians,

[671] Bernard Knox, *The Oldest Dead White European Males And Other Reflections on the Classics* (New York: W. W. Norton, 1994) 66–67.

[672] Daniel J. Boorstin, *The Seekers: The Story of Man's Continuing Quest to Understand His World* (New York: Random House, 1998) 21.

[673] Bruce Thornton, *Greek Ways: How the Greeks Created Western Civilization* (San Francisco: Encounter Books, 2000) 14, 198.

the true precursors of the modern world and all-time successes in the pursuit of the true and the beautiful."[674]

Also, in 2016, *National Geographic* magazine and America's Public Broadcasting System produced a series of historical documentaries on the history and achievement of the Greeks. They introduced their program with this paean:

"[The Greeks] were an extraordinary people born of white rock and blue sea. They invented democracy, distilled logic and reason, wrote plays to plump the deepest recesses of the soul, and captured the perfection of the human form in athletics and art. Quite simply, the Greeks created our world."

National Geographic and PBS praised the Greeks not merely because the Greeks deserved such praise but also because they wished to discover "how their legacy illuminates our present, and will shape our future. The story of the Greeks is the story of us."[675] On August 29, 2018, PBS reported on the "invisible" Athens, its everlasting invention and contribution to democracy.[676]

True, the Greeks are our desired models. This book is proof of that. But unlike the Greeks, we (American, Japanese, European, Russian, and Chinese corporations, industrialized farmers—and armies) keep poisoning and destroying the world. Our factory fishing fleets are desolating the oceans, our timber and mining companies are wrecking forests, rivers, and wildlife, our factory agriculture is unsettling the countryside, poisoning the human and natural worlds, and even engineering plagues. Our petroleum, coal, and natural gas-burning industries are causing climate change, a dire danger engulfing the Earth, wildlife, and civilization. And our wars and nuclear weapons are threatening nuclear apocalypse.

This means "we" are far from being like the Greeks who thought the natural world was sacred: a place for plants, animals, gods, and people for living, dancing, and enjoying life. We need to learn from them. The Tablet, the miniature galaxy and its Cosmos, the world's most ancient computer, is telling us we have to reverse our priorities from a MacBook for private profit to an Antikythera Book largely for the public good. That vision can and must reshape our computers and machinery so we become lovers of the Greeks and lovers of the natural world and the Cosmos.

---

[674] Boardman, *Greek Art*, 9.

[675] "The Greeks," produced by *National Geographic* Studios in association with PBS, June 21, 2016.

[676] "Ancient Invisible Cities: Episode 1: Invisible Athens" by Darius Arya. PBS, August 29, 2018.

In April 1961, the Russians used the Antikythera gear computer to model their spacecraft gear computer for carrying Yuri Gagarin in space. In late August 2016, Stephanos Paipetis, professor of engineering at the University of Patras, addressed an international conference in Olympia about ancient Greece and the modern world. He said he learned from studying ancient Greek technologies. Probably the same thing is happening to NASA. In March 2017, it decided to design a geared machine made from hardened metals, the Automaton Rover for Extreme Environments, for the exploration of the hot planet Aphrodite/Venus. The brain of the Automaton is a device modeled after the Antikythera Mechanism.[677]

Paipetis was certain that ancient Greece is a source of inspiration and learning, especially knowledge that may save the Earth and us from destruction.[678] The Antikythera computer is part of that knowledge.

[677] Jonathan Keats, "Walking with Venus' Wind," *Discovery Magazine*, March 10, 2017.

[678] S. A. Paipetis, "Opening Address" in Ancient Greece and Contemporary World, International Conference, Olympia, Greece, August 28–31, 2016. The proceedings of the conference are now a book: S. A. Paipetis and Chrestos Kechagias, eds., *Ancient Greece and Contemporary World* (Athens: Livanes Publishing for the University of Patras, 2017) 13 (in Greek).

Acknowledgments

It took me several years to write this book. I dug deep into Greek history and the history of science. I followed the news of the scientific studies trying to decode the fragments of the Antikythera Mechanism. To me, this was a voyage of discovery and pleasure. Unexpectedly, I found myself rereading the fragments of philosophers, the works of Aristotle, and other important witnesses of the history of the ancient Greeks. I met lots of interesting people, some of whom played a significant role in discovering the science secrets of this 2,200-year-old computer. I thank them all for their determination to get to the bottom of this extraordinary device, which, at first, looked as if it had arrived from another planet.

I am grateful to Xenophon Moussas for his encouragement. I met him in the summer of 2007. He gave me a guided tour of the Acropolis and the neighboring area where the Athenians had their political and administrative institutions and central market, agora. We stopped at the Tower of the Winds built in the second century BCE. The now gutted stone tower used to house a hydraulic clock, a marvel of architecture and technology. We then went to the National Archaeological Museum and I saw for the first time the fragments of the ancient Greek computer.

I looked at those fragments again in 2009, 2011, 2013, 2015, 2017, and 2018 with the curiosity and delight of a child and the affection and seriousness of an adult. Could it be, I kept saying to myself, that my ancestors, more than 2,200 years ago, first created the model of the society I grew up in, especially the American society that educated me? Yet I could not help feeling embarrassed. Despite the fact I must have been in that museum dozens of times, I had never seen the fragments of the Antikythera Mechanism before 2007. But here I was with Xenophon Moussas who, judging from the way he spoke about those fragments, gave me to understand he had been in love with the Antikythera

Mechanism for several decades. Moussas and I became good friends. He read sections of this book and made useful comments. He also handed me dozens of images of the Antikythera Mechanism.

These images don't belong to Moussas or, for that matter, the Antikythera Mechanism Research Project, an informal organization of scientists who studied the Antikythera device. Most photographs and images are a result of the 2005 imaging of the fragments of the computer at the National Archaeological Museum in Athens. Tom Malzbender, Hewlett-Packard Laboratories, photographed the exterior of the machine. His Polynomial Texture Mapping gave us detailed and revealing images of the inscriptions on the front and back plates. These images may be used, provided the user credits Tom Malzbender and Hewlett Packard. A British company, X-Tek Systems (Nikon Metrology), photographed the interior of the mechanism. Its X-ray Computed Tomography revealed the gears and architecture of the computer. Equipment loaned by X-Tek Systems (now owned by Nikon Metrology) was used to collect the data. Like in the case of Hewlett-Packard, the images (data) of the X-Tek Systems are available to researchers, if they give it credit.

The National Archaeological Museum, which owns the fragments of the Antikythera Mechanism, also owns the rights to still photographs of the device. I am grateful to this museum, the Epigraphic Museum, the Numismatic Museum, the Acropolis Museum, and the Numismatic Collection of the Alpha Bank in Athens. I am using a few images from the Numismatic Museum and the Alph Bank.

I am grateful to Antonios Pinotsis, another physicist at the University of Athens, for the work he has done on the history of Greek science, especially the history of science of Rhodes, an early center of the Greek scientific revolution in the Alexandrian Age. He also read part of this book and his comments were useful. Moreover, I thank Dionysios Kriaris for building models of the Antikythera Mechanism as well as models of major ancient Greek scientific instruments. He is a mathematician living in Athens. I talked to him in the summers of 2013, 2015, and 2018. He explained to me he had built gadgets since his youth. He put his tremendous technical skills into fashioning a small machine that is probably very similar to the Antikythera computer of antiquity. One can see Kriaris's reconstruction of the Antikythera Mechanism in the National Archaeological Museum in Athens. In addition, I am very grateful to him for giving me pictures of his reconstruction of the Antikythera computer and other ancient Greek scientific instruments.

I am thankful to Tony Freeth, Tom Malzbender, and Magdalini Athanasiou for their contributions to the understanding of the Antikythera computer. Athanasiou wrote an incisive Aristotle University of Thessalonike dissertation on the Antikythera computer. Many thanks to the Library of Congress and Wikipedia for expanding our education with images and knowledge about the Greeks. Wikipedia follows a noble tradition with its restoration of museum-locked Greek images to the public domain.

Unfortunately, copyright restrictions hinder research and the presentation of the beautiful, which is an intimate part of Greek science and civilization. Tony Freeth said to me that none of the images of the Antikythera Mechanism are in the public domain. Tom Malzbender, however, confided to me that he wants all images to be in the public domain. Nevertheless, copyright restrictions and the cost of publishing images in color reduced dramatically the number of images in this book, even from the Greek museums.

I reserve my profound gratitude for Georgia Nomikos and her Hellenic non-profit organization, Pan Logos Foundation, for encouragement and support. Eva Sarantea gave me permission for using three of her paintings: the beautiful painting of the Antikythera Mechanism gracing the book cover, a painting of Aristotle, and a painting of Poseidonios. I thank her. Yanna Darilis and Alexandros Hahalis of the New Greek TV in New York interviewed me about the Antikythera Mechanism. I thank them. Darilis and I also share a passion for Alexander the Great.

I benefited from the comments of Christos Evangeliou, John Lenz, Stephen Bertman, Stephanos Paipetis, Robert Garland, Debbie Luican, Graham White, and Sir John Boardman. I thank them.

Science is no one's private asset, being a gift of the gods to all human beings searching for truth. No doubt this book emphasizes the Greek side of science: what the Greeks did with the science they invented. But like the ancient Greek students of science—men like Herodotus, Plato, Aristotle, Eudemos, Pappos, and Proklos—who gave credit to non-Greeks, I, too, give credit to whoever has the most convincing argument and reliable evidence.

I wrote most of the book in Claremont, California, my home since 2008. I am grateful to the Honnold Mudd Library of the Claremont Colleges for opening its rich book collection to me.

Finally, I am very grateful to my wife, Crista. She suggested the title of this book and kept encouraging me, giving me time for writing and reflection.

Bibliography

I. Primary Sources

Anna Comnena, *Alexiad* in *The Alexiad of Anna Comnena*. Tr. E. R. A. Sewter (New York: Penguin Books, 1982).

Apollonios of Perga, *On Conic Sections*. Ed. Evaggelos S. Stamates (Athens, 1975–1976), 4 vols.

Aratos, *Phenomena*. Tr. D. Kidd (Cambridge, 1997).

Archimedes, *Complete Works*. Ed. Kaktos Philological Team (Athens, 2002), 6 vols.

Archimedes in the Middle Ages, Volume I, *The Arabo-Latin Tradition*. Ed. and tr. Marshall Clagett (Madison, 1964).

Archimedes, *On the Sphere and Cylinder, Measurement of a Circle, On Conoids and Spheroids, On Spirals, On the Equilibrium of Planes, The Sand-Reckoner, Quadrature of the Parabola, On Floating Bodies, The Cattle Problem, The Method of Mechanical Theorems*. In Thomas Heath, ed., *The Works of Archimedes* (New York: Dover, 2002; originally published by Cambridge University Press, 1897).

——— *Opera Omnia*. Ed. Iohan Ludvig Heiberg. First published 1910. Corrected and edited by Evaggelos S. Stamates (Stuttgart, 1972), 3 vols.

——— *On the Sphere and the Cylinder*. In *The Works of Archimedes*. Tr. Reviel Netz, Volume I: *On the Sphere and the Cylinder* (Cambridge: Cambridge University Press, 2004).

Archytas of Tarentum: Pythagorean, Philosopher and Mathematician King. Ed. Carl A. Huffman (Cambridge, 2005).

Aristarchos of Samos, *On the Sizes and Distances of the Sun and the Moon*. Ed. Evaggelos Spandagos (Athens, 2001).

Aristotle, *Generations of Animals*. Ed. and tr. A. L. Peck (Loeb, 1943).

———*Historia Animalium*. Ed. D. M. Balme, prepared for publication by Allan Gotthelf (Cambridge, 2002).

———*Meteorology*. Ed. and tr. H. D. P. Lee (Loeb, 1952).

———*Movement of Animals*. Tr. E. S. Forster (Loeb, 1937).

———*Parts of Animals*. Ed. Kaktos Philological Team (Athens, 1994), 2 vols.

———*Parva Naturalia*. Ed. David Ross (Oxford, 1955).

———*Physics*. Tr. Joe Sachs (New Brunswick, 1998).

———*Politics*. Ed. and tr. H. Rackham (Loeb, 1998).

———*Posterior Analytics*. Ed. and tr. H. Tredennick and E. S. Forster (Loeb, 1960).

———*Prior Analytics*. Tr. H. Tredennick (Loeb, 1938).

———*Progression of Animals*. Tr. E. S. Forster (Loeb, 1937).

——— *The Basic Works of Aristotle*. Ed. and tr. Richard McKeon, introduction by C. D. C. Reeve (New York, 2001).

Autolykos of Pitane, *Works*. Ed. Evaggelos Spandagos (Athens, 2002).

Cicero, *The Nature of the Gods*. Tr. Horace C. P. McGregor (Penguin Books, 1972).

——— *The Republic*. Ed. and tr. Clinton Walker Keyes (Loeb, 1928).

——— *Tusculan Disputations*. Tr. J. E. King (Loeb, 1945, 1971).

Codex Theodosianus. Ed. T. Mommsen and P. M. Meyer (Berlin, 1905), 3 vols.

Dio, *Epitome of Roman History*. Tr. E. Cary (Loeb, 1914–1927), 9 vols.

Dio Chrysostom, *Discourses*. Tr. J. W. Cohoon and H. Lamar Crosby (Loeb, 1940), 5 vols.

Diocles. *On Burning Mirrors*. Ed. G. J. Toomer (New York, 1976).

Diodorus Siculus, *The Library of History*. Ed. and tr. C. H. Oldfather et al. (Loeb, 1933–1967), 12 vols.

Diogenes Laertius, *Lives of Eminent Philosophers*. Ed. and tr. R. D. Hicks (Loeb, 1995) 2 vols.

Diophantos, *Arithmetika*. Tr. Thomas L. Heath (Cambridge, 1885 and New York, 1964).

Dioskourides, *Complete Works*. Ed. Kaktos Philological Team (Athens, 2000), 8 vols.

Euclid, *Elements*. Tr. Thomas Heath (Cambridge, 1926 and New York, 1956), 3 vols.

Eunapios, *Lives of Philosophers*. Tr. Wilmer Cave Wright (Loeb, 1998).

Freeth, Tony et al., "Decoding the ancient Greek astronomical calculator known as the Antikythera Mechanism," *Nature*, Vol. 444, November 30, 2006, 587–591.

——— "Calendars with Olympiad display and eclipse prediction on the Antikythera Mechanism," *Nature*, Vol. 454, July 31, 2008, 614–617.

Galen, *On Food and Diet*. Ed. and tr. Mark Grant (New York, 2000).

——— *On the Natural Faculties*. Ed. and tr. A. J. Brock, Loeb, 1916.

——— *On the Properties of Foodstuffs (De alimentorum facultatibus)*. Ed. and tr. Owen Powell (Cambridge, 2003).

——— *On the Sects for Beginners*. Tr. Frede (Indianapolis, 1985).

——— *On the Usefulness of the Parts of the Body*. Tr. Margaret Tallmadge May (Ithaca, 1968), 2 vols.

——— *Prognosis*. Ed. and tr. V. Nutton (Berlin, 1979).

Galileo, *Dialogue Concerning the Two Chief World Systems*. Tr. Stillman Drake (Berkeley, 1967).

——— *Discoveries and Opinions of Galileo*. Tr. Stillman Drake (New York, 1957).

Geminos, *Introduction to the Phenomena*: in *Geminos's Introduction to the Phenomena*. A translation and study of a Hellenistic survey of astronomy by James Evans and J. Lennart Berggren (Princeton, 2006).

Herodas, *The Mimes and Fragments*. Ed. and tr. Walter Headlam and A. D. Knox (Cambridge, 1922).

Herodotus, *The Histories*. Tr. Robin Waterfield (Oxford, 1998).

Heron, *Automatopoietikes* [*Automaton-making*]. Tr. in S. Murphy, 1995: "Heron's Automatopoietikes," *History of Technology* 17, 1–44.

——— *Belopoiika* [*Artillery-construction*]. Tr. E. W. Marsden, *Greek and Roman Artillery: Technical Treatises* (Oxford, 1971) 106–184.

——— *Heronis Alexandrini Opera Supersunt Omnia*. Ed. Wilhelm Schmidt, L. Nix, Hermann Schone and J. L. Heiberg. 5 vols. (Leipzig, 1899, 1903, 1976).

——— *Pneumatics*: in *The Pneumatics of Hero of Alexandria*. Ed. Bennet Woodcroft, tr. Joseph Gouge Greenwood (London, 1851, 1971).

Herophilos, *The Art of Medicine in Early Alexandria*. Ed. and tr. H. von Staden (Cambridge, 1989).

Hesiod, *Theogony, Works and Days, Shield*. Tr. Apostolos N. Athanassakis (Baltimore, 1983).

Hipparchos, *The Geographical Fragments of Hipparchus*. Ed. and tr. D. Dicks (London, 1960).

Hippocratic Writings. Ed. G. E. R. Lloyd, tr. J. Chadwick and W. N. Mann, I. M. Lonie, E. T. Withington (Penguin Books, 1983).

Hippocrates, *Ancient Medicine, Airs Waters Places, Περι διαιτης (Regimen)*. Tr. W. H. S. Jones (Loeb, 1995).

Homer, *The Iliad*. Ed. M. M. Willcock (London, 2001).

Homer, *The Odyssey*. Ed. Kaktos Philological Team (Athens, 1992).

Kallimachos, *Hymns and Epigrams*. Tr. A. W. Mair (Loeb, 2006).
Lucretius, *On the Nature of the Universe*. Tr. R. E. Latham, revised by John Godwin (Penguin Books, 1994).
Pausanias, *Guide to Greece*. Tr. Peter Levi (Penguin Books, 1984), 2 vols.
Philostratos, *Lives of the Sophists*. Tr. Wilmer Cave Wright (Loeb, 1998).
Pindar, *The Odes and Selected Fragments*. Tr. G. S. Conway and Richard Stoneman, ed. Richard Stoneman (Everyman, 1997).
Plato, *Complete Works*. Ed. John M. Cooper and D. S. Hutchinson (Indianapolis: Hackett, 1997).
——— *Gorgias*. Ed. E. R. Dodds (Oxford, 1990).
——— *Euthyphron, Theaitetos, Republic, Phaidros, Timaeus*. Ed. E. A. Duke, Ioannes Burnet (Oxford Classical Texts, 1995).
——— *Phaidon*. Ed. C. J. Rowe (Cambridge Greek and Latin Classics, 2001).
Pliny (the Elder), *Natural History*. Ed. and tr. H. Rackham (Loeb, 1938–1963), 10 vols).
Pliny the Younger, *The Letters of the Younger Pliny*. Tr. Betty Radice (Penguin Books, 1969).
Ploutarchos, *Animals are Rational, The Causes of Natural Phenomena, Convivial Questions, On the Face of the Orb of the Moon, On the Principle of Cold, Whether Land or Sea Animals are Cleverer*, all these works are in *Moralia*. Ed and tr. Frank Cole Babbitt et al. (Loeb, 1928–1976), 16 vols.
——— *Life of Cato the Elder, Life of Marcellus, Life of Pompey, Life of Theseus, Life of Alexander*, in *Lives*. Ed. and tr. Bernadotte Perrin (Loeb, 1967–1994), 11 vols.
Polybios, *The Rise of the Roman Empire*. Tr. Ian Scott-Kilvert, ed. F. W. Walbank (Penguin Books, 1979).
Poseidonios, *The Fragments*. Eds. L. Edelstein and D. Kidd (Cambridge, 1972).
——— *The Translation of the Fragments [of Poseidonios]*. By D. Kidd (Cambridge, 1999).
Price, Derek de Solla, "Gears from the Greeks: The Antikythera Mechanism—A Calendar Computer From ca. 80 B.C.," *Transactions of the American Philosophical Society*, Vol. 64, Part 7 (Philadelphia, November 1974).
Proklos, *Commentary on the First Book of Euclid's Elements*. Tr. G. R. Morrow (Princeton, 1992).
Prokopios, *Secret History (Anekdota)*. Tr. H. B. Dewing (Loeb, 1969).
Psellos, Michael. *Chronographia*. Ed. Vrasidas Karales (Athens, 2004), 2 vols.
——— *Fourteen Byzantine Rulers*. Tr. E. R. A. Sewter (Penguin Books, 1982).
Ptolemaios, *Almagest*. Tr. G. J. Toomer (Princeton, 1998).

——— *Hypotheses*. Tr. B. Goldstein, *Transactions of the American Philosophical Society* 57.4 (Philadelphia, 1967).
——— *Geography*: An annotated translation of the theoretical chapters by J. Lennart Berggren and Alexander Jones (Princeton, 2000).
——— *Optics*. Tr. A. M. Smith, *Transactions of the American Philosophical Society* 86.2 (Philadelphia, 1996).
——— *Tetrabiblos*. Ed. Kaktos Philological Team (Athens, 2001).
Said al-Andalusi. *Book of the Categories of Nations*. Ed. and tr. S. I. Salem and A. Kumar (Austin, Texas, 1991).
Selected Variae of Magnus Aurelius Cassiodorus Senator. Tr. S. J. B. Barnish (Liverpool, 2010).
Seneca, *Epistles*. Ed. and tr. R. M. Gummere (Loeb, 1917–1925), 3 vols.
——— *Natural Questions*. Ed. and tr. T. H. Corcoran (Loeb, 1971–1972), 2 vols.
Sophocles, *Antigone*. Ed. Mark Griffith (Cambridge Greek and Latin Classics, 2003).
Strabo, *Geography*. Ed. and tr. H. L. Jones, Loeb, 1988, 8 vols.
The Greek Sophists. Tr. and ed. John Dillon and Tania Gergel (Penguin Books, 2003).
The Homeric Hymns. Tr. Apostolos N. Athanassakis (Baltimore, 1976).
The Orphic Hymns. Tr. and ed. Apostolos N. Athanassakis (Atlanta, 1977).
Theokritos, *Idylls*. Ed. and tr. A. S. F. Gow (Cambridge, 1950), 2 vols.
Theophrastos, *Complete Works: Causes of Plants, History of Plants, Characters, On Fire, On Fish, Metaphysics, Meteorology, On Odors, On the Senses, On Stones, On Weather-signs*. Ed. Kaktos Philological Team (Athens, 1998), 11 vols.
Thucydides, *The Peloponnesian War*. Tr. Walter Blanco, ed. Walter Blanco and Jennifer Tolbert Roberts (New York, 1998).
Tzetzae, Ioannis. *Historiae*. Ed. Petrus Aloisius M. Leone (Naples, 1968).
Vitruvius, *On Architecture*. Ed. and tr. Frank Granger (Loeb, 1956), 2 vols.
Xenophon, *Oeconomicus*. Ed. and tr. Sarah B. Pomeroy (Oxford, 2001).

II. Selections of Primary Sources

Ancient Greece: Documentary Perspectives. Ed. and tr. Stylianos V. Spyridakis and Bradley P. Nystrom (Dubuque, Iowa, 1985).
Anthology of Classical Myth. Ed. and tr. Stephen M. Trzaskoma et al. (Indianapolis/Cambridge, 2004).
A Presocratics Reader. Ed. Patricia Curd, tr. Richard D. McKirahan (Hackett Publishing Company, 1996).

A Source Book in Greek Science. Ed. and tr. Morris R. Cohen and I. E. Drabkin (New York, 1948).

Byzantium: Church, Society, and Civilization Seen through Contemporary Eyes. Ed. and tr. Deno John Geanakoplos (Chicago, 1984).

Galen, *Selected Works*. Tr. P. N. Singer (Oxford, 1997).

Greek Astronomy. Tr. Thomas L. Heath (First published 1932, reprinted by AMS Press, New York, 1969).

Greek and Roman Military Writers: Selected Readings. Ed. and tr. Brian Campbell (New York, 2004).

Greek and Roman Technology: A Sourcebook. Ed. and tr. John W. Humphrey, John P. Oleson and Andrew N. Sherwood (New York, 1998).

Greek Mathematical Works. Ed. and tr. Ivor Thomas (Loeb, 2002), 2 vols.

Greek Medicine from the Heroic to the Hellenistic Age: A Source Book. Ed. and tr. James Longrigg (New York, 1998).

Greek Science of the Hellenistic Era: A Sourcebook. Eds. and trs. Georgia L. Irby-Massie and Paul T. Keyser (New York, 2002).

Hellenistic Philosophy: Introductory Readings. Tr. Brad Inwood and L. P. Gerson, second edition (Indianapolis, 1997).

Paganism and Christianity 100–425 C. E.: A Sourcebook. Ed. Ramsay MacMullen and Eugene N. Lane (Fortress Press, 1992).

Plato. *Selections*. Ed. Raphael Demos (New York, 1927).

Pliny the Elder. *Natural History: A Selection*. Tr. John F. Healy (Penguin Books, 1991).

Sources for the Study of Greek Religion. Ed. and tr. David G. Rice and John E. Stambaugh (Scholars Press, 1979).

The Greek Anthology. Tr. W. R. Paton (Loeb, 1960–1968), 5 vols.

The Hellenistic Philosophers. Volume 1: *Transactions of the Principal Sources, with Philosophical Commentary*. Ed. and tr. A. A. Long and D.N. Sedley (Cambridge, 1997).

The Presocratic Philosophers: A Critical History with a Selection of Texts. Ed. and tr. G. S. Kirk, J. E. Raven and M. Schofield, second edition (Cambridge, 1983).

The Roman World: A Sourcebook. Ed. David Cherry (Oxford, 2001).

III. Secondary Works

Anastasiou, Magdalini. *The Antikythera Mechanism: Astronomy and Technology in Ancient Greece* (In Greek, doctoral dissertation, Aristotle University of Thessalonike, June 2014).

——— et al. "The Back Dial and Back Plate Inscriptions," *Almagest*, 7–1, May 2016, pp. 138–215.

Angold, Michael. *Church and Society in Byzantium under the Comneni 1081–1261* (Cambridge, 2000).
Artmann, Benno. *Euclid—The Creation of Mathematics* (New York, 1999).
Aveni, Anthony F. *Ancient Astronomers* (Washington, DC, 1993).
Baikouzis, Constantino and Magnasco, Marcelo O. "Is an eclipse described in the Odyssey?" *Proceedings of the National Academy of Sciences of the United States of America*, Vol. 105, No. 26, July 1, 2008, 8823–8828.
Bamm, Peter. *Alexander the Great: Power as Destiny*. Tr. J. Maxwell Brownjohn. (New York, 1968).
Billows, Richard. *Before and After Alexander: The Legend and Legacy of Alexander the Great* (New York, 2018).
Bitsakis, Y and Jones, A. "The Front Dial and Parapegma Inscriptions," *Almagest*, 7–1, May 2016, pp. 69–137.
Boardman, John. *Greek Art*. Fifth Edition. (London, 2016).
Bowra, C. M. *The Greek Experience* (Cleveland, 1957).
Bertman, Stephen. *The Genesis of Science: The Story of Greek Imagination* (Amherst, NY, 2010).
Brunschwig, Jacques and Lloyd, Geoffrey E. R. Eds. *Greek Thought: A Guide to Classical Knowledge* (Harvard, 2000).
Burnet, John. *Early Greek Philosophy*. Fourth edition (London, 1930).
Canfora, Luciano. *The Vanished Library: A Wonder of the Ancient World* (Berkeley, 1990).
Carman, Christian C. and Evans, James. "On the epoch of the Antikythera mechanism and its eclipse predictor," *Archive for History of Exact Sciences*. (2014) 68:693–774.
Carr, Nicholas. *The Glass Cage: Automation and US* (New York, 2014).
Charette, François. "High tech from Ancient Greece," *Nature*, Vol. 444, November 30, 2006, 551–552.
Clagett, Marshall. *Greek Science in Antiquity* (New York, 1963).
Clarke, Arthur C. "Technology and the Limits of Knowledge." In *The Frontiers of Knowledge*. Ed. Brooke Hindle (Garden City, New York, 1975).
Clayman, Dee. *Berenice II and the Golden Age of Ptolemaic Egypt* (New York, 2014).
Cuvier, Georges. *Histoire des sciences naturelles* (Paris, 1841), 5 vols.
D'Angour, Armand. *The Greeks and the New: Novelty in Ancient Greek Imagination and Experience* (Cambridge, 2011).
Dijksterhuis, E. J. *Archimedes*, tr. from the Dutch by C. Dikshoorn, with a new bibliographic essay by Wilbur R. Knorr (Princeton, 1987).
——— *The Mechanization of the World Picture* (Oxford, 1969).
Dodds, E. R. *The Greeks and the Irrational* (Berkeley, 1951).

Dorminey, Bruce. "Antikythera Mechanism May Have Been World's First STEM [Science, Technology, Engineering and Mathematics] Project," *Forbes*, June 30, 2016.

Drachmann, A. G. *The Mechanical Technology of Greek and Roman Antiquity: A Study of the Literary Sources* (Copenhagen, 1963).

Draper, John William. *History of the Conflict Between Religion and Science*, 15th ed. (London, 1881).

Edmunds, Mike. "Does the Antikythera Mechanism have a hidden purpose?" *Daily Mail*, July 20, 2016.

Edwards, Charles Henry Jr. *The Historical Development of the Calculus* (New York, 1979).

Evangeliou, Christos C. *The Hellenic Philosophy: Between Europe, Asia and Africa* (Binghamton, 1997).

——— *Essays and Poems on Aristotle* (Sioux City, IA, 2016), 2 vols.

Evans, James. *The History and Practice of Ancient Astronomy* (Oxford, 1998).

Ferguson, John. *The Heritage of Hellenism* (New York, 1973).

Freely, John. *Aladdin's Lamp: How Greek Science Came to Europe Through the Islamic World* (New York, 2009).

Freeth, Tony. "Decoding an Ancient Computer," *Scientific American*, December 2009, Vol. 301, No. 6, 76–83.

——— "Eclipse Prediction on the Ancient Greek Astronomical Calculating Machine Known as the Antikythera Mechanism," *PLOS ONE*, 9(7), July 30, 2014.

——— et al. "A Model of the Cosmos in the Ancient Greek Antikythera Mechanism," *Scientific Reports*, 11, 5821, March 12, 2021.

Garland, Robert. *Ancient Greece: Everyday Life in the Birthplace of Western Civilization* (New York, 2013).

Georgiou, Aristos. "How Russia's Yuri Gagarin Became the First Man in Space Thanks to Steampunk Contraption," *Newsweek*, April 12, 2021.

Geymonat, Mario. *The Great Archimedes* (Waco, Texas, 2010).

Gibbon, Edward. *The Decline and Fall of the Roman Empire*, ed. Hugh Trevor-Roper (Everyman's Library, 1993), 6 vols.

Gray, Shirley B. "A Centennial Celebration of Two Great Scholars: Heiberg's Translation of the Lost Palimpsest of Archimedes—1907 Heath's Publication on Euclid's Elements—1908," *Notices of the American Mathematical Society*, Vol. 55, No. 7, August 2008, 776–783.

Gribbin, John. *Science: A History 1543–2001* (London, 2002).

Gutas, Dimitri. *Greek Thought, Arabic Culture* (New York, 1998).

Hamilton, Edith. *The Greek Way to Western Civilization* (New York, 1957).

Heath, Thomas. *Aristarchos of Samos* (First published 1913, Elibron Classics, 2007).

Highet, Gilbert. *The Classical Tradition* (New York, 1949).

Irwin, Terence, (ed.) *Classical Philosophy* (Oxford, 1999).

Iversen, Paul A. "The Calendar on the Antikythera Mechanism and the Corinthian Family of Calendars," *Hesperia* 86 (2017) 129–203.

James, Peter and Thorpe, Nick. *Ancient Inventions* (New York, 1994).

Jones, Alexander. *A Portable Cosmos* (New York, 2017).

——— "Historical Background and General Observations," *Almagest*, 7–1, May 2016, pp. 36–67.

Jones, C. P. *The Roman World of Dio Chrysostom* (Cambridge: Mass, 1978).

Kaldellis, Antony. *Hellenism in Byzantium* (Cambridge, 2007).

Kaltsas, Nikolaos et al. (eds.) *The Antikythera Shipwreck: The ship, the treasures, the mechanism* (Athens, 2012).

Kaplan, Sarah. "The World's Oldest Computer is Still Revealing its Secrets," *The Washington Post*, June 14, 2016.

Keats, Jonathan. "What is the Ultimate Machined Object? A New Exhibit of Ingenious Devices Shows How Innovation Works," *Forbes*, July 30, 2019.

Keyser, Paul T. and Irby-Massie, Georgia L. (eds.) *The Encyclopedia of Ancient Natural Scientists: The Greek Tradition and its Many Heirs* (New York, 2008).

Knorr, Wilbur Richard. *The Ancient Tradition of Geometric Problems* (New York, 1993).

Knox, Bernard. *The Oldest Dead White European Males and Other Reflections on the Classics* (New York, 1994).

Kocandrle, Radim and Couprie, Dirk L. *Apeiron: Anaximander on Generation and Destruction* (Cham, Switzerland, 2017).

Lazos, Christos. *Engineering and Technology in Ancient Greece* (Athens, 1998).

——— *Archimedes: The Brilliant Engineer*, fourth edition, in Greek (Athens, 1995).

Leake, William Martin. *Topography of Athens*, second edition, (London, 1841), 2 vols.

Leroi, Armand Marie. *The Lagoon: How Aristotle Invented Science* (New York, 2014).

Lilienthal, David E. *Change, Hope, and the Bomb* (Princeton, 1963).

Lindberg, David C. *The Beginnings of Western Science*. Second Edition (Chicago, 2007).

Lloyd, G. E. R. *Early Greek Science: Thales to Aristotle* (New York, 1970).

——— *Greek Science After Aristotle* (New York, 1973).

——— *In the Grip of Disease: Studies in the Greek Imagination* (Oxford, 2003).

——— *Magic, Reason and Experience: Studies in the Origins and Development of Greek Science* (Indianapolis, 1999).

Long, A. A. *Hellenistic Philosophy: Stoics, Epicureans, Sceptics*, second edition (Berkeley, 1986).

MacMullen, Ramsey. *Christianity and Paganism in the Fourth to Eighth Centuries* (New Haven, 1997).

——— *Christianizing the Roman Empire (A. D. 100–400)* (New Haven, 1984).

——— *Paganism in the Roman Empire* (New Haven, 1981).

Manuel, E. Frank. *Isaac Newton Historian* (Cambridge, MA, 1963).

Marchant, Jo. *Decoding the Heavens* (New York, 2009).

——— "In search of lost time," *Nature*, Vol. 444, November 30, 2006, 534–538.

Moussas, Xenophon. *The Antikythera Mechanism: The Tablet: The First Mechanical Universe: The Most Ancient Computer and Astronomical Instrument* (In Greek, Athens, 2012).

——— "The Antikythera Mechanism and the Pythagoreans," *Socrates*, Vol. 1 (2015) 77–86 (in Greek).

Murray, Gilbert. *The Rise of the Greek Epic* (First published 1907, New York, 1960).

Naddaf, Gerard. *The Greek Concept of Nature* (Albany, 2005).

Netz, Reviel and Noel, William. *The Archimedes Codex* (New York, 2007).

Netz, Reviel et al., (ed.) *The Archimedes Palimpsest* (New York, 2011), 2 vols.

Neugebauer, O. *History of Ancient Mathematical Astronomy*, 3 vols (Berlin, 1975).

Noble, Joseph Veach. *The Techniques of Painted Attic Pottery* (New York, 1965).

O'Leary, De Lacy. *How Greek Science Passed to the Arabs* (London, 1951).

Oleson, John Peter, (ed.) *The Oxford Handbook of Engineering and Technology in the Classical World* (New York, 2008).

Paipetis, Stephanos A. *The Unknown Technology in Homer* (New York, 2010).

Paipetis, Stephanos A. and Ceccarelli, Marco, (eds.) *The Genius of Archimedes—23 Centuries of Influence on Mathematics, Science and Engineering* (New York, 2010).

Paipetis, Stephanoa A. and Kechagias, Chrestos. (eds.) *Ancient Greece and Contemporary World* (Athens, 2017).

Papathanasiou, Maro. *Cosmological and Cosmogonic Thinking in Greece During the Second Millennium BCE* (Athens, 2016) (in Greek).

Pinotsis, Antonios D. "The observatory and the places of observations of Hipparchos of Rhodes," *Dodecanesian Chronicles*, 1994, 169–204 (in Greek).

——— "Why Hipparchos of Rhodes adopted the geocentric system and the models of epicycles and eccentric circles," *Dodecanesian Chronicles*, 1998, 97–137 (in Greek).

——— "The contribution of Poseidonios of Rhodes in Astronomy and Mechanics," *Dodecanesian Chronicles*, 2000, 377–452 (in Greek).

——— "Rhodian astronomers of the 4th and 3rd century B.C.," Proceedings of the 4th Astronomical Conference, 16–18 September 1999, Samos, Greece, 2001, 241–246.

——— "A comparative study of the evolution of the geographical ideas and measurements until the time of Eratosthenes," *Astronomical and Astrophysical Transactions*, Vol. 24, No. 2, April 2005, 127–138.

——— "Comparison and historical evolution of ancient Greek cosmological ideas and mathematical models," *Astronomical and Astrophysical Transactions*, Vol. 24, No. 6, December 2005, 463–483.

——— "The Antikythera Mechanism: who was its creator and what was its use and purpose?" *Astronomical and Astrophysical Transactions*, Vol. 26, Nos. 4–5, August–October 2007, 211–226.

——— "Astronomy in Ancient Rhodes" (Section of Astrophysics, Department of Physics, University of Athens), [2008].

Pollard, Justin and Reid, Howard. *The Rise and Fall of Alexandria* (New York, 2006).

Preus, Anthony. *Science and Philosophy in Aristotle's Biological Works* (New York, 1975).

Price, Derek de Solla. "An Ancient Greek Computer," *Scientific American*, 200 (6) (1959) 60–67.

Renan, Ernest. *Prayer on the Acropolis*, ed. and tr. from the French by Xenophon Dantis (Athens, 1962).

Rice, Rob S. "The Antikythera Mechanism: Physical and Intellectual Salvage from the 1st Century B.C.," USNA Eleventh Naval History Symposium, [1995].

Rihll, T. E. *Greek Science* (Oxford, 1999).

Roebuck, Carl (ed.) *The Muses at Work: Arts, Crafts, and Professions in Ancient Greece and Rome* (Cambridge, MA, 1969).

Rovelli, Carlo. *Reality Is Not What It Seems*, tr. From Italian by Simon Carnell and Erica Segre (New York, 2017).

Russo, Lucio. *The Forgotten Revolution: How Science Was Born in 300 BC and Why It Had to Be Reborn*, tr. from Italian by Silvio Levy (Springer, 2004).

Sambursky, Samuel. *The Physical World of the Greeks*, tr. from the Hebrew by Merton Dagut (New York, 1956).

Sarton, George. *Ancient Science and Modern Civilization* (Lincoln, NE, 1954).

———*A History of Science*, 2 vols (New York, 1970).

Seabrook, John. "Fragmentary Knowledge," *The New Yorker*, May 14, 2007, 94–102.

Shaw, Brent D. "The Divine Economy: Stoicism as Ideology," *Latomus*, Vol. 44, No. 1 (January–March 1985), 16–54.

Shelley, Mary. *Frankenstein or the Modern Prometheus*, ed. M. K. Joseph (Oxford, 1998).

Simmons, George F. *Calculus Gems: Brief Lives and Memorable Mathematics* (New York, 1992).

Siraisi, Nancy G. "Life Sciences and Medicine in the Renaissance World." In *Rome Reborn: The Vatican Library and Renaissance Culture*. Ed. Anthony Grafton (Washington, DC, 1993), 169–197.

Stekas, Kostes. (ed.) *The Antikythera Mechanism: Researchers Unlocking the Secrets of the Most Ancient Mechanical Universe* (Athens, 2016). In Greek.

Stern, Sacha. *Calendars in Antiquity* (Oxford, 2012).

Swerdlow, N. M. "The Recovery of the Exact Sciences of Antiquity: Mathematics, Astronomy, Geography." In *Rome Reborn: The Vatican Library and Renaissance Culture*. Ed. Anthony Grafton (Washington, DC, 1993), 125–167.

Szabo, Arpad. *The Beginnings of Greek Mathematics*. Tr. from the German by A. M. Ungar (Boston, 1978).

Tanes, K., Palyvou, K. and Mpolote, S. (eds.) *Ancient Greek Technology*, in Greek (Thessalonike, 2000).

Tarn, W. W. *Hellenistic Civilization*, third edition revised by the author and G. T. Griffith (Cleveland, 1966).

Thompson, D'Arcy Wentworth. *A Glossary of Greek Birds* (Oxford, 1895).

——— "Natural Science: Aristotle." In *The Legacy of Greece*. Ed. Richard Livingstone (Oxford, 1961) 137–162.

——— *On Aristotle as a Biologist* (Oxford, 1913).

Thornton, Bruce. *Greek Ways: How the Greeks Created Western Civilization* (San Francisco, 2000).

Tzonis, Alexander. *Classical Greek Architecture: The Construction of the Modern* (Paris, 2004).

Vallianatos, Evaggelos. *The Passion of the Greeks: Christianity and the Rape of the Hellenes* (Harwich Port, Cape Cod, 2006).

——— "The Antikythera Mechanism," *Bulletin of Science, Technology and Society* 2009; 29: 506–511.

——— "Archimedes moved the world," *College Quarterly*, Fall 2010, Vol. 13, No. 4.

——— "Deciphering and appeasing the heavens: The history and fate of an ancient Greek computer," *Leonardo*, 45:3, pp. 250–257, May 15, 2012.

——— "Reaching for the Stars: The World's First Computer from the Greeks," *Huffington Post*, August 2, 2012.

——— "Civilization Shift," *Huffington Post*, May 6, 2013.

——— "Storms in the Wine-Dark Sea," *Huffington Post*, May 20, 2013.

——— "Platonism and Secularism in the Thought of Plethon," *Huffington Post*, May 28, 2013.
——— "Archimedes Palimpsest," *College Quarterly*, Spring 2013, Vol. 16, No. 2.
——— "Science Is Thinking About the World in the Greek Way," *Huffington Post*, January 29, 2014.
——— "Archimedes: The Greatest Scientist Who Ever Lived," *Huffington Post*, May 27, 2014.
——— "Greeks Sailing by the Stars," *Huffington Post*, August 26, 2014.
——— "Programming Our Computer-like Brains with Greek Philosophy and Mathematics Leads to Virtue," *Huffington Post*, March 24, 2015.
——— "The Pythagorean Paradigm," *Huffington Post*, May 7, 2015.
——— "The Exceptionalism, Greatness and Immortality of Ancient Greece," *Huffington Post*, 20 July 2015.
——— "Archimedes' Mirrors in the Mojave Desert," *Truthout*, July 18, 2016.
——— "Leisure and the Perfected Life," *Huffington Post*, February 12, 2017.
——— "The Beautiful and the Good in Greek Civilization," *Hellenic Insider*, March 6, 2018.
——— "Olympic Dreams," *Helios*, March 23, 2018.
——— "The Greek Cosmos," *Helios*, April 4, 2018.
——— "Searching for Aristotle in Lesbos," *Helios*, July 7, 2018.
——— "Greek Dancing," *Helios*, September 29, 2018.
——— "How I Fell in Love with Greek Art," *Counterpunch*, January 4, 2019.
——— "Vultures Over Greece and Macedonia," *Counterpunch*, February 5, 2019.
——— "The Lure of the Past," *Counterpunch*, February 12, 2019.
——— "Literature and Theater During War: Why Euripides Still Matters," *Counterpunch*, February 19, 2019.
——— "On Home Ground in a Distant Land," *Counterpunch*, March 4, 2019.
——— "Robots in the Vast Memory Palace of Myth," *Counterpunch*, March 11, 2019.
——— "Human Flourishing (Eudaimonia)—an Antidote to Extinction?" *Counterpunch*, April 25, 2019.
——— "Time Machine: Re-Visiting the Golden Age of Greek Science," *Counterpunch*, April 29, 2019.
——— "Dare to Know: The Paradox of Greek History," *Counterpunch*, May 3, 2019.
——— "My First Cousin Odysseus," *Counterpunch*, July 26, 2019.
——— "What We Owe Homer," *Counterpunch*, August 9, 2019.
——— "Power and Tragedy," *Counterpunch*, September 11, 2019.
——— "The Evolution of Pandora," *Counterpunch*, February 7, 2020.
——— "Herakles in the Age of Climate Chaos," *Counterpunch*, February 18, 2020.

——— "Plague and Civilization," *Counterpunch*, April 13, 2020.
——— "War and Power in Classical Greece: Lessons for Superpowers and the World," *Counterpunch*, May 18, 2020.
——— "The Golden Age of Greek Science," *Greek Reporter*, June 6, 2020.
——— "Frankenstein Chimeras," *Counterpunch*, June 24, 2020.
——— "A Just and Talented Government for Our Hazardous Age," *Counterpunch*, July 6, 2020.
——— "Knowledge, Virtue and Power," *Counterpunch*, July 13, 2020.
——— "The Meaning of the Battle of Salamis," *Counterpunch*, August 14, 2020.
——— "Another Renaissance?" *Counterpunch*, October 9, 2020.
——— "Hellas Reborn?" *Counterpunch*, October 16, 2020.
——— "Thebes of Myth and History," *Counterpunch*, December 25, 2020.
——— "Freedom or Death: How Adamantios Koraes Transformed Greek Thought into the Greek Revolution of 1821," *Hellenic News of America*, March 22, 2021.
——— "Free Besieged," *Hellenic News of America*, March 23, 2021.
——— "The Cosmos of Greek Science," *New Greek TV*, April 25, 2021 (in Greek).
——— "The War on the Classics," *Counterpunch*, June 7, 2021.
Vernant, Jean-Pierre. *Myth and Thought Among the Greeks* (New York, 2006).
Vlastos, Gregory. *Plato's Universe* (Parmenides, 2005).
Wells, Colin. *Sailing from Byzantium: How a Lost Empire Shaped the World* (New York, 2007).
White, Andrew D. *A History of the Warfare of Science with Theology in Christendom*, first published in 1896 (New York, 1955), 2 vols.
White, Lynn Jr. *Medieval Religion and Technology* (Berkeley, 1978).
Wiener, Philip P and Aaron Noland. *Roots of Scientific Thought* (New York, 1957).
Wilford, John Noble. "Discovering how Greeks Computed in 100 B.C.," *The New York Times*, July 31, 2008.
Wills, Garry. *Bomb Power: The Modern Presidency and the National Security State* (New York, 2010).
Wilson, N. G. "Archimedes: The Palimpsest and the Tradition," *Byzantinische Zeitschrift* 92 (1999) 89–101.
———. *Scholars of Byzantium* (Baltimore, MD, 1983).
Wright, M. T. "The Antikythera Mechanism reconsidered," *Interdisciplinary Science Reviews*, 2007, Vol. 32, No. 1, 27–43.
Zerefos, Christos S. and Vardinoyiannis, Marianna V. (Eds.) *Hellenistic Alexandria: Celebrating 24 Centuries* (Oxford, 2018).

Zeruneith, Keld. *The Wooden Horse: The Liberation of the Western Mind, from Odysseus to Socrates*. Tr. from the Danish by Russell L. Dees (New York, 2007).
Zhmud, Leonid. *The Origin of the History of Science in Classical Antiquity*. Tr. from the Russian by Alexander Chernoglazov (New York, 2006).

Index

Lightning Source UK Ltd.
Milton Keynes UK
UKHW020628051022
409964UK00011B/929

9 781627 343589